THE PROBIOTIC PLANET

CARY WOLFE *Series Editor*

*(continued on page 333)*

# THE PROBIOTIC PLANET

## Using Life to Manage Life

*Jamie Lorimer*

**posthumanities 59**

UNIVERSITY OF MINNESOTA PRESS

MINNEAPOLIS | LONDON

Chapter 3 draws on materials first published in "Probiotic Environmentalities: Rewilding with Wolves and Worms," *Theory, Culture, and Society* 34, no. 4 (2017): 27–48. A version of chapter 5 was first published as "Parasites, Ghosts, and Mutualists: A Relational Geography of Microbes for Global Health," *Transactions of the Institute of British Geographers* 42, no. 4 (2017): 544–58. Chapter 6 draws on elements of Jamie Lorimer and Clemens Driessen, "From 'Nazi Cows' to Cosmopolitan 'Ecological Engineers': Specifying Rewilding through a History of Heck Cattle," *Annals of the American Association of Geographers* 106, no. 3 (2016): 631–52. Chapter 6 also draws on elements from Clemens Driessen and Jamie Lorimer, "Back-Breeding the Aurochs: The Heck Brothers, National Socialism, and Imagined Geographies for Nonhuman Lebensraum," in *Hitler's Geographies: The Spatialities of the Third Reich*, ed. P. Giaccaria and C. Minca, 138–60 (Chicago: University of Chicago Press, 2016).

Published by the University of Minnesota Press
111 Third Avenue South, Suite 290
Minneapolis, MN 55401-2520
http://www.upress.umn.edu

Printed in the United States of America on acid-free paper

The University of Minnesota is an equal-opportunity educator and employer.

28 27 26 25 24 23 22 21          10 9 8 7 6 5 4 3 2

Library of Congress Cataloging-in-Publication Data
Names: Lorimer, Jamie, author.
Title: The probiotic planet : using life to manage life / Jamie Lorimer.
Description: Minneapolis : University of Minnesota Press, [2020] |
Series: Posthumanities ; 59 | Includes bibliographical references and index. |
Identifiers: LCCN 2020020376 (print) | ISBN 978-1-5179-0920-8 (hc) | ISBN 978-1-5179-0921-5 (pb)
Subjects: MESH: Probiotics | Environmental Health | Health Behavior | Environmental
   Microbiology | Public Policy
Classification: LCC RA784 (print) | NLM WA 30 | DDC 613.2—dc23
LC record available at https://lccn.loc.gov/2020020376.

To MAGALI, AMELIE, AND LOUIS
for foraging, fermentation, and unwavering love and friendship

# CONTENTS

# PREFACE

THE MANUSCRIPT of *The Probiotic Planet* was completed at the end of 2019 and the book was published at the end of 2020. I write this preface in April 2020 in the midst of the global Covid-19 pandemic. The United Kingdom and many other countries are in lockdown. Universities, schools, and most workplaces are closed. Travel and general mobility are tightly restricted, and citizens are social distancing: keeping apart to try to restrict the spread of the disease. The elderly and the already unwell are dying in greater numbers and a profound economic recession is predicted. A general atmosphere of anxiety, uncertainty, and frustration pervades personal and public discourse.

One on level, it is an unfortunate time to be publishing a book that maps and encourages a probiotic enthusiasm for microbes and other forms of life. We are living through an amplification of the modern, antibiotic approach to managing life. Hands, surfaces, and volumes of air are awash with antimicrobial chemicals, people are panic-buying personal protective equipment to secure their lungs and bodies, and the pressures for distancing and isolation enact boundaries and territories around bodies, homes, and nations. Resources are dedicated to antiviral drugs and vaccines to accelerate the development of immunity. Such a response is understandable, for Covid-19 is a new and lethal pathogen that we must make time to tame and live with.

But as the pandemic hopefully eases and we come to better understand how it emerged and spread, how it was known and experienced, and how such pandemics might be averted in the future, I wager that we might benefit from the types of probiotic thinking and action that are mapped in this book. A probiotic approach involves using life to manage life; intervening in the ecological dynamics of a system to deliver desired functions and services. Often probiotic approaches use keystone species, or species with disproportionate agency relative to their ecological abundance, to manage ecosystems. Probiotic enthusiasts tend to stress the importance of biological abundance and diversity for delivering the resilience of their target system—from the body to the farm to the nature reserve—and for transforming it to a desired stable state.

Covid-19 is no keystone species. It is a virus, and thus on the margins of the menagerie of forms commonly included in the category of life. As a virus it is also anomalous in its pathogenicity. There is a myriad of viruses, many of which are undescribed, and only a tiny subset is harmful to humans and animals. An even smaller number have the ability to jump species. Covid-19 certainly needs fighting, but it would be a shame if the necessary push for antiviral control results in an amplified fear of nonhuman life and collateral damage to ecological diversity and health.

Probiotic thinkers advocate what Heather Paxson terms a "post-Pasteurian" approach to hygiene and the wider management of nonhuman life. A post-Pasteurian approach requires a nuanced taxonomy of life-forms and a more selective armory of control methods than the modern antibiotic (or Pasteurian orthodoxy). It retains a vigilant control over specific pathogens, but it also permits the existence of those organisms that restrict the spread of disease and encourages those that deliver desired functions and services. A post-Pasteurian would certainly go after Covid-19, but she would also look at a full range of ecological knowledges and interventions to understand social practices and to enable viral control.

Many people in the urban, Western world live lives cut off from encounters with unruly organisms. We might garden, keep a pet, or ferment, but we are rarely threatened by the vitality of the nonhuman world. With the arrival of Covid-19, many have had to learn how to think like a virus, tuning into the threats posed by a highly mobile infectious agent and adjusting everyday regimes to try to exclude it. Engaged publics are learning about the science of "contact tracing" and its revelations about international, local, and then more intimate domestic movements and infections. Public health officials sketch the viral mobilities enabled by a cough and a sneeze. They encourage us to space ourselves and to choreograph our movements to prevent transmission. We speculate as to how the virus persists on different surfaces and visualize the microscopic landscapes it inhabits. Anxiously we learn to gauge the temporalities of infection through practices of self-isolation and the experience of illness.

As newly forged microbial citizens, Western publics are developing a degree of ecological awareness that is unprecedented since the advent of antibiotics and the successful eradication of infectious disease. And as with other infectious diseases the ability to dodge the virus and the experience of infection is strongly stratified by socioeconomic status. The virus is laying bare stark inequalities in health care, housing, and economic support.

It is also becoming clear that there is no single way of thinking like a virus, in the same way as there is no single model of probiotic knowledge. As I write, disagreements abound among expert epidemiologists as to the reproduction rate of the virus and its trajectory. Rifts are emerging between deductive mathematical modelers and inductive public health researchers keen to test empirical realities. On the margins of the scientific establishment, maverick figures are advocating off-label and untested antibiotic solutions that find favor among desperate politicians. Covid-19 exposes the powerful strand of anti-Pasteurian expertise and health care that takes issue with mainstream science and medicine. Some populist leaders downplay or deny the etiology and the magnitude of the pandemic to justify business as usual. Alternative experts frame the virus as divine judgment and offer conspiracy theories linking the virus to familiar bogeymen. The internet is awash with retailers peddling alternative cures. In this book I explore how the uncertainties caused by the absence of nonhuman life permit rich epistemic diversity. With Covid-19 we see how the uncertainties and insecurities posed by microbial excess open comparable space for denial, doubt, and debate.

At the time of writing, the ecological drivers of the emergence of Covid-19 are not fully known. One hypothesis links the disease to "wet markets" selling live animals in China and a convoluted multispecies transmission pathway from bat to pangolin to pig to human. Comparable zoonotic diseases like SARS and H5N1 have been linked to the hot spots created by industrial livestock agriculture with its concentration of animal bodies, waste, and antimicrobial chemicals. In either case, it is clear that the intensities and accelerated connections of the globalized economy and the globalized food system will have played key roles in this pandemic. In disentangling the pathogenesis of Covid-19, it is likely that probiotic thinkers will point to how the novel, simplified, and unstable ecologies of the food system created the ideal "dysbiotic" conditions for disease emergence. They will advocate interventions that de-intensify production to introduce a greater deal of ecological diversity, stability, and control. They may well find that the exuberant application of antibiotic chemicals leaves pathogenic legacies as skin, bodies, and domestic ecologies recover in the aftermath of continued chemical scouring.

The different ways in which nations and citizens are responding to the pandemic scrambles the political and economic orthodoxy in a fashion much more dramatic than in any of the cases I describe in this book. Many of the core functions of modern economies have been suspended. In a fundamental shift in policy after years of austerity, the U.K. government is now providing a basic income and housing the homeless. A furloughed workforce is volunteering in large numbers to care for those in need, and citizens appear carefully in public to applaud a newly discovered category of "key workers": nurses, carers, supermarket staff, and delivery drivers. This compelling moment of communal solidarity is fragile and conditional. But a crisis precipitated by ecological dysbiosis and nonhuman vitality is making history in ways inexplicable to humanist models of social theory. It remains to be seen whether this newfound recognition in powerful circles of human vulnerability to nonhuman life shifts social relationships and the management of the nonhuman world. Too many microbes are precipitating planetary change, but what new worlds will this pandemic create?

# LIFE IN THE ANTHROPOCENE

*The Antibiotic Anthropocene*

In 1983, ecologists working for the Dutch government introduced herds of back-bred cattle and horses into a nature reserve just north of Amsterdam. The animals were left to de-domesticate themselves in the hope that they would restore the naturalistic grazing that had been missing since the rise of modern agriculture. The aim was to simulate the ecological conditions that would have characterized Northern Europe at the end of the last ice age, with the cattle and horses standing in for the extinct aurochs and tarpan. The project was promoted as an act of rewilding, and it has inspired a wider movement for rewilding Europe with large herbivores, predators, and other ecologically significant species.

In 2004, a British immunologist deliberately infected himself with *Necator americanus*, a species of human hookworm. He wanted to explore the old friends hypothesis, which proposes that the rise in autoimmune, allergic, and inflammatory disease in the West is caused by the absence of some microbes with which humans coevolved. Having demonstrated that it was safe to host the worms, which remain common in tropical areas, he began a clinical trial to test their therapeutic potential for training and exercising the human immune system. Popular interest in the human microbiome and this ecoimmunology has inspired a network of DIY helminth users, who raise, share, sell, and host a variety of worms for the purposes of biome restoration.

Rewilding and biome restoration are exemplary of a probiotic turn that is underway in the management of life in some parts of the WEIRD (Western, Educated, Industrialized, Rich, and Democratic) world.[1] I use the adjective "probiotic" in an expansive sense to describe human interventions that use life to manage life, working with biological and geomorphic processes to deliver forms of human, environmental, and even planetary health.[2] Going probiotic goes well beyond a preference for live yogurt; it links a range of efforts that aim to change the composition of biophysical systems to modulate the rhythms and intensities of their ecological interactions. These projects look for tipping points within these ecologies and manage their dynamics so that they stay within desired boundaries. They control, restore, and enhance the functioning of ecosystems to secure a desired set of systemic properties—in these two cases, biodiversity and immunity.

This book wagers that this probiotic turn can be detected across a range of ecological scales and science and policy domains that are generally conceived in isolation. Further microscale examples include the wider use of probiotic microbes and biotherapeutic organisms (like leeches and maggots), and the adoption of prebiotic diets and lifestyles designed to foster their emergence and survival.[3] It is evidenced in emerging health and hygiene practices that nurture microbial life on the body and in the built environment. These develop targeted approaches that differentiate between good and bad germs and that manage human microbial exposure across the life course.[4] They extend to the long-standing approaches of permaculture, organic farming, and biological pest control.[5] Going probiotic involves the selection and engineering of organisms for bioremediation and waste management, alongside the revived interest in fermentation in food processing.[6] The probiotic interest in rewilding is part of a wider rethinking of the management of biophysical systems, including forms of natural flood management and coastal realignment,[7] localized methods for weather modification,[8] and nature-based, planetary-scale schemes for geoengineering through afforestation, ocean seeding, or solar radiation management.[9]

Advocates position these probiotic innovations as antidotes to the pathologies caused by modern, antibiotic ways of managing life. Like the term "probiotic," I use "antibiotic" in an expansive sense to mean much more than a class of pharmaceuticals that restricts the growth of specific microbes. Being antibiotic describes systematic efforts to secure the Human through the control of unruly ecologies. It involves efforts to eradicate, control, rationalize, and simplify life that are common across landscapes, cities, homes, and bodies. It describes the scientific and political developments in the nineteenth and twentieth centuries that led to economic growth, food surpluses, and disease eradication. In going antibiotic, bodies have been purified, cities sanitized, and some natural environments made safe, productive, and orderly. Many human lives have been extended and made better. But in recent decades, scientists and citizens have in many cases considered this antibiotic approach to be excessive; obsessions with purity, division, simplicity, and control lead to blowback and the emergence of new pathologies.[10] Modern modes of managing life and the earth may be disturbing and intensifying natural processes, helping drive the planetary transition into the Anthropocene.

Environmentalists were among the first to flag the pathologies of modern natural resource management.[11] They demonstrated how the rationalization of environmental processes (like reproduction, grazing, predation, flooding, and fire) can undermine the functionality of ecological systems and cause extinctions. Agriculturalists warned that the simplification, standardization, and intensification of modern farming can lead to soil degradation, can cause the death of pollinators, and can create hot spots for disease emergence.[12] Meanwhile, doctors have raised concerns that modern antibiotic drugs, diets, health care, and lifestyles are creating widespread dysbiosis in the collection of microscopic organisms that live in, on, and around the human body known as the human microbiome.[13] Missing microbes might account for recent increases in allergic, inflammatory, and autoimmune diseases.[14] There is great anxiety that the use and abuse of antimicrobial chemicals accelerates the evolution and spread of drug resistance.[15] On a planetary scale, earth systems scientists caution that the earth is approaching critical boundaries as a result of the Anthropocene's great acceleration in fossil fuel consumption.[16] These scientists suggest that the planet is close to tipping out of Holocene conditions and moving into a new state that might become pathological to all but the hardiest forms of life.

These diagnoses of antibiotic blowback offer a profound critique of modern dreams of mastery and progress. They have been met by denial, obfuscation, and deferment, particularly when they threaten the political economic order and lack simple technological solutions. Subsequent popular responses have tended to fall in two opposing directions. One aims for a techno-optimistic amplification of the status quo: one final effort to achieve our Enlightenment destiny as a God species capable of ecological planetary control.[17] Better technology, better government, and more science and reason will save the day through an accelerationist vision of ecological modernization.[18] The other looks backward, promoting a return to idealized premodern or paleo pasts. We will be taken back to Nature through less technology, less government, and less science and reason. As Bruno Latour observes, seen this way, an otherwise heterogeneous collection of policies and practices consistently array civilization and its futures along a linear development trajectory. We are offered either more of the same or a retreat into history.[19]

While these popular twentieth-century responses to the pathologies of antibiotic ways of managing life certainly inform the probiotic turn, it is also being shaped by more contemporaneous anxieties that call into question some of the certainties of modern and antimodern environmentalism as well as health care. For example, those involved in rewilding must come to terms with how the diagnosis of the Anthropocene undermines the idea of a stable earth—an earth that provides the background to human history and is a source of timeless value. Although this argument predates the political event of the Anthropocene, conservationists have come to realize that there really is no longer a nature to which we can return. Similarly, although on a different scale, the figure of the human as an unraveling holobiont[20]—a multispecies chimera that is kept alive, sane, and rational by its microbes—unsettles the modern idea of health as the absence of microbes and of the Human as a mind in a vat.[21] The probiotic turn is influenced by the more-than-human intellectual zeitgeist catalyzed by the Anthropocene and microbiome, which questions these modern ontological reference points and the "more of the same" or "back to nature" programs they inform.

This zeitgeist is also evidenced in a common challenge, made by prominent advocates of going probiotic, to reductionist scientific epistemologies that model life as adhering to universal and mechanistic laws on which firm predictions can be made. Probiotic approaches that use life to manage life pull at the binary and linear models of resource management that are founded on them.[22] Probiotic thinking must thus grapple with the contemporary "postepistemological" or "posttruth" condition of a world marked by uncertainty and complexity, the slipperiness of which has compounded a growing distrust in science among some publics and in some parts of politics.[23] The probiotic diagnosis of the false promises of modern ways of managing life, as well as a growing awareness of their inherent pathologies, has weakened the modern settlement between science and politics, and between science and its publics.[24] Here objective and rational scientists generate knowledge outside of society and in advance of politics. Such knowledge is provided to policy makers, who receive it with gratitude and deference, then use it to shape the publics in its image. This has always been a straw man depiction, but the end of Nature and the Human, as well as the loss of certainty, have led to the proliferation of knowledge claims and controversies, the formulation of new modes of expertise, and the proliferation of novel forums in which publics engage (and disengage) with science and technology.[25]

The probiotic turn is happening at a contemporary juncture in which common anxieties about ecological dependency, dysbiosis, and precarity profoundly challenge modern approaches to health and environmental management, animate a range of ecomodernist and reactionary alternatives, and prompt a far-reaching consideration of the futures of progress, prosperity, and multispecies survival. I do not want to overstate my case. The probiotic turn is not an emic category used by the actors whose stories I report; no one talks of going probiotic in such terms. It is not synchronized across policy domains. Some of these practices are novel; others have longer histories. Nor is it occurring everywhere with the same intensity; it has a geography. Many of these practices are marginal and contested; they also differ in important ways. Nonetheless, in this book, I hope to demonstrate that they represent a coherent and important shift in how life is conceived and managed in parts of the WEIRD world.

To be clear from the outset, this book is not a celebration of the triumph of ecological thinking. The probiotic turn is not a panacea for planetary health. No single technological or governmental solution is going to fix the current political and ecological juncture. Nor is there only one way of going probiotic. Instead, in this book, I have two initial aims. The first is to specify the probiotic turn, identifying the common characteristics of probiotic approaches to managing life and providing rich empirical illustration of their practice, history, and variety. The second is to critically analyze the different manifestations of probiotic thought and practice. The probiotic turn has promise, but it requires close critical differentiation. There are many different ways to go probiotic, each of which has different beneficiaries.

## *What Does It Mean to Go Probiotic?*

The probiotic approach to managing life has six general characteristics; these are composed of a common scientific ontology, mode of biopolitics, approach to knowledge production, relational geography, temporality, and political ecology. These are the respective subjects of chapters 2 to 7, and I will summarize them below. I focus this exposition on the cases of rewilding and biome restoration with which I opened this chapter, and which are my principal empirical reference points.

The probiotic turn is founded on a common scientific conceptual framework in which the world and its constituent parts form a nonlinear but nonetheless self-regulating system—or, better, a system of nested systems. Borrowing from the work Donna Haraway, Bruno Latour, and Isabelle Stengers, I describe this as a Gaian ontology, after the influential work of James Lovelock and Lynn Margulis.[26] Probiotic scientists, thinking like Gaia, are preoccupied with evolutionary and ecological relationships as well as the ways living organisms shape and are shaped by their environments. These scientists think in cycles; they concern themselves with feedback loops with the potential to modulate or amplify ecological change. They focus on questions of symbiosis and symbiogenesis—the reciprocal, coevolutionary relationships and dependencies between life-forms—as well as the ways these configure the structure, intensities, and dynamics of ecologies, climates, and geomorphologies. This science stretches and subverts the neo-Darwinian figure of the bounded human individual, emphasizing both its microbial composition and its planetary entanglements.

In my examples, probiotic science is especially concerned with keystone species—that is, the subset of organisms with disproportionate ecological importance resulting from their location within ecological networks. Some species, like wolves, beavers, or earthworms, are capable of configuring entire landscapes through their effects on the food chain or their ecosystem engineering. In the absence of this engineering, or the top-down regulation performed by their ecological interactions, ecologies tip into degraded and dysfunctional conditions. Ecologists and immunologists describe dysbiotic human populations marked by missing microbes and a planet haunted by extinct or diminished keystone species as a result of antibiotic modes of managing life. Research attention focuses on mapping the shape or topology of ecological interaction networks, identifying keystone species and tipping points, then designing and testing experiments to revert ecologies to their desired composition and functionality. Here the planet and its component systems adhere to multiple stable states, few of which are provident for human life. Gaia is neither vengeful nor beneficent but rather is agitated and in transition. Humans figure centrally in these novel ecosystems, naturalized as a "hyperkeystone species" capable of both disaster and remedy.[27]

This Gaian ontostory of nested systems shaped by symbiotic relations informs a common set of management practices that are geared toward nurturing and modulating ecological dynamics to secure the delivery of desired ecological functions and services. In chapter 3, I explain how this approach involves the strategic deployment of keystone species within and beyond human bodies, and targeted interventions to create ecological conditions conducive to their flourishing. These mechanisms range across nested scales, from the nature reserves that are the targets of rewilding to the human bodies that are subject to biome restoration. Probiotic governance is targeted at the conduct of individual humans and animals; it also involves interventions into populations of keystone species. However, the ultimate aim is to transform the dynamics of the ecologies with which they are entangled, working from the bodies of animals out to the planetary concentration and circulation of atmospheric gases. In using life to manage life, probiotic governance differs markedly from the command-and-control logics of modern antibiotic approaches to human and environmental health.

Informed by Michel Foucault and those who have further developed his work, I suggest we understand probiotic governance as a novel mode of environmental biopower that works on and through human subjects to modulate the dynamics of socioecological systems.[28] This involves shaping the knowing conduct of human actors through what I term holobiont governmentalities. This concept describes how the scientific figure of the human as a holobiont is enrolled to legitimate new projects and practices of probiotic subject formation that recalibrate human encounters with animals, microbes and the wider environment. These interventions seek to enable ecological change. Taboo modern elements like dirt, rot, damp, and mess are given salutary probiotic potential. Probiotic governance also involves symbiopolitics (a term I borrow from Stefan Helmreich): the application of keystone species or anthropogenic surrogates to modulate the intensities of symbiotic relationships to deliver ecological functions and services.[29] These two forms of power are discursively and materially connected, but they involve different methods and targets for governmental action. The first targets the self-aware behaviors of the human host; the second focuses on the dynamics of nested ecosystems. Both come to shape the lives of the humans and animals that are governed as and through populations of keystone species. Probiotic governance has similarities with other efforts to institutionalize systems biology—for example, to deliver biosecurity or resilience.[30]

Drawing on Heather Paxson's work on the biopolitics of raw-milk cheese production, I suggest we see the probiotic turn as a post-Pasteurian recalibration, rather than a rejection, of antibiotic management practices.[31] Probiotic governance involves a nuanced differentiation between beneficial and unruly organisms and the "controlled decontrolling of ecological controls" to put them to work.[32] This symbiopolitics requires a different set of knowledge practices to those that predominate modernist science. A Gaian ontology of novel ecosystems offers deeply uncertain and complex foundations on which to legitimate ecological management decisions. There is no singular Nature to which scientists or politicians can make recourse; publics, in all their emergent heterogeneity, can be caught up in the conduct and analysis of any probiotic experiment. In chapter 4, I explore how, in grappling with this reality, probiotic knowledge makers have shuttled between the scientific truth spots of the laboratory and the field site. They have observed natural experiments in which keystone species have gone missing or made their own self-willed return. They have also designed experiments in nature, managing clinical and countryside settings to monitor change and test hypotheses. They have arrived, somewhat uneasily, at a model of real-world (or what Clemens Driessen and I have previously termed "wild") experiments that involve working closely with unruly ecologies and publics.[33]

The diverse scientists involved in these wild experiments dedicate much of their time to mapping ecological interactions and nonlinear processes.[34] They develop their ways of thinking in cycles, flows, and feedbacks. They refine apparatus for detecting the ecological anachronisms and vacuums left by the absence of coevolved ghost species, and they seek data and models to disclose how ecosystems are wired and rewired by keystone species. But some of these scientists, and the citizens with whom they interact, also seek to hack their focal ecologies, conducting open-ended experiments, sometimes at the edge of the law, to see what happens when keystone species are introduced.[35] They learn to be affected by their subject organisms, developing craft skills and experience to understand novel ecological interactions.[36] This hacking values the epistemic potential of surprises, developing public methods to nurture and apprehend unpredictable and nonanalog ecological dynamics. The probiotic turn is marked by an uneasy constellation of these mapping and hacking practices that generates a lively knowledge politics. Advocates disagree over what the wild is, and how it might be known, valued, and managed.

There are stark and unequal geographies to the probiotic turn in terms of where its science is developed and applied, and in the distribution of its benefits and impacts, which I trace in chapter 5. Going probiotic involves developing mutually beneficial relations with species and processes, many of which (like wolves or helminths) have previously threatened human existence or health, and which have been subject to concerted programs for control or eradication. These keystone species can be understood as pathobionts, or organisms that are not inherently good or bad but whose role is configured by their ecological context. They can be parasites (who depredate their hosts), ghosts (whose absence causes ecological dysbiosis), and mutualists (whose presence benefits their hosts). The possibility of going probiotic with such organisms is strongly configured by political, economic, and ecological relations. The situations in which designed mutualistic (or Mutualism 2.0) relations are feasible have a common geography that is patchy and uneven, and interdependent on relations elsewhere. Probiotic relations are predominantly confined to specific parts of the temperate, white WEIRD world in which life was already tightly controlled and the resources for controlled de-controlling are readily available.

There are parts of the non-WEIRD world that have yet to be made subject to antibiotic ecological relations, where people may live in preantibiotic and mutualistic relations with keystone species. There is a complex geography here, but these areas tend to be rural, poor, and Black, and they tend to involve traditional or indigenous ecological relationships. But the political-ecological intensities experienced by people and places in many less affluent parts of the world may also tip relations so that keystone species become parasitic, leading to deleterious accumulations, dysbiosis, or even death—for example, in the accumulation of helminth infections among those living in plantation political ecologies. Given the magnitude of antibiotic global management in both the macrobiome and the microbiome, the majority of the world is haunted by some degree of ghostly relations. Dysfunctional, "trophically degraded"[37] ecologies and human "epidemics of absence"[38] are now the norm. But the human experience of this haunting is strongly configured by compounding exposure to other parasites and by their differential ability to access artificial surrogates for missing keystone species, like immuno-suppressant drugs or pest-control products. Many people thus experience a doubled dysbiosis of lost old-friend microbes and amplified exposure to (drug-resistant) crowd infections.[39] The possibility of going probiotic is thus largely an elite experience, and one that reflects and is maintained by longer histories of global exchange and exploitation. The probiotic turn cannot be read as a linear universal trend in a progressivist history of world development.

If the probiotic turn has a common, patchy geography, then it also evidences a common temporality. The stories told by advocates of going probiotic are founded on a history that frames the present as a critical inflection point, one occurring at a moment of crisis and engendered by the pathologies of modern, antibiotic approaches to managing life. Probiotic proponents present themselves as seers, visionaries able to look back in time in order to diagnose the problems of the present, dismiss rival contemporaneous solutions, and herald new restorative futures. That is, they offer future pasts.[40] On the one hand, their ecological politics is often profoundly retrospective, advocating rewilding, biome restoration, back-breeding, and de-domestication, for example. They look to past baselines to identify desired ecological conditions. They tell revisionist histories that challenge the linear and triumphalist narratives of Western agriculture and health care, tracing ecological unraveling through the premodern and then modern periods. On the other hand, such seers are prospecting into the future and are anticipating novel ecologies. They seek functional landscapes that will emerge along unexpected pathways. The quality of these desired future pasts is not directly indexed to the authenticity of historical reenactment. Instead it involves a recalibration of human ecologies and their keystone species. The historical reference points for these future pasts range from the postpaleo to a postpastoral, both of which signal the centrality of agriculture and domestication to contemporary anxieties about when the wild was, and how it was lost. As I trace in chapter 6, this retrospective periodization permits some dark, reactionary, and ecofascist models of going probiotic.

The final common dimension to the probiotic turn is a consistent political economy marked by a drive to value the ecological work done by keystone species in economic terms. This involves a range of discursive, legal, and financial practices that depart from established methods of bestowing value. Certain organisms are afforded higher value by virtue of the types of ecological work they do. Keystone species shift from being pests (like hookworms) or resources (like cattle or beavers) and come to the fore as idealized ecosystem engineers: skilled and well-networked middle managers that coordinate the activities of others in a nonhuman division of labor. They figure as a source of nonhuman intelligence, a model for biomimicry and the creation of artificial surrogates. Keystone species thus become lively commodities involved in ecological work, and exchanged and sold in different economic formations.[41]

This naturalization of nonhuman work involves a range of proprietorial practices through which a once open source resource (like abundant worms or beavers) is first turned into common property, then privatized as a source of nonhuman intellectual property. Sometimes keystone species become natural capital; other times, they remain in common. Yet rarely are they permitted to live in an unalienated relationship with their own means of (re)production. This process instigates what Jessica Dempsey terms an "ecological–economic tribunal for life," one in which the future survival of a species or other ecological component is indexed to its functional utility.[42] But this framing is partial and incomplete, and is resisted by a range of human and nonhuman actors let loose by a "ticklish Gaia."[43] The nonhuman elements of Gaia proliferate in ways that caution against the hubris of some of the more ambitious dreams of probiotic planetary control.

Taken together, these six dimensions provide us with a first approximation of the common characteristics of going probiotic. They offer a means of distinguishing it from its antibiotic antecedents, and they serve as useful foci for critically differentiating ways of going probiotic; the overview provided above masks a great deal of political and ecological heterogeneity. To lay out the conceptual underpinnings for my analysis of this difference, I must make a detour into social theory.

## Why Gaia, and Why Now?

This book is not a traditional work of science studies that reports a history of probiotic ideas or the coproduction of probiotic science and society. It does a bit of both of those things, but it is also interested in the manifestations of the probiotic turn in social theory. It tracks a triangular relationship between probiotic science, its social context, and an emerging body of probiotic social theory. More specifically, one of my aims is to critically examine the convergence between strands of new materialist and environmentalist thinking in the social sciences and the Gaian science associated with the probiotic turn. This convergence comes in several forms. The most prominent comprises the collaborations among social theorists and some of the microbiologists, immunologists, ecologists, and earth systems scientists involved in the probiotic turn. The outcome is an increasingly normative body of ecologized social theory that aims to diagnose and address the problems caused by modern antibiotic modes of biopolitics, or ways of conceiving and managing life.

French anthropologist Bruno Latour has been collaborating and writing with earth systems scientists (and even sought an audience with James Lovelock, the purveyor of the Gaia hypothesis, now a hundred years old).[44] He defends Gaia as scientific framework and has developed and promoted a synoptic version of Gaian thinking in books on how science and politics might be conjoined in the "new climatic regime."[45] Similarly, feminist science studies scholar Donna Haraway has long admired the work of Lynn Margulis, and has maintained close links with those involved in developing the field of ecological evolutionary developmental biology, to which Margulis's science gave rise. Haraway promotes Margulis's science in several of her essays as well as in her collaborations with biologists Scott Gilbert and Margaret McFall-Ngai.[46] Finally, anthropologist Anna Tsing has collaborated with Haraway and some of these scientists while also working with ecologists central to rewilding, like Jens-Christian Svenning.[47] These transdisciplinary collaborations have been enabled by a series of high-profile and well-resourced research programs that have taken the Anthropocene and the microbiome as generative provocations for new forms of academic, artistic, and literary endeavor.[48]

Reading across this work, we can establish that prominent figures see a turn to Gaian science as a solution to two key challenges that social theory faces in the early years of the twenty-first century. First, Gaian thinking offers a palatable liberal and ecological political ontology—or ontostory[49]—for rematerializing theory after the idealist excesses of the cultural turn. Second, it provides common epistemic ground for rebuilding alliances with the natural sciences after the science wars, and in the face of rising science denialism. I will take these two points in turn. In the aftermath of eugenics, the Holocaust, and the environmental determinism of some colonial scholarship, much Western social theory disavowed materialist accounts of the world that sought explanations for, and justifications of, social patterns in biological, climatological, or geomorphic phenomena. In a now well-told history, influential strands of critical theory presented the social as a realm of signs, symbols, and discourse peopled by disembodied human minds. Culture was separated from nature, and critical attention focused on how the latter was used to construct or naturalize the former.

Special attention was paid to the late twentieth-century rise and influence of neo-Darwinian sociobiology. This depiction of selfish and violent individual actors, struggling to survive and reproduce in a world without top-down regulation, was taken as patriarchal, neoliberal theory parading as biology.[50] In contrast, Haraway, Latour, and others make an alliance with the ontostory of Gaia and of symbiosis because it provides a markedly different model of social and multispecies relations. The postmodern synthesis in biology, which is gradually superseding the modern synthesis of Darwinian natural selection and Mendelian genetics, places greater emphasis on reciprocal and mutualistic interactions, marked by tolerance, dialogue, and communication. We learn of the lateral (rather than vertical) exchange of biological materials, and the environmental context (or niche) of an organism is afforded much more importance in shaping and regulating behavior.[51] The selfish individual is subsumed within a wider ecology. When coupled with the planetary models of earth systems science, this communitarian, self-regulating model entangles the human in the earth, challenging the reductionism and anthropocentrism of some forms of social Darwinism.[52] As Haraway puts it, "The order is re-knitted: human beings are with and of the earth and the biotic and abiotic powers of this earth are the main story."[53] It is not hard to see why this science provides an appealing worldview for liberal left academics concerned with social inequality and ecological degradation. It offers Latour grounds to ecologize politics, bringing them "down to Earth" to "face Gaia."[54] Writing in *Science* with earth system scientist Timothy Lenton, he argues,

> Earth has now entered a new epoch called the Anthropocene, and humans are beginning to become aware of the global consequences of their actions. As a result, deliberate self-regulation—from personal action to global geoengineering schemes—is either happening or imminently possible. Making such conscious choices to operate within Gaia constitutes a fundamental new state of Gaia, which we call Gaia 2.0. By emphasizing the agency of life-forms and their ability to set goals, Gaia 2.0 may be an effective framework for fostering global sustainability.[55]

Latour is careful to note that Gaia is not an organized totality with teleology, and is neither vengeful nor providential in its inclinations toward life. But Latour's return to earth and his investment in an earth systems science framework for global governance are indicative of the rise of a wider "general ecology" among prominent thinkers and political theorists, the ranks of which include new materialist illuminati like Jane Bennett, William Connelly, Timothy Morton, and Cary Wolfe.[56]

For Latour in particular, Gaian thinking also offers new epistemic grounds for social theorists to collaborate with scientists. The potential planetary changes it makes tractable provide a clear imperative for why they should do so. Latour comes to Gaia after a long and distinguished career dedicated to placing science in its social context, tracing the practices, networks, and power games through which specific accounts of the natural world come to prominence. Though he was far less of an idealist than some of his science studies colleagues, his interventions helped critique a powerful political epistemology of natural science (as outside of society and politics) and revealed its social construction. Latterly, and somewhat remorsefully, Latour has come to observe how such methods are mirrored by those who doubt or deny the science of climate change. Critique, for Latour, "runs out of steam" when it is aligned with these "obscurantist" projects, shorn of any epistemological or materialist grounds.[57]

Going on the offensive (and drawing on both Haraway and Stengers), Latour proposes a new model of "composition" in which social and natural scientists collaborate with each other, and with publics, to develop new forms of knowledge and new models of deliberative democracy. The indeterminacy of Gaia here offers grounds for epistemic pluralism and political action.[58] More concretely, Haraway and Tsing have used Gaian thinking to ground interdisciplinary research programs that foster "arts of living on a damaged planet."[59] On the basis of long histories of friendship and social interaction, trust and mutual respect, they have worked with natural scientists to develop and refine new "arts of noticing":[60] knowledge practices that enable the types of composition envisaged by Latour. Tsing describes these as "transdisciplinary mutualisms,"[61] or experiments that cut across the spaces of knowledge production, spanning the field, the laboratory, and the studio, and folding together disparate representational idioms including opera, film, craft, and writing.[62] In keeping with the Anthropocene zeitgeist, Haraway and Tsing also engage with science fiction as a way of bridging epistemic difference.[63]

This rapprochement between critical social theory and science is historically striking and important. It is also politically vital in the context of the problems by which it is propelled. In this book, I share these figures' enthusiasm for transdisciplinary collaboration while thinking critically about the terms of reengagement and the ends to which such Gaian science might be put.

## Going Probiotic Otherwise

The probiotic turn in science, politics, and new materialist theory has been greeted with rather more skepticism by other social theorists, and it is to their criticisms that I now turn because they help identify some of the criteria I will use to differentiate ways of going probiotic. Three issues stand out here: scientism, quietism, and hubris.

Sympathetic critics of the new materialist turn in the social sciences have expressed concern about the epistemic grounds on which some prominent thinkers have engaged with the model ecosystems offered by the Gaian natural sciences. For Heather Paxson and Stefan Helmreich, model ecosystems are "contemporary tools that scientists . . . use to describe desires for exemplary ways of studying human entanglements with nonhuman agencies."[64] Models do important epistemic work, but they also operate "in a prescriptive sense, as tokens of how organisms and human ecological relations with them *could, should,* or *might* be."[65] Paxson and Helmreich are concerned that new normative desires to ground social and ecological projects in biological science might occlude scientific uncertainty and heterogeneity, thus causing us to lose sight of the practices through which scientific knowledge is generated and legitimated. They caution that some "new materialist tactics often veer toward universalizing metaphysical claims about the nature of 'matter' as such and also, at times, take scientific truth claims about the world at face value—a move that we consider a step backward."[66] In a specific example, they cite a review by Maureen O'Malley of a book by Myra Hird, who develops Margulis's science toward a microbial ethics, worrying that "by removing from the picture how the science was done as well as the ongoing revisability and contestability of microbiological findings, we are left with the sense that there is a fount of straightforwardly produced and accepted knowledge from which we can drink."[67]

Bruce Braun voices similar criticisms about the work of Jane Bennett, noting a surprising lack of reflexivity among some new materialists and a tendency toward scientism:

> In looking to the natural sciences the (new materialist) literature too often takes "science" to speak in one voice, or in what often amounts to the same thing, draws selectively from the natural sciences in order to find the ideas and concepts it needs, ignoring science's heterogeneity and side-stepping vibrant internal debate over models and paradigms. . . . There is an irony here, for even as many new materialists propose an ontology that is non-deterministic and non-teleological, they often deploy a very different epistemological position when it comes to the emergence of their ideas, which are viewed as universal rather than particular, and necessary rather than contingent: the world

is marked by indeterminacy and contingency, except when it comes to theories of indeterminacy and contingency![68]

These criticisms are not directed at Haraway or Latour, whose training in science studies informs more transparent and nuanced accounts of their epistemic alliances with Gaian science.[69] But the risk of scientism serves as a useful warning for the alliance with science I make in the analysis of the probiotic turn that follows—one I hope to address by paying close attention to the practices of probiotic science (chapter 4) as well as the geographical, historical, and political economic contexts in which this knowledge is produced (chapters 5 to 7, respectively).

A second and connected set of criticisms accuses probiotic social theorists and scientists of a political quietism: that an end of critique and alliance with science risks downgrading the role of the theorist to speak truth to power, to challenge the status quo, and to imagine and advocate for transformative alternatives. This case is made most forcefully by a group of geographers and political scientists concerned with what David Chandler terms the "ontopolitics of the Anthropocene": emerging frameworks of knowledge and governance that fold the uncertain, planetary science of disturbed, or ticklish, earth systems into the management of human and nonhuman life.[70] Chandler, Sara Nelson, Stephanie Wakefield, and others have identified two tendencies in this work.

The first, and most significant, examines the governance of environmental and economic systems that are framed as on the verge of tipping points, where the potential for collapse is folded into the governance of the present in regimes for biosecurity and environmental security. They trace a long history in which policy makers have borrowed concepts from ecology and systems biology to forge a "resilience dispositif"—a powerful mode of adaptive governance, enabled by the Gaian science of resilience thinking, the primary aim of which is to secure the (often unjust and unsustainable) status quo in the face of coming disasters, be they climatic, economic, or microbiological.[71] Here a particular strand of new ecological thinking is understood to naturalize neoliberal modes of social and ecological organization.[72] Chandler and others do not direct their critique at Latour and his compositionalist model; indeed, Latour himself does not make explicit use of resilience as an organizing concept in his high-level reformulation of global governance for Gaia 2.0. However, but there are commonalities worthy of further investigation.

Second, Chandler and Wakefield identify a contrasting tendency in the ontopolitics of the Anthropocene to affirm life amid the blasted ruins of landscapes destroyed by capitalist modernity.[73] They argue that this genre is best exemplified in the writings of Tsing on the political ecologies of global matsutake mushroom collection and exchange.[74] Tsing's work focuses on ecologies that have already been tipped into novel and degraded political and ecological conditions by colonial and capitalist development. Such ecologies are marked by extinction as well as by the proliferation of undesired forms of life, like pathogens and invasive species. But these ecologies also evidence the precarious, multispecies survival strategies of those living amid the ruins, which sometimes enable socioecological "resurgence"—a term Tsing differentiates from resilience, and that she takes to describe the emergence of desired ecologies and livelihoods.[75] In spite of its critique of neoliberal capitalisms, Chandler and others take issue with this affirmation of ruins. They suggest it too willingly gives up on progress, science, and the possibilities of reworking modernity. It diminishes human potential and struggles to think beyond capitalist economic models.

Out of this critique of resilience and ruins, a range of thinkers has sketched a third model of ontopolitics, one in which a politicized engagement with Gaian science (especially resilience thinking) helps critique the status quo, attending in particular to the political and economic relationships that come to configure the unequal experience of ecological change across a wide range of scales. This work stokes up critique (to rework Latour's metaphor), developing an experimental political ecology that can "transform" the unequal, proprietary, and anthropocentric character of antibiotic modes of managing life.[76] The aim is to find and nurture just and sustainable ways of responding to what Isabelle Stengers calls "the intrusion of Gaia" into social theory: salvaging some of the tools of modernity to build better communal futures along trajectories that are "orthogonal" (after Latour) to popular alternatives of ecological modernization or reactionary retreat.[77] The most vibrant strands of this work are found in novel feminist and eco-Marxist thinking that engages with how the material properties of animals, ecologies, and geophysical systems configure neoliberal capitalism to both enable and constrain alternative futures.[78]

The third and final criticism of the traffic of Gaian thinking into social theory comes from the new materialists, voiced by theorists like Clare Colebrook, Nigel Clark, Elizabeth Povinelli, and Kathryn Yusoff, who are concerned with "geopower" and who write from indigenous, feminist, and postcolonial perspectives.[79] In different ways, they caution that a celebration of symbiosis risks maintaining the anthropocentrism of modern theory as well as its preoccupations with human survival and the management of life. They suggest that the probiotic turn in social theory does not fully acknowledge the "radical asymmetry" of geological processes (we need them much more than they need us), the vast longevity of earth's history, and the magnitude of disruption that will follow from the now unavoidable end of Holocene conditions.[80] They caution that aspirations for biopolitics (however affirmative or ecological) are hubristic and ill-fated.

Their work develops a much darker, sometimes apocalyptic environmentalism that at the very least provides a necessary corrective to some of the bright green probiotic dreams of the techno-optimist eco-modernists that will be encountered in this book. Often written from the perspective of those on the front line of Anthropocenic antibiotic change, their writings offer profound philosophical and empirical lessons in "learning to die in the Anthropocene," where learning to die involves less the literal end of life and more a willingness to experiment with alternative modes of becoming otherwise as the planet passes into, and perhaps beyond, the coming disaster.[81] I engage with these ideas throughout my analysis, attending to situations in which the aftermath of antibiotic interventions or the unpredicted outcomes of probiotic experiments give glimpses of alternative—sometimes inhuman—political ecological futures.

Taken together, this body of probiotic social theory, including both its advocates and its skeptics, provides the conceptual tools I use to differentiate and critique the probiotic turn in chapters 4 to 7 of this book. I return to my evaluation of the relationships between Gaian science, contemporary ecological crises, and probiotic social theory in the conclusion. The ways of going probiotic that I suggest hint at the many ways there are of facing Gaia—and the many Gaias that one might face. I focus in particular on how approaches to going probiotic differ with regard to the hoary themes of humanism, progress, and justice. Through this parsing of probiotic diversity, I evaluate the epistemic and critical potential of a transformative model of Gaian social theory that makes a careful, political, and humble alliance with Gaian science.

## Methods

This book makes some rather grand claims, which makes it necessary to provide details about my methods and evidence. The arguments I develop are informed by nearly a decade of empirical research. Some of this I did alone; some I conducted with PhD students, postdoctoral researchers, and colleagues. I am a human geographer by training, but even as a geographer, I deploy a somewhat heterodox methodology and disciplinary praxis. Geography has given me a diverse and generous discipline that tolerates a wide range of empirical interests and epistemic commitments. Being a geographer, I spend nearly as much time conversing and collaborating with natural scientists as I do with social scientists. Teaching geography in an institution dedicated to small group tutorials, I read my colleagues' work and help students make sense of their writings. Although I am not trained as an ecologist or an earth systems scientist, I spend a lot of time around them. My interest in the microbiome has taken me outside of geography and into the companionship of a network of microbiologists. I helped found the Oxford Interdisciplinary Microbiome Project, and I led an interdisciplinary research group exploring the microbiome of the domestic environment. I was taught microbiology by scientific friends and have had to teach it to the publics involved in these research activities. In different ways, these institutional commitments have led me to partially inhabit the worlds of Gaian scientists, who give me space to ask questions, develop hypotheses, and road test some of the ideas in this book. I intend this book to be read as a conversation with this science rather than a distanciated critique. There is too much at stake to forego transdisciplinary collaboration.

I have deployed a more formal set of methods for tracing the probiotic turn in science. I read science journals to track general trends, and I have set e-mail alerts to notify me of themes of interest. I follow key scientists and science journalists on social media and read popular science books. Chapter 2 is based on a more systematic reading of the scientific literature on earth systems science and the microbiome, with specific focus on rewilding and ecoimmunology, the two examples of the probiotic turn that feature most prominently in this book. While I can't claim to fully understand the technical details and methods of these fields, my reading identifies their key ontological principles, concepts, and metaphors. I read this literature guided by existing work in the history of science—especially on the histories of ecology, immunology, and conservation biology—to better understand their precursors and the significance of the trends I trace. I look for disagreement and dissent; I also note the absence of conversations across what would seem to be related domains.

My analysis was further developed through interviews with over fifty microbiologists, immunologists, and ecologists whose work I was reading. I spoke to them on the phone or visited them in their laboratories, offices, and field sites. I asked for clarification of key methods and concepts, gained unpublished insights, and was also able to rehearse my argument of a probiotic turn. Several of my informants were kind enough to demonstrate their research, and I witnessed probiotic science in action in the context of rewilding in situ, in the laboratory, and in clinical experiments with microbes. I developed my understanding of the social worlds and epistemic character of probiotic science by attending several conferences and workshops, sometimes presenting my own work on rewilding and participatory microbiology. In the chapters that follow, I tend to cite these scientists' published papers, but my analysis is strongly shaped by my firsthand conversations and field observations.

However, the probiotic turn is not only a scientific event. Researching its public dimensions required a further set of methods, tailored more explicitly to my case studies, which I discuss in more detail in the following chapter. In both examples, I spent time accessing the popular literature and visual culture of rewilding and biome restoration, following how they were presented and discussed in legacy and social media, and on film. For biome restoration in particular, I joined a set of social media support groups and spent time witnessing the social interactions of advice, support, and conspiracy that characterize this virtual space. Through this virtual ethnography, I was able to identify at least twenty key informants among the DIY communities who are using worms and human feces to treat a range of health conditions. Speaking to further participants and reading their accounts, I learned how to raise a worm colony at home, and how to know when one's worms are doing well. I spoke to those who organize the online networks that made these microbes widely available, and I traced the legal and regulatory challenges they face in distributing their organisms across a range of national jurisdictions.

Understanding the social dimensions of rewilding took me and my collaborators to a range of field sites and archive locations. Part of the analysis presented in this book draws on work I undertook with Clemens Driessen on the history and politics of rewilding in the Netherlands. Clemens carried out extensive archival research in Germany tracing the back-breeding of Heck cattle. He also conducted interviews with those involved in their contemporary reintroduction for rewilding Europe. I draw on these materials in chapter 6. I also visited several of the best-established examples of rewilding in Germany, the Netherlands, and the United Kingdom, including the Knepp Wildland Project in Sussex and the various beaver reintroduction projects in Devon, Scotland, and Bavaria. I spoke with those leading these projects as well as with various local supporters and opponents. I also sought out examples of self-willed rewilding—like the wild boar in Europe—by soliciting interviews with enthusiasts for their return and reading popular coverage of the legal, political, and economic challenges they pose.

## A Guide to Reading

I anticipate that this book will have a range of readers who will come to it with different backgrounds and interests. It uses a diverse range of social and natural scientific terminology, not all of which will be familiar to everyone. I have provided a glossary of key terms at the end of the book. *Probiotic Planet* is designed to be read from start to finish. The book's first part provides an overview of the probiotic turn, and the second offers a critical appraisal. I would encourage everyone to read chapter 1, which provides a detailed overview of rewilding and biome restoration, and which introduces the specific case studies that feature in the chapters that follow. Chapter 1 also provides a more substantial discussion of the differences between an antibiotic and probiotic approach, and it gives more detail on the probiotic turn as it manifests in other domains of knowledge and action, including food, agriculture, and waste and environmental management. This chapter justifies why we might take rewilding and biome restoration as indicative of wider trends across these other domains. Chapters 2 to 7 then work through the six characteristics of the probiotic turn. Chapter 2 summarizes the common elements of probiotic Gaian science, chapter 3 covers the governance practices this informs, and chapter 4 looks at the knowledge practices from which it is developed. Then chapter 5 maps the unequal geography of the probiotic turn, chapter 6 traces its temporalities, and chapter 7 turns to its political economy. The conclusion returns to the three broad aims of the book and draws out its contributions.

# 1

## THE PROBIOTIC TURN
### Rewilding and Biome Restoration

THIS BOOK focuses on two case studies of going probiotic: rewilding in environmental management and biome restoration in human health care. This chapter provides an overview of these two areas of science as well as their policy and practice. It introduces the key characters, sites, ideas, and events that figure in the chapters that follow, gathering the background and empirical information required to make sense of the analysis. Here I situate rewilding and biome restoration within the wider collection of examples and trends that I refer to as the probiotic turn, fleshing out some of the broad claims made in the Introduction about a shift from antibiotic to probiotic modes of managing life. As far as is possible, I use the discourse of my research subjects as they explain what it means to rewild and restore, and as they narrate the origins, trajectories, and discord within their respective movements. I also deploy a fair amount of editorial license in curating the narrative that follows. I want to establish the common characteristics and trends that are the subject of more detailed discussion in the second section of the book. Where appropriate, I note the contested nature and heterogeneity of probiotic thought and practice, flagging the tensions that are the subject of the third section of the book.

### Rewilding

"Rewilding" is a capacious and contested term, the remit of which reaches well beyond environmental management to encompass a host of social, psychological, and spiritual practices. The boundaries and thus the definition of rewilding are subject to much debate.[1] Here I focus primarily on what is known as trophic rewilding: an approach to ecological restoration that uses species introductions to restore ecological interactions and to promote self-regulating and biodiverse ecosystems.[2] This involves the deliberate return of a small number of keystone species—that is, organisms with disproportionate ecological significance relative to their abundance, which are valued because they restore ecological processes and functions. Trophic rewilding has come to prominence only in the last twenty years, shifting the focus of conservation from preventing extinctions and maintaining species composition to managing processes. It also shifts the ecological benchmarks that guide conservation, looking beyond the predominant premodern archetypes to consider prehistorical landscapes, and looking forward to the novel ecologies emerging in the Anthropocene. Finally, rewilding expands the geographical scale of conservation from discrete protected areas toward landscape- or regional-scale networks as well as the management of ecological connectivity. The general history, character, and regional variations in rewilding have been well reported in the literature.[3] Here I focus on the set of rewilding case studies that are the subject of analysis in this book.

My primary interest is in developments in European conservation, and I start with the flagship example of the Oostvaardersplassen (OVP) in the Netherlands. The OVP is a 5,600-hectare nature reserve located just north of Amsterdam. It forms part of the Flevoland, the largest artificial island in the world, which was reclaimed from the sea in 1968. The OVP is on a polder that lies below sea level, and most of it is kept dry through continuous pumping. It was initially earmarked for the development of an oil refinery, but plans changed after the 1970s oil crisis, and it was abandoned. The site is on a flyway for migratory birds, which colonized the site and began to make use of its unmanaged habitats. This new ecology came to the attention of Dutch bird watchers, including paleoecologist and civil servant Frans Vera. Vera was especially interested in greylag geese and the ways in which their grazing seemed to keep the OVP open, preventing the growth of trees and thus the expected ecological succession toward a closed-canopy forest.[4]

Vera and his colleagues at Staatsbosbeheer (the Dutch government organization responsible for forestry and the management of nature reserves) helped secure the designation of the OVP as a nature reserve.

In 1983, they introduced thirty-two Heck cattle into the reserve, which were joined by eighteen Konik ponies in 1984, and fifty-two red deer in 1992 (Figure 1). These large herbivores were encouraged to de-domesticate themselves, breeding and finding new behaviors outside of regimes of human care and management. Their populations initially grew, before stabilizing and fluctuating depending on food availability. By 2013, there were about 3,000 deer, 1,000 horses, and 350 cattle.[5] The herbivores were deployed as naturalistic grazers that would simulate the ecological activities of their prehistoric ancestors: Heck cattle serve as a surrogate for the now-extinct aurochs *(Bos primigenius)*, while the Konik ponies stand in for the tarpan *(Equus ferus ferus)*.

The OVP was promoted as an experiment to test a radical hypothesis, developed by Vera, that the climax community of the paleoecology of Europe was not characterized by a closed-canopy forest, as is commonly believed.[6] Inspired by his observations of the geese as well as by visits to African savanna, he proposed a past Europe that resembled a shifting mosaic of forest–pasture landscapes that were kept open by

*Figure 1.* The large herbivores at the Oostvaardersplassen, taken in the early 2000s. Source: GerardM, Wikimedia commons.

grazing and browsing. Much European ecology and nature conservation is founded on the idea that low-intensity, premodern agriculture opened up the primeval forest, diversifying the landscape and creating biodiversity. An elaborate policy architecture currently aims to maintain or simulate such agricultural practices, stalling or reversing processes of agricultural intensification. Vera's work challenged this orthodoxy, suggesting that it was the herbivores, not the farmers, that opened the forest. He claimed that biodiversity survived in spite of, rather than as a result of, farming. He hoped that the de-domesticated herbivores at the OVP would act as ecosystem engineers, restoring the past grazing regimes that generated diversity and created functioning and resilient ecologies.

The Heck cattle that Vera selected had been back-bred in Germany in the 1930s by Heinz and Lutz Heck, two brothers who ran Munich and Berlin zoos, in parallel attempts to re-create the aurochs. Lutz worked closely with the National Socialist elites. His primary patron, Hermann Göring, planned to reintroduce his back-bred aurochs for hunting in the occupied territories of Eastern Europe. Lutz also promoted the animals to Konrad Meyer—who was responsible for Nazi spatial planning—as tools to restore what he perceived to be the degraded cultural ecologies of the European steppe. Most of the Hecks' original animals were killed in World War II. However, some survived, dotted across various nature reserves. By the 1980s, their provenance had largely been forgotten. However, their descendants had developed a reputation for hardiness, and it was this that drew Vera's attention.

Since the mid-1990s, Dutch and German rewilding enthusiasts have embarked on new aurochs back-breeding projects. They are dissatisfied with the small size and heterogeneity of Heck cattle (and are somewhat queasy about their dark past).[7] For example, those involved with the Tauros Programme (formerly known as the TaurOs Project) are using data and tools provided by the sequencing of the aurochs genome, alongside the genomes of other cattle breeds. This genetic information is combined with the expertise of animal breeders, archaeologists, an artist, and ecologists to select the cattle that will be combined to create a bigger, more sexually dimorphic, but nonetheless dog- and people-friendly "Aurochs 2.0."[8] These so-called Tauros cattle are promoted as a keystone species for rewilding (Figure 2). The parallel, but less advanced,

*Figure 2.* A sketch of Tauros cattle as a keystone species. Illustration by Jeroen Helmer; reprinted with permission.

Uruz Project proposes the de-extinction of the aurochs through genome editing and in vitro reincarnation.

Back at the OVP, Vera and colleagues presented their experiment as both a restorative form of rewilding and as a model of "new naturing" or "nature development" *(Natuurontwikkelings).*[9] The OVP was to be one node in a wider "ecological main structure" *(Ecologische hoofdstructuur)*: an ambitious national spatial plan to link together the country's nature reserves and to connect them to neighboring sites in Germany. This would involve the acquisition of corridors and the construction of so-called ecoducts: green bridges to allow safe passage across major roads. They hoped that in the future the OVP would be linked to the neighboring Veluwe forest to create an expansive Oostvaarderswold into which the expanding populations of large herbivores might disperse.

As a showcase for Vera's thinking, the OVP has helped inspire a wider enthusiasm for rewilding. The last two decades have seen a growing number of similar rewilding projects across Europe, using large herbivores as tools for naturalistic grazing. Reintroduced cattle, horses, and deer have been joined by bison, moose, and beaver alongside carnivores, including wolves, lynx, and brown bears. This growth has been facilitated by Rewilding Europe, a Dutch NGO with ambitious plans to rewild one million hectares of land in Europe.[10]

Rewilding Europe has sought to shift the legal and regulatory frameworks for European conservation by lobbying to change the agroenvironmental subsidy regime of the European Union's Common Agricultural Policy. Specifically, they want to shift the focus from the preservation of rare species toward ecological processes and the creation of functional landscapes. They also aim to remove the obligations on landowners to keep habitats in a condition amenable to swift reversion to agricultural production. Rewilding Europe has seized the political and ecological opportunities presented by an expanding Europe and a growing sense that subsidized agriculture is no longer affordable or desirable in many areas where agriculture is marginal. The globalization and intensification of agriculture and an aging population have led to the cessation of farming and to land abandonment in marginal parts of Europe. Rewilders have sought to encourage the wildlife comeback

*Figure 3.* Rewilding Europe map. Illustration by Jeroen Helmer; reprinted with permission.

enabled by these shifts and by decreases in hunting.[11] To bring about the vision shown in Figure 3, they translocate animals, including the back-bred Tauros cattle, and promote ecotourism as a means to revitalize rural economies.

Many of the rewilding projects promoted by Rewilding Europe are taking place in existing public nature reserves and national parks. They are funded by state lotteries, the European Commission Nature–LIFE program, and the World Wildlife Fund. But they have also bene-fited from the growing interest in rewilding among a network of private landowners and philanthropists. A collection of wealthy industrialists—including Paul Lister (furniture), Anders Povlsen (clothing), Lisbet Raus-ing (packaging), Doug and Kristine Tompkins (clothing), and Hansjörg Wyss (medical technology)—have all purchased land, provided fund-ing, and helped lobby in support of rewilding.[12] Povlsen has become the largest landowner in Scotland, and he and Lister own large estates where they are conducting projects for reforestation and for predator conservation and reintroduction. In 2018, Rausing invested $30 million to support eight landscape-scale rewilding projects across Europe.[13] In the past twenty-five years, the Tompkinses have spent over $170 million on rewilding, purchasing large tracts of land in Argentina and Chile.[14]

The OVP has inspired the most developed example of rewilding on private land in the United Kingdom: the Knepp castle estate in Sussex, which comprises a Regency castle, a Repton-designed deer park, and a 1,415-hectare former intensive dairy farm. Knepp is owned and managed by Sir Charlie Burrell and his wife, Isabella Tree. In the midst of the 1990s dairy crisis in Europe, they began to take their land out of production and transition toward a naturalistic grazing project that uses hardy local breeds of cattle, horses, and pigs, alongside several species of deer. The estate is run as an ecotourism venture that offers safaris, glamping, and high-end corporate hospitality. It works as a rewilding–agriculture hybrid: the cattle, pigs, and deer are slaughtered and their meat sold at a pre-mium. The primary aim is the cost-effective restoration of natural pro-cesses like grazing and decomposition to create mosaic forest–pasture landscapes. They have also changed the drainage system to rewet the land to generate wetland habitats, slow and store water, and mitigate the risks of downstream flooding. The project is advised by Frans Vera, and Charlie Burrell is the chairman of Rewilding Britain, an influential NGO advocating rewilding.[15] Knepp has become a U.K. rewilding showcase, akin to the OVP in the Netherlands.

In addition to these large herbivores, much attention in Britain has focused on the beaver. Beavers were once native but went extinct in the wild in the seventeenth century.[16] Experience in Europe (especially in Bavaria) and in North America has demonstrated the role of beavers as ecosystem engineers; they are keystone species capable of reorganizing drainage regimes at catchment scale through their dam-building activities.[17] Beavers are now promoted for their ability to slow the flow of water through a catchment, thereby preventing flooding and drought, and improving water quality (Figure 4).[18] Beavers have been reintroduced to Britain from Bavaria for many years, initially by private landowners into small zoos and wildlife parks. These beavers bred well and proved hard to enclose. They gnawed through fencing and escaped during flooding events. These self-willed returnees have been assisted by the so-called beaver bomber, a citizen enthusiast who has been proactively

*Figure 4.* Beaver dam at Bamff, Scotland. Photograph by author.

introducing beavers into ecologically suitable locations, sometimes with the tacit support of local landowners.[19]

The appearance of beavers in new locations forced the devolved parts of the English and Scottish governments to commission official trials and then to regulate in favor of their residence. Beavers had become so politically palatable by 2018 that Michael Gove, then the secretary of state for the environment, presided over an official reintroduction in the Forest of Dean (Figure 5). Beavers are promoted as cheap, efficient, hardworking managers that deliver protection to the upland parts of river catchments in parts of Britain that have been blighted by flooding. Bavarian experts have introduced U.K. beaver enthusiasts to a range of technologies to protect trees and other infrastructure. These include fencing, tree guards, and "beaver deceivers," which silently lower the level of water behind a beaver dam so that the beaver does not notice (Figure 6).[20]

Finally, agricultural cessation and land abandonment, coupled with diminished hunting and active releases by animal activists, have facilitated self-willed processes of rewilding, including reforestation and the growth and dispersal of populations of keystone species. For example,

*Figure 5.* Derek Gow reintroducing a beaver into the Forest of Dean in England, in the company of the secretary of state for the environment, Michael Gove. Photograph by David Broadbent. Courtesy of Forestry England.

a

b

*Figure 6. (a)* A beaver deceiver at Bamff. Image shows the protected upstream entry point. *(b)* The fenced-in outflow. Photographs by author.

wolves are now present across the majority of European countries.[21] There is even a breeding pair in the Netherlands—the continent's most densely populated (sizable) nation—although they have not taken up residence at the OVP. Meanwhile, wild boar populations have grown and proliferated across the landscape, with boars even taking up residence in European cities like Berlin and Barcelona.[22] This unplanned rewilding is creating a series of political and biosecurity challenges. Wolves threaten livestock and are still subjects of popular fear.[23] Meanwhile, wild boar eat crops and attack dogs. Boars and their ticks have become a vector for the African swine fever virus. This disease comes from warthogs and has become endemic in Russia and post-Soviet states. It causes high mortality and has no treatment. Mobile infected boars threaten domestic pig farming and are subject to draconian control regimes in countries like Denmark.[24] Comparable problems are associated with the growth of deer populations and an associated rise in crop damage and cases of Lyme disease, which is transmitted by deer ticks, though the latter association is contested.[25]

These European developments intersect with and depart from enthusiasm for rewilding elsewhere in the world. For example, there is a rich parallel history of rewilding in North America, where the initial focus was on the three Cs: cores, corridors, and carnivores. Carnivores figure more prominently than grazers in America, the most famous example being the gray wolf *(Canis lupus)*. Wolves were eradicated from most of North America by the 1920s, populations of herbivores like elk *(Cervus canadensis)* boomed, and significant changes in vegetation were noted. Wolves were reintroduced to Yellowstone National Park in 1995, and as their populations grew, those of elk reduced by over 50 percent, leading to an increase in new-growth aspen and willow on the valley bottoms and in populations of beaver and bison.[26] Biologists suggest that these shifts relate in part to direct predation but more significantly to changes in elk behavior. Wolves create a "landscape of fear" that causes elk to change their grazing patterns; they no longer venture into deep thickets and open areas out of fear of being attacked.[27] This understanding of animals' affective atmospheres in the presence and absence of predation has come to shape rewilding interventions worldwide.

The continental ambitions of Rewilding Europe are matched by proposals for rewilding North America, which include introducing keystone species from outside the continent in cases where they have gone extinct and a surrogate is available.[28] This Pleistocene rewilding looks primarily to Africa and imagines importing species like the African elephant and the cheetah to perform grazing, predation, and other functions carried out by long-extinct ecological equivalents.[29] Such acts of ecological surrogacy have informed the translocation of tortoises for seed germination and dispersal in the rewilding of island ecologies in Mauritius and the Galápagos Islands, and the use of feral horses in parts of Central America.[30] Perhaps the most ambitious proposal for rewilding with surrogates is located in Siberia, where Sergei and Nikita Zimov run Pleistocene Park, an experiment in rewilding tundra landscapes.[31] They propose the de-extinction of the mammoth as a keystone herbivore capable of restoring grazing regimes that would prevent the anticipated melting of permafrost and the subsequent release of their frozen stored carbon. The mammoth is promoted as "the proper tool to fight global warming."[32]

## From Antibiotic Relations to Blowback to Probiotic Relations in Rewilding

Rewilding advocates like Frans Vera, Wouter Helmer (Rewilding Europe), and Ronald Goderie (Tauros Programme) present rewilding as a critical inflection point, one that is occurring at a moment of crisis that is engendered by the pathologies of modern, antibiotic approaches to environmental management. They tell revisionist histories that challenge the celebration of the rise of Western agriculture. For example, one narrative emphasizes how rewilding departs from the human–cattle–environment relations associated with modern agriculture in Europe. It begins with the long coevolution of aurochs, hominids, and their landscapes that moved in synchrony with the advance and retreat of the ice sheets. It traces the arrival of domestic cattle in Western Europe in about 6,000 BCE with the spread of Neolithic (sedentary, pastoral, agricultural) relations. It charts the historic decline of the aurochs and its extinction in 1627 resulting from hunting, competition with farming, and exposure to the diseases of domestic cattle. The narrative follows the rise of cattle breeds and the deleterious effects of the development of modern beef and dairy industries.[33]

The subsequent diagnosis of antibiotic blowback is twofold. It focuses on the effects of intensive livestock systems on biodiversity, water quality, and climate, as well as the poor animal welfare associated with productivist breeding, veterinary care, and animal enclosure. We learn of dysfunctional landscapes bereft of a myriad of species and the loss of ecological abundance. We hear of polluted rivers and disastrous emissions. We are introduced to suffering industrialized organisms that reach their apotheosis in the Holstein-Friesian turbo cow.[34] Rewilders take issue with the deficiencies of traditional conservation that seeks to subsidize low-intensity forms of such agriculture. They present rewilding as redemption—a salvatory, probiotic response. We are told that rewilding generates a modern simulation of prehistorical ecological relations in which happy cattle flourish while working to generate biodiversity, ecosystem services, tourism revenue, and (in some cases) organic meat.

Such promises are contested and controversial, as I explore in the chapters that follow. For example, some paleoecologists doubt the veracity of Vera's wood pasture hypothesis and the degree to which it is supported by the pollen record.[35] They defend the closed-canopy model. Other European conservationists are concerned about the open-ended character of rewilding experiments, and what it might mean for the rare species dependent on low-intensity agricultural landscapes.[36] Farmers ask where food will be grown in these wild futures, and they take issue with the messy, unproductive character of rewilded landscapes. Some are concerned that Vera's thinking undermines the rationale for the agroenvironmental subsidies that keep much low-intensity farming viable.[37] Finally, those concerned with animal welfare contest the treatment of the large herbivores, like those at the OVP that were initially left to die of starvation in harsh winters, and suggest that these animals should be subject to the welfare protections normally afforded animals enrolled onto scientific experiments.[38]

## Biome Restoration

Biome restoration is the proactive management of the human microbiome for the purposes of restoring health. The human microbiome is the collection of organisms found in association with the human body.[39] It comprises bacteria, fungi, archaea, viruses, and a few eukaryotic animals. It is sometimes referred to as the multibiome to acknowledge those human-associated organisms, such as helminths, that are visible to the naked eye.[40] While the presence of specific microbes has long been connected to the incidence of specific diseases, biome restoration approaches the microbiome as a dynamic ecology nested within the human body, which is reconceived as a holobiont, or superorganism. It takes health as the outcome of the ongoing interactions between the microbiome and the host, in which the absence of specific microbes—and the functions they perform—can be as pathological as their over-abundance.[41] This is a more recent, less developed, but better resourced field of science compared to rewilding. It has come to the fore in the last decade as a result of the growing speed and affordability of genome-sequencing technologies, which make visible the myriad microbes that cannot be cultured by traditional laboratory methods. As microbiome science matures, it is transitioning from a traditional natural historical interest in describing what species and genes exist to exploring ecological relationships. As I explore in the next chapter, with this rise of an eco-logical approach, microbiologists increasingly borrow concepts, methods, and metaphors from their macroecology colleagues.

Biome restoration primarily involves immunologists and micro-biologists, as well as networks of expert patients who undertake their own DIY therapeutic experiments. It is concerned with a wide range of microbial organisms, but my primary focus in this book is the therapeu-tic use of helminths. Helminths are parasitic worms that live in human and other animal bodies (Figure 7). More than 350 species of helminths have been associated with humans, of which only a small number have been used for helminth therapy. In 2004, British parasite immunolo-gist David Pritchard infected himself with *Necator americanus*, a species of human hookworm.[42] The life cycle of *N. americanus* requires that eggs excreted in human feces be deposited in warm, moist soil (Figure 8). Once the larvae hatch, they must crawl through exposed skin before they can travel through the blood vessels, heart, lungs, mouth, and throat to return to the gut. Successful reproduction therefore requires a suitably warm climate and soil, frequent human defecation, and opportunities for skin contact.

*N. americanus* is one of several helminth species that used to be a common part of the human microbiome.[43] It remains so in many rural areas of the Global South, where a serious infection can be fatal.[44] *N. americanus* struggles to survive in temperate climates; it has largely

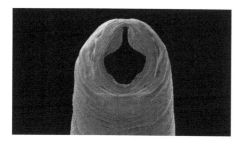

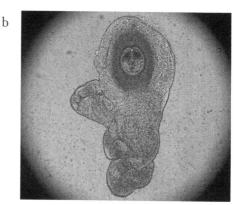

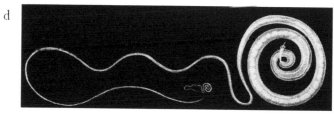

*Figure* 7. The four species of helminth used in helminth therapy: *(a)* human hookworm *(Necator americanus)* larvae; *(b)* rat tapeworm *(Hymenolepis diminuta)* cysticercoids; *(c)* pig whipworm *(Trichuris suis)* ova; *(d)* human whipworm *(Trichuris trichiura)*. Illustrations provided by John Scott and William Parker.

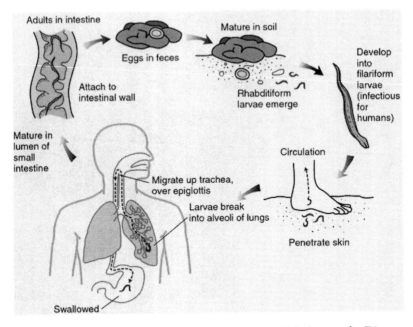

*Figure 8.* The life cycle of *Necator americanus*. Source: U.S. Centers for Disease Control and Prevention.

been eradicated from parts of the Global North thanks to twentieth-century public health campaigns and general improvements in sanitation. Footwear and flush toilets prevent reinfection. Pritchard, working in Papua New Guinea in the late 1980s, became aware of epidemiological evidence that linked patterns in the rise of allergic and autoimmune diseases—like hay fever and asthma—to the absence of *N. americanus*. He wanted to conduct clinical research to see if returning hookworms could ameliorate these conditions. He sourced hookworms from his field site and infected himself to demonstrate that it was safe to take the worms. Pritchard and fellow immunologists in the United States, the United Kingdom, and the Netherlands have since conducted clinical trials testing the therapeutic potential of *N. americanus* and a few other helminths for diseases including asthma, inflammatory bowel disease, multiple sclerosis, rheumatoid arthritis, and autism.[45]

This work is supported by extensive laboratory experiments using mouse models. These use strains of mice that have been selectively bred to express disease conditions that simulate human equivalents. Mice are kept in tightly managed ecological conditions that might be sterile or gnotobiotic, such that the mice are born and develop without a microbiome. But a growing awareness of the ways in which the microbiome shapes organism development and health has driven a shift to wild mouse models.[46] Self-described wild immunologists have been dirtying up the ecological conditions of their laboratories and have started using different strains of mice sourced from the wild; from feral, urban settings; or from domestic pet shop strains.[47] These clean or dirty mice are then challenged with murine helminths that serve as surrogates for those commonly found in, or recently lost from, human bodies.

In framing the rationale for their research, several of these immunologists advance a model of evolutionary or Darwinian medicine.[48] Helminths are known to have parasitized hominids since the Pliocene (5.3–2.6 million years ago).[49] Parasitologists suggest that N. americanus likely evolved into its current form as a result of a long history of antagonistic relations with small mammals, our hominid ancestors, and then modern humans. Theories of parasite evolution suggest that over time, successful parasites tend to become less aggressive and deadly, as those that kill or significantly harm their host run the risk of not completing their life cycle.[50] Shifting away from common martial metaphors, these ecoimmunologists suggest that during their long coevolutionary history, N. americanus has learned to train, modulate, or calibrate the human immune system to achieve host tolerance.[51] They propose that hookworms can communicate with the commensal bacteria that are in contact with the human gut. This enables them to disguise their presence and suppress the host's normal immune response.[52] As a result, many contemporary infections with N. americanus are unknown and asymptomatic; many people can tolerate a moderate worm burden without obvious consequences.

William Parker, a surgeon at Duke University and a prominent advocate of the therapeutic use of helminths, presents some worms as keystone species, akin to the wolf or the beaver, arguing that they exert a disproportionate ability to engineer our internal ecologies relative to their abundance in the body.[53] Graham Rook, a British immunologist, presents N. americanus and a small subset of other microbes as beneficial (or at least mutualistic) old friends, a term he differentiates from crowd infections like cholera, typhoid, and malaria.[54] He explains that the crowd infections emerged later in human evolution as part of the novel microbial ecologies generated by the agricultural revolution; these reached unprecedented infection intensities with the rise of large cities and global transportation networks. Rook, Parker, and others suggest that in the absence of helminths and other old friends, the microbiome of the gut goes awry. The immune system becomes poorly calibrated, generating a generalized state of dysbiosis that triggers a range of inflammatory, allergic, and autoimmune diseases.[55] Set in the context of its long history of incorporation into the human body, *N. americanus* come to figure as a ghost. It is a former keystone species whose demise drives self-reinforcing conditions of autoimmunity and inflammation.

In 1989, David Strachan linked the rise in inflammatory disease in the West to the excesses of modern hygiene and the deleterious ecological and immunological consequences of antibiotic lifestyles.[56] His so-called hygiene hypothesis has since been nuanced by immunologists into the biome depletion or biodiversity hypotheses.[57] These hypotheses, and the news of trials with helminths, generated popular interest among patients with inflammatory, allergic, and autoimmune diseases. Many early adopters were dissatisfied by the state-of-the-art treatments made available to them by their doctors. The causes of inflammatory diseases remain poorly understood, and therapies rarely do more than deal with symptoms. Such patients found themselves chronically dependent on immunosuppressant drugs with declining efficacy and debilitating adverse effects. Although clinical trials with helminths have proved inconclusive, a growing number of often desperate people have begun to obtain their own worms and to conduct DIY experiments to explore the worms' therapeutic potential to treat a range of conditions. As with the case of AIDS and other autoimmune diseases in the 1980s, these users are willing to act outside and ahead of the official regulatory apparatus for drug development and delivery; some have developed close personal and professional relationships with the clinicians who are pioneering new work and conducting mouse model research.[58]

One early adopter was Jasper Lawrence, a British entrepreneur with asthma. After learning of Pritchard's trial, Lawrence traveled to Cameroon to walk in latrines in the hope of picking up *N. americanus*. He was infected by the wrong worm, but after a further visit to Central

America, he was able to cultivate his own colony of *N. americanus*. As his asthma improved, he learned to isolate, incubate, and count *N. americanus* larvae so that he could reinfect himself. In 2007, he started a company, Autoimmune Therapies, that sells worms and provides support to patients willing to travel to his clinic, which was then located in Northern Mexico. Helminth therapy has since grown in popularity. Upward of seven thousand people now purchase *N. americanus* and three other species from at least four commercial providers.[59] They give advice and support, and share their positive experiences through a wiki and in online forums.[60]

Much of this online activity is coordinated by John Scott, a retired British headmaster with a range of autoimmune diseases. Scott participated in Pritchard's first clinical trial and was one of Lawrence's first customers. Prominent figures in this community of users echo the discourse of their scientific forebears and present helminth therapy as a form of biome restoration,[61] in which helminths are enrolled as "gut buddies" or "colon comrades"[62] that work to train and exercise the immune system. But DIY helminth therapy is controversial. It is illegal to bring live *N. americanus* into the United States, and in the absence of clinical data supporting its efficacy, advocates are not permitted to advertise any health benefits. Scientists and doctors in Europe and North America have expressed doubts about whether the therapy works, and they are concerned about the risks to users of an unmanaged infection and to the public of a potential accidental environmental reintroduction.[63] Much correspondence and exchange in the online hookworm underground is anonymous.[64] Many users deploy e-mail encryption and cryptocurrencies, and must often travel to sympathetic jurisdictions to secure reinfection.

Those using helminth therapy look to a future in which worms are legal. They continue to expand the range of helminth species used for therapy, and they speculate about the possibilities of further domesticating their worms, either through selective breeding or by deliberate genetic engineering.[65] Mainstream clinical immunological research is skeptical of this model of "bugs as drugs," but it anticipates a new range of immunosuppressant molecules that might be revealed in work with mouse models and from the recent decoding of the hookworm genome.[66] Pritchard and his colleagues hope to synthesize (and patent) these as "drugs from bugs" for delivery in pill form. They have been able to obtain some interest from the pharmaceutical industry. Critics of this molecular model, like Parker, suggest it will be impossible to chemically recapitulate the ecological work done by a whole animal. There is a conspiratorial tenor to the discussions of these developments in the helminth therapy support groups. In the wake of the failure of a well-publicized, badly designed phase 2 clinical trial assessing infection with *Trichuris suis* ova,[67] some protagonists suspect "Big Pharma" of sabotaging the future of live-organism therapeutics to shore up established investments in lucrative immunosuppressant drugs.

Urbanization, improved sanitation, and the distribution of anthelminthic drugs continue to decrease *N. americanus* infection rates worldwide, but global *N. americanus* control remains a distant prospect. *N. americanus* is endemic in many rural areas of the Global South as well as in some poor parts of the United States (Figure 9).[68] Hookworms remain a cause of debilitating disease. Antihelminthic programs have been reinvigorated under the banner of Global Health, with new funding from the Gates Foundation. Such programs are directed toward

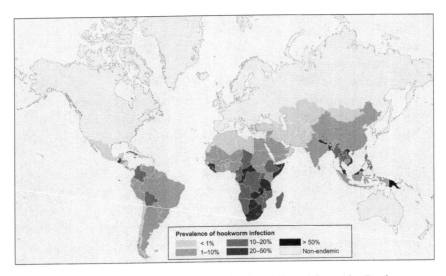

*Figure 9.* Map of global hookworm distribution. Adapted from Alex Loukas, Peter J. Hotez, David Diemert et al., "Hookworm Infection," *Nature Reviews Disease Primers* 2 (2016): 16088.

school-based deworming using out-of-patent or freely provided deworming drugs. There is hope that the decoding of the hookworm genome might also inform new rounds of anthelminthic drug innovation, and significant investments have also been made into the development of hookworm vaccines. These interventions aim to prime the human immune system to attack hookworm larvae and make the gut a less hospitable location. Researchers think it unlikely that such interventions will be able to engender sterilizing immunity and achieve permanent exile of hookworms, but hope remains that infection intensities can be downmodulated without requiring expensive—and politically and ecologically challenging—sanitary interventions.

The rise of helminth therapy must be understood in the context of the wider scientific, clinical, and popular interest in the microbiome and biome restoration. The decreased cost and increased speed of genetic sequencing have enabled scientists to identify the roles played by microbes in a wide range of bodily systems, including metabolism and cognition. Beneficial microbes are the subject of an ever-expanding range of laboratory and clinical research. The best-established translation of this research involves fecal microbiota transplantation (FMT)—that is, the replacement of dysbiotic gut bacteria via donated stool. The emergence of FMT mirrors that of helminth therapy. It was pioneered by a small number of clinical researchers before catching on among a now-vibrant DIY community that has pioneered FMT techniques outside (and well ahead) of the traditional research pipeline. Biobanks were founded to make healthy stool available; the breakthrough moment came with the successful completion of a clinical trial to treat long-term infections with antibiotic-resistant *Clostridium difficile*.[69] Regulatory authorities in Europe and North America licensed its use, thus catalyzing wider trials to test the efficacy of FMT for other gut-related conditions.

The case of FMT has reinvigorated established interest in biotherapeutics and provoked a renewed excitement among clinicians and the pharmaceutical industry in microbial therapeutics. The consumption of probiotic and prebiotic health supplements continues to grow; meanwhile, commentators predict that a new wave of probiotics will emerge from microbiome research.[70] These move from the administration of single strains of microorganisms, as with drinks like Yakult, or the use of a cocktail of bacteria, as is commonly found in probiotic pills. Instead, the microbial therapeutics company Seres Therapeutics (and its competitors) hope to develop ecobiotics—that is, functional microbial ecologies reverse engineered from data gathered by the sequencing of healthy gut ecologies and assembled from lab strains to be delivered in pill form. In order to secure the investment required to fund the necessary trials of these bugs as drugs, Seres Therapeutics and other companies that work with microbial therapeutics have requested that the U.S. Food and Drug Administration grant new investigator drug status to FMT. This would give the company that successfully tested an FMT surrogate at a phase 2 trial a monopoly on its production. It would also require providers and recipients of live FMT to cease operations—a course of action resisted by many clinicians and DIY users on the grounds that it will greatly enhance the cost of what is now a nearly free therapy.[71]

Although the efficacy of many of these products is contested or remains to be proven, these medical developments—and the wider microbiome zeitgeist—have been accompanied by the rise of a wide array of guides and gurus promoting personalized microbiome sequencing and diagnostics, as well as new modes of microbiome management. The microbiome has been picked up by the paleo and wild fermentation movements, for example.[72] It has inspired several high-profile microbiologists to write popular science books reappraising modern approaches to health and hygiene, and proposing elaborate means for recalibrating exposure to microbes. Attention has focused in particular on diet and exercise, avoidance of antibiotics, vaginal birth, breastfeeding, and childhood exposures to soil and domestic animals. These works tend to extol the virtues of dirt and the benefits—especially for children—of recalibrated forms of dirty play.[73]

In many cases, these guides offer a holistic disavowal of the pathologies of modern urban living, linking immunological issues with psychological (and even spiritual) problems associated with nature-deficit disorder.[74] For example, in his best-selling book on rewilding, George Monbiot suggests that risky encounters in remote rewilded places and amid wild people will help tackle the ecological boredom of modern urban life.[75] Similarly, microbiologist Jeff Leach locates his personal microbiome rewilding project among Hadza hunter-gatherers in Tanzania. After undergoing FMT from a Hadza man of his own age, he advocates "primitive" dietary, hygiene, and cultural practices as the best means for recalibrating dysfunctional Western bodies and lifestyles.[76] Monbiot and Leach offer popular hot takes on an emerging holistic understanding of the drivers of microbial dysbiosis and on the links between human and environmental health. This science is starting to inform ambitious research and planning initiatives that aim to map, monitor, and eventually manage the microbial and macrobial ecologies of the built environment. These focus in particular on recalibrating hygiene practices (broadly conceived) that seek to develop healthy and functional microbial urban environments.[77]

### From Antibiotic Relations to Blowback to Probiotic Relations in Biome Restoration

In a similar fashion to rewilding, advocates for biome restoration present their work as a radical departure from modern approaches to health and hygiene. The narratives offered by Graham Rook and William Parker take us through the evolutionary history of human–helminth–microbiome relations to make sense of the present. We learn that humans as prehistoric hunter-gatherers coevolved with many species of helminths and bacteria, with evolution favoring the emergence of mutualistic or old-friend relations. Infections with species like hookworm increased in sedentary agricultural situations, especially those with poor sanitation, where shit is used as fertilizer. Hookworms and other bacterial crowd infections become especially pathological in the intensities of colonial plantations and slum ecologies in which weakened human bodies are regularly reinfected with new worms or bacteria. The near eradication of hookworms and other old friends from North America and Europe by the middle of the twentieth century is taken to exemplify the excesses of antibiotic approaches to public health. Missing microbes lead to blowback, manifesting in "epidemics of absence" of autoimmune and inflammatory disease, and the rise of drug-resistant bacteria.[78] As I explore in more detail in chapters 5 and 6, this critique implicates a wide array of health, hygiene, and lifestyle practices that are central to twentieth-century ideals of what it means to be modern and developed. As with rewilding cattle, the careful use of *N. americanus*, healthy stool, and other keystone species is proposed as part of a holistic program for probiotic restoration that would recalibrate human–microbe relations and ameliorate the dysbiotic ecologies generated by antibiotic relations.

## A Wider Probiotic Turn

In the chapters that follow, I draw out the common characteristics of rewilding and biome restoration, focusing in part on their scientific foundations, modes of managing life, and associated knowledge practices. I explore their geographies, temporalities, and ways of valuing the work done by animals, noting both commonalities and internal heterogeneity within these ways of going probiotic. However, the significance of my argument rests on the premise that we can take rewilding and biome restoration as indicative of a wider probiotic turn, which I think is underway in parts of the WEIRD world. I will spend the rest of this chapter legitimating this claim by situating rewilding and biome restoration among a range of examples of going probiotic in other knowledge and policy domains. I briefly touch on some of these examples in the chapters that follow; however, they offer ample scope for future research.

We can first position rewilding and biome restoration alongside longer-established practices in agriculture that seek to avoid or mitigate the antibiotic excesses of modern intensive approaches to food production. Probiotic alternatives involve the conservation or introduction of beneficial organisms that deliver ecological services necessary for growing crops and raising animals. Examples include practices associated with permaculture and organic farming, like biological pest management.[79] This approach uses the diseases and predators of agricultural pests in place of chemical pesticides or culling to control their populations. Examples include slug-killing nematodes and bird-scaring hawks. Further, reinvented traditional practices of composting and recycling of animal manure work in place of synthetic fertilizers and provide prebiotic nutrients that feed desired soil microbes.[80] Probiotic interventions also characterize the recent interest in soil care promoted as a solution to the crisis facing soil in industrial farming systems.[81] Probiotic microbes and prebiotic diets are also used to boost farm animal productivity by shifting gut ecologies to improve digestion and metabolism, thereby restoring dysbiotic ecologies after or alongside the use of antibiotics as growth promotors.[82]

Probiotic practices are also on the rise in Western food processing, where there has been a rediscovery and reinvention of fermentation practices in the production of beer, cheese, and other foodstuffs, expanding the rather narrow roles afforded bacteria in modern food production.

Writers have explored the novel social and human–microbial relations associated with the rise of raw-milk cheese, kombucha, miso, and kimchi production.[83] These novel fermentation practices have come to the fore as part of the local and naturalistic turn in high-end Western food cultures, best exemplified by the rise of the New Nordic food movement in Scandinavia.[84] In less salubrious circumstances, such probiotic approaches are well established in waste management, where beneficial microbes are deployed to break down unwanted food and sewage, with the aim of reducing landfill and incineration. This process is known as biodegradation. Beneficial microbes have also been developed for cleaning industrial pollution, a practice known as bioremediation, for example to tackle oil spills.[85] There is a growing interest among domestic and personal hygiene companies in developing probiotic cleaning and beauty products.[86] The idea here is to use bacteria capable of breaking down detergent-resistant biofilms, thus reducing the use of toxic chemicals, and seeding skin and surfaces with a mix of species that will grow to prevent colonization by pathogens.

Rewilding and biome restoration can be understood alongside the rise of new approaches to managing the dynamics of abiotic processes. For example, the Dutch approach to nature development that shaped the OVP experiment also informs new approaches to water management.[87] Modern techniques that rationalize and accelerate the movement of water in river catchments are giving way in some places to more naturalistic techniques. These aim to slow the flow by removing upstream drainage, making river patterns more complex, and allowing water back onto a floodplain.[88] The aim of this rewetting is to maintain water on the land to attenuate downstream flood surges, or in drier areas to hold up water to prevent drought. These projects also aim to enhance biodiversity in and around the river, thus improving water quality.[89] This work is sometimes enabled by the reintroduction of beavers.[90] Such thinking extends to the coast. It informs practices of managed retreat through the removal or abandonment of hard flood defenses and the development of more naturalistic models for moving sediment. One example is the Dutch Sand Motor (*Zandmotor*) project, in which sand is not dumped on the shore but rather fed into dynamic water systems offshore to modulate coastal flows.[91]

These naturalistic models of water management are mirrored in new approaches to fire management in temperate forestry. Modern forestry techniques often sought to prevent fire by suppressing fire dynamics. New approaches, developed, for example, by the U.S. National Park Service, involve regimes of controlled burning that aim to simulate prehistoric cycles.[92] Such approaches to forest, river, and coastal management are examples of what is known as ecosystem-based disaster risk reduction, which has been defined as "the sustainable management, conservation and restoration of ecosystems to provide services that reduce disaster risk by mitigating hazards and by increasing livelihood resilience."[93] This approach aims to move from hard or structural engineering solutions to working with natural processes. Such interventions often use animals and vegetation as a means to prevent or mitigate possible disasters. Examples include planting forests to secure volcanic ash, sand, or mobile soil, or to redirect or slow rockfalls and avalanches.[94] They extend to the creation of oystertecture—that is, littoral oyster beds that protect against storm surges.[95]

These local- and regional-scale interventions can be considered alongside nascent proposals for planetary-scale geoengineering, in which biological and geological agents are deployed to modify atmospheric composition and to modulate earth system dynamics. We have already encountered such claims in the proposal by the Zimovs to de-extinct the mammoth and to return it as a keystone species for grazing and browsing the Siberian tundra. The claim is that the restoration of mammoths (and other large herbivores) would help prevent the anticipated melting of permafrost, the release of frozen methane, and the subsequent runaway climate change.[96] This represents a high-profile example of an emerging suite of nature-based climate solutions that take issue with the hard engineering solutions proposed for geoengineering and instead reframe some of the elements of rewilding and ecological restoration as tools for keeping the earth system within planetary boundaries.[97] Less naturalistic examples of such probiotic geoengineering include proposals to seed oceans with bacteria to accelerate processes of carbon sequestration.[98]

## Probiotic Planet

Figure 10 provides a summary of the different examples of going probiotic across a range of domains of knowledge and practice, from the micro scale to the planetary scale. In this book, I wager that we can take these diffuse developments as a coherent and important shift in how life is being conceived and managed in parts of the WEIRD world, and that we can take rewilding and biome restoration as indicative of these wider trends.

In making this wager, I have looked for similarities across these domains rather than seeking examples in which antibiotic modes of managing life remain ascendant. I acknowledge that there is a risk of both overselling my claim and of losing sight of important differences within and between the case studies I present. I conclude this chapter with some important caveats that begin to identify the themes that organize the critical analysis I perform in the second half of the book. The probiotic turn is marginal in its political and ecological reach. Although it is gathering momentum, it is still far from the mainstream in Western health and environmental policy, and it is absent from many parts of Global South. The probiotic turn has also been asynchronous in its origination and development across different policy domains. It begins with food and agriculture; moves into conservation and environmental management; and becomes established in discussions of health and hygiene. The causes of this asynchrony are beyond the scope of this chapter, but they no doubt relate in part to the progressive miniaturization (from macroecology to microecology) of technologies for detecting and explaining ecological change.

The probiotic turn also has a stark geography that maps onto a range of current and historic patterns of inequality. The ability of any one individual or landowner to go probiotic is premised on the existence of a range of technological, political, and economic factors. There are many people in the world who have yet to see the benefits of antibiotic approaches to managing life. Many experience the blowback of antibiotic approaches without having the means to redress or restore dysbiosis. In some cases, the ability to go probiotic—for example by taking fertile land out of production—is premised on subjecting other humans and their ecologies to forms of antibiotic management. We need to consider whether the probiotic turn might be a zero-sum game; we need to examine the different political ecological models through which the probiotic turn is being enabled.

There are also important differences between probiotic thinking and practice inside and outside the human body. Although there are clear

|  |  | *Antibiotic model* | *Blowback* | *Probiotic alternative* |
|---|---|---|---|---|
| *Micro* | Health | Antibiotic drugs, caesarean sections, low rates of breast feeding | Loss of microbial diversity. Dysbiosis and epidemics of absence: rise in allergic, inflammatory, and autoimmune disease. Antibiotic resistance and new pathogens. | Probiotic and prebiotic supplements, new birthing practices, helminth therapy, fecal microbiota transplant. BIOME RESTORATION. |
|  | Hygiene | Water purification, urbanization, antimicrobials, limited contact with animals, smaller families, clean living |  | Probiotic cleaning products, the promotion of dirt, and contact with domestic animals |
|  | Diet | Pasteurization, ultra-processed food, demise of live food |  | Raw, live, fermented, and paleo diets and forms of food processing |
| *Macro* | Agriculture and forestry | Use of pesticides and herbicides, artificial fertilizers, antibiotics in livestock, intensive management systems | Biodiversity loss, invasive species, zoonotic and other diseases, loss of soil fertility, loss of crop diversity, loss of ecosystem services, pesticide resistance, inability to adapt to climate change | Biological pest control, organic and permaculture systems, bioremediation, animal probiotics |

| | | | |
|---|---|---|---|
| Conservation | Preservation of low-intensity agricultural systems in small nature reserves | Habitat fragmentation, loss of adaptive capacity, trophic downgrading, extinction | REWILDING, back-breeding, de-domestication. De-extinction. |
| Environmental management | Rationalization of environmental processes and disturbance regimes, e.g., in rivers, coasts, fire, pest control | Loss of resilience, increased risks of natural disaster. Extreme events. | Rewetting, managed retreat, naturalistic erosion, and fire management |
| Planetary management | Resource use that exceeds planetary boundaries: carbon, nitrogen, water | Climate change: global warming, extreme weather, sea level rise, positive feedbacks | Geoengineering and nature-based climate solutions: ocean seeding, forest planting, rewilding the tundra and other systems |

*Figure 10.* Table of probiotic approaches to managing life showing antibiotic antecedent and associated blowback.

commonalities in the ecological science that informs biome restoration with hookworms and rewilding with cattle, the ecology of a nature reserve is different than the ecology of the human body. These differences also relate to contrasts between the perception and governance of life in here and life out there. In simple terms, it is much more common to make space for awkward natures out there than it is for people to make bodies a home for nonhumans whose proximity might engender some degree of discomfort. Hookworms become gut buddies because they restore gut functions, but this corporeal generosity is rarely extended to increasingly rare but nonetheless lethal helminths, like the Guinea worm.[99]

Despite these caveats, I hope to demonstrate that the probiotic turn represents a coherent shift. Although its traction on practical action is modest, it is significant for its intellectual consistency across a range of normally disparate disciplines and fields of science-led management. This probiotic turn evidences the claim made by historians and social theorists of the rise of a "general ecology" across twenty-first-century responses to the Anthropocene.[100] A close examination of the probiotic turn helps specify and permit critique of the types of ecology and environmentalism that it brings into existence.

# 2

# THINKING LIKE GAIA
## The Science of the Probiotic Turn

I have lived to see state after state extirpate its wolves. I have
watched the face of many a newly wolfless mountain, and seen
the south-facing slopes wrinkle with a maze of new deer trails.
I have seen every edible bush and seedling browsed, first to
anaemic desuetude, and then to death. I have seen every edible
tree defoliated to the height of a saddlehorn. . . . I now suspect
that just as a deer herd lives in mortal fear of its wolves, so does a
mountain live in mortal fear of its deer. . . . So also with cows.
The cowman who cleans his range of wolves does not realize that
he is taking over the wolf's job of trimming the herd to fit the
range. He has not learned to think like a mountain. Hence we
have dustbowls, and rivers washing the future into the sea.

—Aldo Leopold, *Sand Country Almanac*

ALDO LEOPOLD wrote this lament in 1947, toward the end of a long pioneering career as an ecologist and founder of the science of wildlife management in North America. Leopold had grown increasingly dissatisfied
with established approaches to game and forest management premised on
varmint (predator) extermination and clean (tightly controlled) production. Retreating to a cabin on a small degraded farm in Midwest sand
country, he proposed his famous land ethic. He encouraged wildlife
managers to "think like a mountain," developing a holistic appreciation
of the interdependencies between species and the landscapes they inhabit.
He wrote lucidly about what ecologists now describe as a trophic cascade,
with top-down regulation performed by predation. He appealed for a
model of conservation that could live with species like wolves and coyotes.

In this chapter I explore the ecological science that informs the probiotic turn. I trace how the scientists involved in rewilding and biome

restoration conceive the relations and interdependencies within and between the different ecological scales at which they work. Many of these figures may never have encountered Leopold, but all, in their different ways, are informed by the Gaian thinking of James Lovelock and Lynn Margulis, two scientists who might be understood as inheritors of Leopold's commitment to holistic, ecological, and relational science. I call this chapter "Thinking Like Gaia" to capture what I see as a reworking of Leopold's exhortation to think like a mountain. I suggest that those working at the forefront of probiotic ecology expand Leopold's holistic view of dynamic natural systems from the scale of the landscape down to the scale of the microbiome and out to the scale of the planetary. The strands of Gaian thinking that are prominent in ecoimmunology and earth systems science, and the social theory that borrows from them, are centrally concerned with the interrelationships between organisms (taken as individuals, as ecologies, and as aggregate life) and the abiotic environment. For Latour, Haraway, and others, this science thinks like Gaia. Here Gaia is ticklish, intrusive, and unruly; Gaia exists in a metastable, sympoietic world with no necessary beneficent or vengeful commitment to humanity.[1]

Thinking like Gaia involves a commitment to a common ecological ontology, a scientific account of how the world works that comes to shape and ground the ontostories of probiotic social theorists like Haraway, Tsing, and Latour. This chapter maps this ontology, identifying, connecting, and synthesizing concepts that are found in common across the seemingly disparate sciences that support rewilding and biome restoration. As with the previous chapter, my narrative seeks to stay faithful to the accounts of key practitioners. I focus on the five sets of concepts that are most prominent within the scientific literature and from my interviews. I highlight those emblematic of the wider shifts in scientific thinking associated with efforts to diagnose and address the pathologies caused by antibiotic approaches to conceiving (and then managing) life.

I have taken the necessary liberty of uniting the domain specific ideas of rewilders and biome restorers under the conceptual umbrella of the postmodern synthesis, as it is presented in so-called eco-evo-devo (ecological–evolutionary–developmental) biology. Scott Gilbert defines eco-evo-devo as "the scientific programme that incorporates the rules governing the interactions between an organism's genes, development and environment into evolutionary theory."[2] Eco-evo-devo is strongly associated with the work of Margulis and her radical theory of the symbiotic origins and evolutionary dynamics of life, which influenced Lovelock and shaped their joint writings on Gaia.[3] In its attention to context, mutualisms, and planetary precarity, the postmodern synthesis departs from the individualism, reductionism, and anthropocentrism of the modern synthesis. Not all of my participants are involved in eco-evo-devo; many are not even aware of its existence. However, in the final section of this chapter, I show how it offers a useful framework for understanding the intellectual significance of the proliferation of Gaian thinking and for mapping the flows of ecological concepts between the micro and the macro that are central to the probiotic turn.

## Symbiosis

The *Oxford English Dictionary* defines symbiosis as "an association between two different organisms which live attached to each other . . . and contribute to each other's support." Symbiotic relationships tended to be treated as rare and unimportant events in the modern synthesis of neo-Darwinian biology, with its preoccupations with competitive relations between individuals. However, this marginalization has been challenged by a series of developments that happened in the different branches of biology and ecology caught up in the probiotic turn. I focus on three here. The first challenge comes from Margulis's radical theory of the symbiotic origins of eukaryotic (or multicellular) life.[4] Margulis argues that life evolved when one bacterium incorporated another, thereby domesticating it to create the mitochondria common to all eukaryotic life-forms.[5] She calls this process symbiogenesis. Subsequent work has confirmed the microbial (bacterial and viral) origins of a significant portion of the genetic diversity of animals. It has also revealed the ability of bacteria to share genetic components laterally, without reproduction.

The diminished cost and increased speed of high-throughput DNA sequencing has helped map the human microbiome—or, better, the multibiome. Most of these microorganisms are commensal or even beneficial mutualists. This work challenges the prevalent pathological associations of microbes (as germs) and implicates them in the successful development and everyday functioning of animal bodies.[6] In this process, these microbes shape animal behavior and mood in ways that are understood to help secure their own futures. The microbiome is presented as an animal organ, akin to the heart or the lungs, and the biological individual is refigured as a porous, symbiotic holobiont, a "eukaryotic organism [host] plus its persistent symbionts."[7] The past primacy of the individual is being displaced by a focus on relations of interdependency.[8]

William Parker, a prominent immunologist working with hookworms, explains, "What we figured out, and what we got ridiculed for at first, but now everybody accepts, is the idea that the immune system is actually supporting the growth of the microbiota and spends most of its time, effort and energy devoted to supporting that symbiotic growth."[9] Certain microbes, like hookworms, are now understood to play vital roles in calibrating the infant immune system as well as enabling metabolism and ensuring the development of the brain. In accounting for the developmental and salutary role of microbes, Graham Rook develops the concept of an "environment of evolutionary adaptedness"; he suggests that for humans, this is the "hunter-gatherer environment."[10] He acknowledges that the hunter-gatherer lifestyle and environment were heterogeneous, but he argues that three types of old-friend microbial organisms would have been abundant in all cases. He terms these the commensals, the pseudo-commensals (found in water and mud, but not reproducing in the human body), and the helminths. He suggests, "Over millions of years a state of Evolved Dependence might have developed, where the induction of appropriate levels of immunoregulation by the 'old friends' has become a physiological necessity. In other words, some genes involved in setting up appropriate levels of immunoregulation are located in microbial rather than mammalian genomes."[11] As Rick Maizels and Ursula Wiedermann, immunologists working with helminths, put it, "We carry the evolutionary imprint of microbes throughout the immune system."[12]

Symbiotic relations have also become a central concern in work on the theory of niche construction—that is, "an evolutionary idea that emphasizes the capacity of organisms to modify their environments and thereby act as codirectors of their own evolution and that of other species."[13] The concept of heredity beyond vertical genetic transmission has a long and contested history in evolutionary biology; indeed, it fell out of fashion with the rise of neo-Darwinian biology.[14] But in the postgenomic present, scientists now talk of an ecological inheritance that is passed down between generations of organisms.[15] This inheritance is partly embedded within the local ecological context within which an organism develops. It involves both abiotic factors (like the hydrological effects of a beaver dam) and biotic relationships with other species (like those between the hookworm and its host). Ecological inheritance configures the evolved dependencies outlined above (e.g., hookworms to train the immune system), and works both inside and outside of animal bodies.

Symbiotic relations have become more prominent through the growing interest in ecology in the role of interaction webs in configuring the structure and function of ecosystems. A shift has been taking place in ecological theory since at least the 1970s, in which an increased importance is afforded top-down regulation over bottom-up supply-side pressures.[16] This theory—which is central to the rise of rewilding—argues that the composition and dynamics of an ecology are shaped not only by physical processes and the productivity of producers and consumers at the bottom of the food chain, but also by the regulatory pressure applied by a select group of strongly interactive keystone species, the most famous example being the wolves of Yellowstone National Park and their ecology of fear.

Microbiologists Heather Filyk and Lisa Osborne note the recent passage of this approach into work on the multibiome, where the focus is on how the interkingdom cross talk between highly interactive organisms influences microbial colonization and ecological structure.[17] For example, helminth researchers have explored how communication between microbes and their hosts influences "local and systemic immune homeostasis, health, and disease."[18] They suggest that "similar to other ecosystems, the intestinal community is dynamic, responsive, and regulated by interactions between distinct biological entities."[19] Margaret McFall-Ngai and colleagues suggest that these symbiotic relationships occur within nested ecosystems that cut across and link the familiar biological scales of the body and its environment.[20] This model of scalar relationships resonates with the influential concept of panarchy from complex systems theory. Philosopher Josef Keulartz explains how panarchy "refers to the dynamic interactions among system scales. A system at a particular scale will usually be comprised of smaller subsystems as well as being nested within larger systems."[21]

## Keystone Species

Within this broader interest in symbiotic relations, attention has focused in particular on a relatively small number of strongly interactive keystone species, including a subset termed "ecosystem engineers."[22] A keystone species is an organism that exerts disproportionate influence on an ecology relative to its abundance or body mass. The term originates in studies conducted in the 1960s by marine ecologists (like Robert Paine) of the top-down regulatory influence of predators (starfish, otters) on populations of consumers (mussels, urchins) and subsequently on producers (seaweed) within a food web.[23] Like Leopold, Paine and his colleagues were concerned with the environmental effects of human hunting—in this case of otters—on ecological functioning. Subsequent work on the topology (or spatial arrangement) of species interactions has focused in particular on the role of apex predators, like the wolf, in regulating populations of herbivores, like elk or deer, and shaping the abundance and diversity of plant species. The keystone metaphor has since proliferated to encompass a wider set of ecological relations, with ecologists claiming keystone herbivores, keystone parasites, and keystone mutualists.[24]

The trophic cascade is a central concept in this ecology of asymmetric interactions. It describes "the propagation of impacts by consumers on their prey downward through food webs."[25] The interest here is in the disproportionate ability of some species to shape and modulate the flows of energy within their ecology.[26] Ecologists have described a range of processes through which these cascades occur, including how the "affective atmospheres"[27] associated with species interactions come to shape behavior. For example, attention has focused on the landscape or ecology of fear created by changes in the behavior of prey species in the presence of predators.[28] Work on the wolves introduced to Yellowstone National Park highlights how the vigilance and avoidance behaviors of elk affect their grazing, with landscape-scale consequences for the distribution and abundance of different plant species.[29] The argument, which is still contested, is that it is the place-specific fear of death, rather than absolute increases in mortality rates, which comes to reconfigure ecological interaction networks.

An ecosystem engineer is a specific type of keystone species that is capable of creating, maintaining, or destroying a habitat.[30] All organisms change their habitat to some degree, so this label requires a calculation of the magnitude of impact. Ecosystem engineers are defined as "agents of system state change."[31] Tim Caro explains that ecosystem engineers tend to be divided into two types: "autogenic engineers that change environments via their own physical structures (i.e., their living and dead tissues), and allogenic engineers that change the environment by transforming living or non-living materials from one physical state to another via mechanical or other means."[32] Corals are a classic example of an autogenic engineer and beavers of an allogenic engineer. Darwin notes the ecosystem engineering abilities of earthworms, and subsequent work has identified a great diversity of engineering techniques.[33] These include grazing and browsing (as enacted by the cattle, horses, and deer at the OVP), seed dispersal (by introducing tortoises to oceanic islands or feral horses to Central America), dam making (by beavers), and disturbance regimes (enacted by boars or elephants). The concept of an ecosystem engineer is a vital conceptual precursor to the concept of niche construction.[34]

The concepts of keystone species and ecosystem engineers continue to have traction as they are imported into the science that informs biome restoration, although much less is known about the ecological composition, structure, and dynamics of the human multibiome. Microbiologists have long been interested in beneficial or probiotic bacteria and the roles they play in delivering human health. Specific strains of bacteria have been promoted and used for therapeutic purposes alongside the early efforts by gastroenterological surgeons who pioneered FMT to assist postsurgical patient recovery. The science behind probiotic supplements has an extensive, albeit checkered, history.[35] There has been a new wave of scientific interest in probiotic bacteria with the advent of metagenomics, the development of mouse models of the human microbiome, and the growth in data on normal and pathological gut microbial ecologies.[36] These developments have enabled scientists to model the topology of interspecies ecological interactions in the microbiome and to identify a small number of strongly interactive keystone species that exert a disproportionate influence on gut ecologies.[37] Interest in the ecological structure of the microbiome is especially prominent in research seeking to explain the microbial mechanisms associated with successful FMT. Scientists studying how healthy human stool restores the gut microbiome have identified the bacterial compositions that can provide desired forms of ecological functionality—in this case an ability to establish a stable ecology, and to outcompete and exclude pathogenic flora.[38]

Studies of the gut multibiome have been extremely bacteriocentric, with relatively little attention being paid to ecological roles of microeukaryotes like helminths.[39] Nonetheless, William Parker and colleagues have proposed that hookworms and some other helminths might act as keystone species within gut ecologies. They develop the hypothesis that humans have an evolved dependency on microbes so that a small subset of old-friend microbes might exert a disproportionate influence on the ecology of the human gut and on the human immune system.[40] Scientific interest in helminths as old friends has focused on the role of these organisms in first training and then modulating or exercising the human immune system.[41] Extensive laboratory studies using mouse models have begun to reveal the mechanisms through which worms communicate directly with their host and modify the commensal bacteria of the gut in order to achieve immunosuppression.[42]

Immunologists suggest that helminths communicate with their host and its microbes by releasing a wide range of molecules. Some of these chemicals interact with host immune cells to down-modulate the standard aggressive immune response. Other chemicals encourage the production of mucus on which worms feed. A further set stimulates the host to release antimicrobial and probiotic products that create conditions favorable to a desired subset of commensal bacteria. Helminths enhance this ecosystem engineering through host manipulation by communicating directly with the gut bacteria.[43] They are able to sense the presence of different bacteria to locate and orientate themselves within the gut so that they take up residence in the right location.[44] Worm secretions further encourage the selective growth of desired commensals, which help them feed and assist with masking their presence from the host immune system.[45] A complex three-way network of symbiotic relationships is presented in which hosts, helminths, and some commensal bacteria are mutually dependent. Although immunologists have yet to draw on the concept, some helminths, like hookworms, appear able to engineer gut ecologies toward their own ends.

## Tipping Points and Regime Shifts

Rewilding and biome restoration conceive of ecologies as having shifted into an undesirable state as a result of the excesses of antibiotic modes of managing life. They are founded on a common understanding of the dynamics of ecological change that draws on concepts and metaphors from systems biology, especially those concerned with the resilience of socioecological systems. Resilience thinking has become popular across a wide range of policy domains, and it has both passionate advocates and critics. I focus here on a few key concepts as they feature in the science of rewilding and biome restoration.

The first common idea is that an ecology, be it a temperate nature reserve or the human gut, can have multiple stable states. There is no single timeless balance of nature. Many different ecological conditions are possible. As such, there is no necessary linear trajectory along which an ecology will develop or change, and there is no universal final destination. The model of symbiotic ecologies regulated and engineered by keystone species is therefore founded on a nonlinear and nonequilibrium model of ecological change or succession. In the terms of resilience thinkers, the different possible stable states of an ecology are known as strong attractors. These are the conditions toward which an ecology will tend in the presence of common abiotic and biotic influences.

Ecological disturbances have been generally pathologized in the equilibrium model that informed antibiotic modes of immunology, natural resource management, and the traditional approaches to nature conservation that concern rewilders. Infection, pests, floods, and fire all threatened social and ecological productivity. In contrast, a probiotic approach emphasizes the generative importance of disturbances for ensuring the functional integrity and ongoing stability of an ecosystem Wolves scare deer, cattle open up the forest, and helminths exercise the immune system, and in so doing, each maintains a degree of homeostasis within its ecology. Homeostasis describes the "dynamic processes that return the system to a balanced state after perturbation."[46] In rewilding and biome restoration, ecologies are understood to be metastable—that is, they are stable provided they are subjected to no more than small disturbances.

Yet these ecologies are also characterized by tipping points or thresholds. These are the points at which an ecology shifts from one state to another, pushed out of one basin of attraction with the potential of shifting toward another stable state. This may happen as a result of changes to the topology of the interaction webs within an ecological system—for example, when the loss of a keystone species like a wolf or a hookworm shifts the passage of energy within the trophic cascade. As geographer Steve Hinchliffe and his colleagues suggest in their work developing this thinking in the context of agricultural biosecurity, such topological twists shift the intensities within an ecosystem, potentially triggering complex, nonlinear changes in composition and function.[47] In this reading, ecological change is emergent from or immanent to an ecosystem, rather than something that is visited upon it from the outside. Ecologist Daniel Botkin offers a musical metaphor, suggesting that ecologies are marked by "discordant harmonies"—that is, multiple, interacting trajectories of potential change.[48] Environmental anthropologist Anna Tsing offers a comparable depiction of ecologies as "polyphonic assemblages."[49]

This thinking is evident in the theory of ecological succession advanced by Frans Vera in his wood pasture hypothesis.[50] Vera challenges the Clementsian (after Frederic Clemens) model, in which ecological succession follows a linear trajectory toward a stable archetype: the closed canopy forest in the case of Northern Europe. He instead suggests that the ecosystem engineering of large herbivores leads to a continuous, nonlinear, but nonetheless cyclical rhythm of ecological change. The animals' grazing and browsing maintains open areas within the forest, and the regeneration of trees species like oak is dependent on what have become known as Vera cycles (Figure 11): the growth of saplings within unpalatable or impenetrable scrub. For Vera, the stable high forest climax community—beloved of many Western European conservationists—results from the absence of the large herbivore disturbance regime. It is evidence of an ecology that has crossed a tipping point into a different stable state and is missing its regenerative disturbance regime.

Tipping points and regime shifts are central to the adaptive cycle of socioecological change proposed by resilience theorists. As the philosopher Josef Keulartz explains, in this account, "ecosystems tend to move through cycles of change. These changes are not entirely predictable, but often follow a pattern in which four phases are commonly observed. Generally, they move from a rapid *growth* phase through to a *conservation*

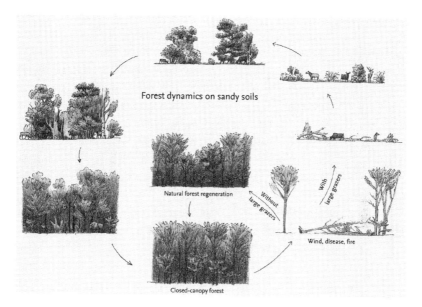

*Figure 11.* The Vera cycle. Illustration by Jeroen Helmer; reprinted with permission.

phase in which resources are increasingly unavailable, locked up in existing structures, followed by a *release* phase that quickly moves into a phase of *reorganization* and hence into another growth phase."[51] This nonteleological model of ecological change suggests that there are many different ecological combinations that could emerge in any given context. For example, there are lots of normal, functional human multibiome compositions that ensure eubiosis, or "the microbiota in a disease-free host."[52] Similarly, several different but normal, functional ecologies can emerge in any biogeographical region. Those involved in rewilding and biome restoration focus on ecologies subject to such severe disturbance regimes that they have undergone "catastrophic regime shifts,"[53] crossing thresholds and tipping into new conditions with deleterious consequences for valued forms of human and nonhuman life.

In the macrobiome, ecologists have focused on the trophic downgrading that results from the (largely anthropogenic) loss of keystone species. One strand of this work links the prehistoric loss of megaherbivore[54] ecosystem engineers (like elephants, rhinos, or mammoths) to declines in biodiversity, and to regional and global changes in biogeochemistry and climate. The loss of these animals changes the distribution of seeds and nutrients within ecosystems. It alters the distribution of vegetation, leading to changes in carbon dioxide and methane cycles (more vegetation, less farting) as well as shifts in global albedo (more vegetation leads to less light reflection).[55] A second strand has focused on the impacts of the more recent declines in the populations of large carnivores (like wolves, lynx, and lions).[56] Together, this work suggests that in the absence of the trophic cascades performed by these keystone species, "ecosystems convulse through harsh transitions to simpler alternative states."[57]

Similar concerns with catastrophic regime change drive those seeking biome restoration. Here, missing microbes are understood to have triggered contemporary epidemics of absence.[58] The widespread rise of allergic, autoimmune, and inflammatory disease is associated with the loss of the roles played by old-friend microbes in both training the human immune system and regulating the composition of the gut microbiome. Immunologists and microbiologists speak of a generalized condition of dysbiosis that results from the excesses of modern health, hygiene, and lifestyle practices. Meanwhile, Graham Rook and others have expressed concerns that the absence of old-friend microbes might also enable colonization by pathogenic crowd infections, including bacterial strains with drug resistance.[59] However, the mechanisms through which the absence of worms cascades through a gut immune ecology are still poorly understood. Others have even suggested that the coexistence of helminths and crowd infections might actually enhance disease risk, as the ability of worms to suppress the immune system might facilitate infection.[60]

A nascent body of scientific work is beginning to link catastrophic changes in the microbiome with equivalent changes in the macrobiome. This research offers a "biodiversity hypothesis" for the rise in inflammatory disease.[61] It proposes that the loss of wildlife (both inside and outside the human body), coupled with modern lifestyles that reduce human exposure, helps account for the absence of old friends and their immune-regulating functions.[62] This writing, which links immunologists and conservationists, tends to pathologize the disconnections associated with urban life and draws attention to the benefits of exposure to soil and agricultural animals associated with a rural lifestyle.[63]

Those involved with rewilding and biome restoration are especially concerned with the positive feedback loops associated with regime shifts, as well as with the ease with which their focal ecological system might be tipped back into a desired stable state. This ease of transition is known as hysteresis. In the context of biome restoration, William Parker and Jeff Ollerton suggest that "elements of a vicious cycle are in place, since immune hypersensitivity due to loss of helminths enhances the tendency of the immune system to interact adversely to a normal microbiome, and alteration of a normal microbiome by aberrant immune activity is likely to further destabilize the immune system."[64] Others have noted the irreversibility of some of the effects of biome depletion as a result of the influence of missing microbes on the development of the human body—for example by generating type 1 diabetes or autism. Similar concerns are expressed about situations in which the missing keystone species has gone extinct and cannot be replaced with a surrogate, as in the case of saber-toothed tiger or woolly rhinoceros, or where the ecology has shifted to such a degree that a returning keystone species would not be able to survive or would come into immitigable conflict with human land uses, as in the return of wolves, hippos, or elephants to lowland Britain.

Finally, in theorizing what prevents regime change (both catastrophic and recuperative), those involved in rewilding and biome restoration focus on a common set of systemic properties that confer resilience on an ecology. Here resilience is understood as the ability of the system to withstand and recover from shocks. It is linked to biological abundance and diversity, and understood in terms of both species and functions. Importance is afforded functional redundancy (lots of species with an unclear ecological role) in the context of systems understood to adhere to complex, nonlinear dynamics, and in a science marked by high levels of epistemic uncertainty. It is important to note that the resilience of a micro- or macroecology is not necessarily valued as a normative good. Prominent rewilding scientists, like Jens-Christian Svenning, caution that there are plenty of depleted or dysfunctional ecologies that show high levels of resilience to socioecological change.[65] Instead, quoting a much-cited paper by Steve Carpenter, such scientists interrogate the "resilience of what to what" and value resilience only when it is the property of a desired ecology or enables the transition toward a desired future condition.[66]

## Dysbiosis and Resurgence

Fully functional ecologies of the kind desired by those involved in re-wilding and biome restoration are extremely rare. Much of the scientific thinking associated with the probiotic turn has therefore emerged from studies of ecologies that have undergone catastrophic regime changes. A great deal of attention has thus focused on situations marked by forms of dysbiosis. This term has become common in work on the microbiome. It has been defined in several different ways, with a common reference to "any perturbation of the normal microbiome content that could disrupt the symbiotic relationship between the host and associated microbe and result in diseases."[67] Dysbiosis is tied to changes in ecological composition, especially the loss of keystone taxa or beneficial commensals, and a general loss of biodiversity. Some functional definitions of dysbiosis are now emerging.[68] Dysbiosis is not often used to describe trophically downgraded or otherwise dysfunctional macroecologies, but I borrow this term as a general descriptor to identify common explanations of the character of dysfunctional ecologies.

One common conception of dysbiosis emerges from a shared engagement in rewilding and biome restoration with both paleoecology and evolutionary biology. There is a consistent reference in these fields to the idea of "ecological memory," or how contemporary ecologies are haunted by the absence of keystone species.[69] Ecologists talks of "ghosts from the past"[70] that leave an "extended legacy"[71] or "ecological inheritance"[72] in the present. They worry about ecological anachronisms—that is, ecologies whose current (seemingly dysfunctional) dynamics can only be understood with reference to the now-absent and past presence of a megaherbivore or predator.[73] In an example from his best-selling book on rewilding, George Monbiot suggests that tree species like hawthorn, which make up hedgerows in places like the United Kingdom, evolved their ability to withstand heavy browsing and frequent destruction in the presence of now-extinct elephants. In a mournful passage, he reflects, "We live in a shadowland, a dim, flattened relic of what once was, of what there could be again . . . our ecosystems are the spectral relics of another age, which, on evolution's timescale, is still close."[74] Michael Soulé cautions that there is a time lag to this unraveling in which the "functional extinction of species interactions often occurs well before the species themselves have completely disappeared."[75]

In a similar fashion, William Parker and Graham Rook attribute the rise in allergic and inflammatory disease to an evolutionary mismatch between the human immune system and the modern microbiome to which it is exposed.[76] They offer a "fish-out-of-water model of disease"[77] in which the evolved dependence of the immune system on the hunter-gatherer environment means it goes awry in the absence of the calibrating and symbiotic effects of the old-friend microbes. This idea of the evolutionary mismatch between modern, sedentary urban life and humans' Paleolithic past is a central component of a wider field of medical, psychological, and social theory associated with what has become known as the paleo movement. The mismatch—without or without microbial mediation—has been used to account for a wide range of pathologies, including obesity, stress, depression, addiction, and violence.[78]

Like extinct megaherbivores or absent predators, some helminths have come to figure as ghosts whose absence haunts dysbiotic gut ecologies, although research on the immunomodulatory abilities of helminths and the autoimmunity associated with their absence has drawn attention to the importance of the timing of their going missing. Researchers have suggested that the ecological memory of human-evolved dependence on helminths might be transient, and that the gut microbiome and immune system might be able to adapt. Immunologists speculate about an epigenetic mechanism that might account for this lag.[79] In a similar fashion, rewilding scientists developing an "ecological memory-rewilding framework" to guide interventions suggest that the strength and transience of ecological memory is likely linked to the speed of abiotic and biotic turnover in an ecosystem.[80]

Another common strand of thinking associated with these discussions of dysbiosis relates to the concept of blowback that I introduced in chapter 1. In particular, ecologists working between the macro and the micro scales of ecology have noted how the absence of the top-down, regulatory role of keystone species can lead to what Anna Tsing has called the "proliferation" of a small number of invasive species or pathogens.[81] In this context, microbiologists have developed the concept of the pathobiont: an organism that can switch between a mutualistic and a pathogenic relationship, often as a result of significant ecological disturbance.[82] For example, antibiotics reduce microbial diversity and remove colonization resistance, allowing some otherwise commensal but drug-resistant bacteria—like *Clostridium difficile*—to proliferate in the lower intestine.[83] Likewise, research on the microbial drivers of colon cancer has identified the role played by pathogenic "alpha bugs," a term that describes a small number of commensal bacterial species that come to configure dysbiotic microbial ecologies in patients with low-fiber diets that in turn induce a cancer-causing immune response.[84] In chapter 5, I explore the hookworm as a pathobiont, whose immunological relations are configured by the intensities of its political ecological relations.

While the concept of the pathobiont is not commonly used in the macrobiome, ecologists suggest that otherwise mutualistic species can become invasive as a result of trophic downgrading. Terborgh and Estes describe a "trophic shunt" enacted by "organisms that . . . short-circuit the hierarchical food chain by usurping primary productivity at low trophic levels."[85] For example, as Aldo Leopold observes in the epigraph to this chapter, the loss of apex predators—like wolves—can lead to increases in their herbivore prey—like deer—with deleterious consequences for plant diversity,[86] whereas the absence of apex predators can enable the expansion in populations of smaller (or meso) predators—like coyotes or foxes—that they would previously have controlled. This process of mesopredator release may lead to declines in populations of their prey. One example is the increase in the predation of beavers by coyotes, which had been released from predation by the wolves' near absence in much of North America.[87]

Ecologists also suggest that invasive species proliferate more easily in heavily disturbed ecologies, especially those marked by trophic downgrading. This is the case for many common animal and plant species dubbed pests and weeds. As with microbial pathobionts, these organisms are not inherently deleterious to their host ecology but become so once its structure and dynamics are altered in ways conducive to their proliferation. Many invasive species reproduce quickly and have evolved to make use of newly disturbed conditions. Some of these figure as "transformer species" that can engineer new, stable, and resilient ecologies.[88] Sometimes the new ecologies they create are marked by significantly lower levels of biological diversity than the ecologies they have replaced, though the past tendency to pathologize all nonnative species has been criticized and tempered, in part in response to the challenges of adapting to the novel ecosystems of the Anthropocene.[89] Nonetheless, a common argument is that deleterious changes to ecological function and the emergence of pathogens result from changes to the character and intensity of ecological interactions. As Steve Hinchliffe, Anna Tsing, Rob Wallace, and others have argued, pathogenesis is immanent to intensive and antibiotic modes of ecological management like plantations or factory farms, which become disease hot spots.[90] Disease and dysbiosis are inherent problems, not necessarily the result of an encounter with an exotic, foreign, or otherwise premodern other. Indeed, premodern, antibiotic, and probiotic modes of management all have the potential to create such disease hot spots.

Behind all of this work on dysbiosis is a shared commitment to understanding and facilitating what Tsing terms ecological resurgence, or "the remaking of livable landscapes through the actions of many organisms."[91] Tsing differentiates resurgence from proliferation, equating the former with the convivial premodern human–environment relations of the Holocene and the latter with the intensive and globalized relations of the Anthropocene, exemplified for her in the political ecologies of agricultural plantations. In the context of the probiotic turn, this resurgence is commonly figured as the self-willed outcome of the ecological ability of keystone species to reconfigure dysbiotic ecologies through the return of their regulatory effects.[92] Rewilders describe this as the (often unexpected) emergent properties of a landscape ecology.[93] Wolves redress the trophic cascade caused by their absence, beavers engineer aquatic ecologies, and hookworms exercise the immune system and foster eubiotic microbial ecologies. Here resurgence is the transformative enhancement of ecological structure and functions—like diversity, immunity, and resilience—through interventions that change the antibiotic status quo. This valorization of the self-willed landscaping abilities of transformer and keystone species creates ambivalence and discord among ecologists involved with both rewilding and biome restoration. It raises challenging biopolitical questions as to which forms of life are let or made to live and die to ensure multispecies flourishing.

## Anthropocene Humanisms

As responses to the excesses of antibiotic modes of managing life, both of these fields of science acknowledge the profound ecological, even geological, influence of human activities. Ecoimmunology and conservation biology are crisis disciplines. They focus on human–environmental entanglements, which are conceived through a range of ontologies that depart from the modern nature–society and human–nonhuman binaries that have been so central to antibiotic modes of managing life. For example, while they make frequent reference to paleoecology, those developing "science for a wilder Anthropocene"[94] place their interventions within novel ecosystems: ecologies so inflected by human actions that they are not analog to what has come before.[95] They acknowledge the importance of human infrastructure in configuring the contemporary biogeographies of species distribution and movement, and they attend to the legacies—and potential—of antibiotic technologies for managing life, like chemicals, fences, and genetic engineering.

Likewise, the recognition of the microbial origins of human life and of our symbiotic dependence on microbes for core bodily functions, including cognition, has decentered the modern figure of the rational Human as a disembodied mind in a vat. Thinking the human as holobiont flags the porosity and precarity of the body and the risks posed by the unraveling of the human microbiome. It draws attention to the ways in which humans and other life-forms are inadvertently shaped by a host of modern health, hygiene, and lifestyle practices such that no component of the microbiome remains unaffected by human activity. In this way, and to differing degrees, both rewilding and biome restoration have begun to link their science to the concept of the Anthropocene and the wider intellectual zeitgeist that it has named and propelled.[96] Ecologists involved with rewilding draw the hypothesized advent of this new epoch back from the modern (eighteenth to twentieth centuries) origin points that are popular in the geosciences. The significance of coal or nuclear energy is downplayed to give preference to the planetary effects of the extinction of the megaherbivores (100,000 to 10,000 BP), subsequent declines in predators, and widespread trophic downgrading.[97] The discussion of the Anthropocene and its contested periodization are less pronounced in work on the microbiome, although immunologists involved in biome restoration share a comparable interest in identifying the moment of the historical fall to which the origins of dysbiotic relations can be traced.

As such, the science of rewilding and of biome restoration operate ontologically "after nature" in the sense described by Marilyn Strathern.[98]

Like Heather Paxson's probiotic cheese makers, these immunologists and conservation biologists are "at once post-nature, recognizing that there is no pristine natural world outside human cultural activity, and also ever in pursuit of some kind of remade nature as a ground for appropriate human action."[99] There is significant and contested heterogeneity to the types of remade natures that come to ground the probiotic projects that feature in this book. Here I will foreground the most prominent position afforded people in the novel ecologies of rewilding and biome restoration, which I call "enlightened anthropocentrism," following Josef Keulartz.[100]

The diagnosis of the Anthropocene has been taken by many environmentalists as a caution against the hubris of modern dreams of planetary management and a portent of apocalyptic futures with the potential to unravel the hospitable conditions of the Holocene. Those involved in rewilding tend not to share these concerns. Instead, rewilding is promoted as a redemption narrative in which optimistic, beneficent environmentalists turn to face the future and seize the opportunities presented by the event of "the age of Man."[101] Such accounts mobilize the long history of (largely) inadvertent human planetary influence to present humans as potential ecosystem engineers capable of large-scale niche construction. In some prominent accounts, humans figure as enlightened "hyperkeystone species" with the potential for facilitating a functional future of mutualistic coevolution.[102] Popular science writer Ed Yong describes the human hyperkeystone as follows: "We are the influencer of influencers, the keystone species that disproportionately affects other keystone species, the ur-stone that dictates the fate of every arch."[103] In keeping with the tenor of the North American ecomodernist movement, some rewilding manifestoes propose that humans (aggregated in the universal "we") embark on ambitious, continental-scale schemes for environmental management. Such schemes would involve "decoupling" people from nature and setting aside large areas for new wildlands.[104]

This model of enlightened anthropocentrism also characterizes the position afforded people in biome restoration. While the identification of the human as a symbiotic holobiont has been taken by some social theorists to undermine models of the competitive, individual human subject and to challenge the modern "masters of our own destiny" narrative, this is not how the microbiome is commonly conceived

by immunologists working with helminths. The theory of biome depletion does recognize the anthropogenic nature of "modern dysbiotic
drift"[105] and its associations with a wide range of scientific and technological developments previously heralded as archetypally modern:
sanitation, antibiotics, urbanization. But these are not understood to
challenge the logics of human exceptionalism. Instead, the prevalence of
human impacts on the microbiome is read as evidence of human power
and a portent of the promise of more rational and ecological modes of
microbiome management.[106] Ecoimmunology after nature operates cog-
nizant of the Anthropocene signature in the microbiome, but in search
of a naturalistic ontology on which to ground new projects of anthropocentric world making.

## Gaian Thinking in Context

It is well beyond the scope of this book to fully situate these concepts
in the history of ecological ideas, or to disentangle the flow of ecological concepts into and between conservation biology, immunology, and
microbiology. But it is worth taking the time to place Gaian thinking
into an intellectual context. Scientific concerns with ecological changes
caused by the anthropogenic loss of species have a long and varied provenance. Since at least the late nineteenth-century writings of Élie
Metchnikoff and Sergei Vinogradsky, microbiologists have developed
ecological understandings of bacteria, including an interest in the salutary roles of beneficial microbes and concerns that their absence leads
to dysbiosis.[107] Ecological research on bacteria and their viruses (or bacteriophages) was especially prominent in the Soviet Union (Russia,
Georgia, and Poland) in the early part of the twentieth century, and
phages and probiotics were used to treat citizens and Soviet soldiers in
World War II.[108] Ecological research remained prominent in the post-
war microbiology of figures like Rene Dubos and Joshua Lederberg, but
receded (at least in Europe and North America) during what the historian
Scott Podolsky has described as "the antibiotic era."[109] Since the production of penicillin in 1942, successive waves of so-called wonder drugs
have helped control infectious bacteria and diminished the importance
of research on the ecological origins of disease. But the decline in new
drugs and the growth of concerns about antibiotic resistance since the

1980s, coupled with the advent of metagenomics and concerns for missing microbes in the 2000s, have all placed microbial ecology firmly back on the agenda.[110]

Likewise, as this chapter's epigraph makes clear, scientific concerns about what is now known as trophic downgrading have a longer history, in some cases preceding Leopold's cautionary observations. In North America, opposition to the science-led agriculture and forestry of settler colonialism catalyzed a romantic tradition in ecology that was driven by a combination of elite hunters and naturalists encompassing figures like John Muir, Teddy Roosevelt, and Henry Thoreau.[111] In Europe, an earlier pastoral tradition of poets, writers, and natural historians, including Robert Burns, John Clare, and William Wordsworth, initiated an ecological concern with the loss of species that found scientific expression in the early twentieth-century writings of figures like Charles Elton, Julian Huxley, and Arthur Tansley—figures who were instrumental in founding the institutions of European and international conservation.[112] Meanwhile, ecologists working within and against the institutions of European colonialism in Africa and South Asia frequently commented on how indigenous and colonial hunting, trapping, and pest control led to landscape-scale change for species as diverse as elephants, tigers, and tortoises.[113]

Conservation biology, microbiology, and immunology have all come to embrace ecology in response to different social anxieties about environmental or human health. Scientists have turned to ecology to forge mission-oriented crisis disciplines that aim to name, popularize, and address socioecological problems.[114] As I trace in chapter 4, the compressed temporality of this crisis framing, and the normativity of mission-orientation, shape their epistemic practices. This coproduction of ecological science and social concerns about environmental and then human health has had important consequences for the history of ideas. Historians of science have diagnosed various "ages of ecology" across different disciplines. Donald Worster argues that the first Earth Day in 1970 marked the advent of the age of ecology in environmental management, as the ecological science of figures like Leopold went mainstream and was heralded as providing a blueprint for planetary survival.[115] Others suggest that this age comes later, with the rise of anxieties about climate change.[116] In comparison, the early 2000s have been identified as a golden age for microbial ecology, and historian Alfred Tauber has claimed that an ecological turn is underway with the twenty-first-century rise of ecoimmunology.[117] Tauber links this ecological enthusiasm to the wider influence of "the ether of environmentalism": "the collective experience that the earth is facing an environmental crisis, if not a catastrophe."[118]

The asynchronous timings of these ages of ecology hint at the unequal flow of ecological ideas between micro- and macrobiology. The origins of ecological science lie in thinking with larger plants and animals whose dynamics are slower, are easier to observe, and more readily excite popular sensibilities. While Lovelock and Margulis both shaped Gaian thinking, the subsequent traffic in probiotic concepts has been somewhat unequal, with microbiologists and immunologists borrowing more frequently from their macroecology colleagues. It is only with the twenty-first-century advent of metagenomics that microbiologists could begin to map the types of ecological network interactions that had become familiar to plant and animal biologists in the later decades of the twentieth century. Only very recently have they been able to test the utility of concepts like niche construction, trophic cascades, and keystone species in the microbiome. Indeed, the idea of helminths as ecological engineers has been most comprehensively discussed by wildlife biologists studying the immune systems of wild animal populations, perhaps because they are more acquainted than immunologists with this branch of ecological theory.[119] The ether of environmentalism infused conservation biology well before it intoxicated microbiology and immunology, so the science of plant and animal restoration has served as the primary concept generator for the probiotic turn. Conversations across this scalar divide are only just beginning, and we might expect future conceptual traffic as Gaian thinking takes stronger hold.

In his classic history of ecological ideas, Donald Worster identifies four axes of philosophical difference to help us to begin to specify and contextualize the type of ecology involved in thinking like Gaia: reductionist–holistic, Arcadian–imperial, equilibrium–nonequilibrium, and compositionalist–functionalist (Figure 12).[120] I only sketch these here; they need further research and elaboration. The first is a shift away from the reductionist approaches that characterized much twentieth-century biology, toward more holistic approaches. In microbiology, this is shown by the return to long-standing interests in microbial ecology

*Figure 12.* Four conceptual axes for specifying the type of ecology that informs probiotic science.

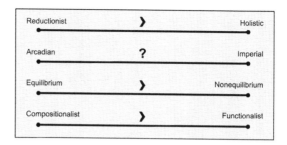

with symbiosis and in the new focus on symbiogenesis, which have culminated in the postmodern synthesis. Margaret McFall-Ngai describes biologists in "future shock" as they struggle to deal with the wealth of data made available by new sequencing technologies, trying to adapt their models to grapple with a much wider range of relevant ecological relations.[121] A holistic turn in immunology is evidenced by the figure of the holobiont, the growing appreciation of coevolved mutualistic relations between microbes and their host, and of the role of the microbiome in both enabling and constraining antibiotic resistance. This holistic understanding challenges reductionist models that link disease to microbial presence, foregrounding the pathological potential of microbial absence and dysbiosis. The holistic character of rewilding is evidenced by its focus on ecological interactions, trophic cascades, and top-down regulation, shifting away from the reductionism of both endangered species conservation and productivist models of single resource management. As I explore in chapter 4, this holism drives an emphasis on specificity, complexity, contingency, and emergent properties over the derivation of general mathematical principles.

We can orientate holistic probiotic science in relation to the differences Worster identifies between the Arcadian and imperial traditions in ecology. For Worster, the Arcadian view "advocated a simple, humble life for man with the aim of restoring him to a peaceful coexistence with other organisms," whereas the imperial tradition aimed "to establish, through the exercise of reason and by hard work, man's dominion over nature."[122] Here there is more ambiguity as to where to position probiotic science, which relates in part to the different forms and degrees of anthropocentrism that we will encounter in the chapters that follow. On the one hand, the prevalent model of enlightened anthropocentrism folds the human as holobiont into a probiotic planetary ecology. Microbiologists and immunologists take issue with neo-Darwinian model of individualistic competition and with imperial metaphors of immunity as all-out Hobbesian warfare. Cold War–esque depictions of the martial and antibiotic defense of the pure human self are being recalibrated toward more historical, dialogic, and communitarian understandings commensurate with the human as a porous and entangled holobiont.[123] Likewise, some rewilders promote themselves as hyperkeystone species, capable of instigating nature-based solutions to anthropogenic planetary dysbiosis. On the other hand, some figure the human as a God species with the potential for rational planetary stewardship. While Lovelock's Gaia has been folded into popular models of Arcadian environmentalism, his science and the emerging model of Gaia 2.0 has the earth as humans' dominion. However, this imperialism is not universal, as I explore in chapter 7 and in the conclusion.

It is easier to place probiotic science on the third of Worster's axes. The ecology of rewilding and biome restoration departs from the equilibrium models that are prominent in the Arcadian tradition. Common interests in sympoiesis, tipping points, and metastability situate probiotic science within the "ecologies of chaos" that Worster argues came to prominence in the second half of the twentieth century with the rise of nonequilibrium systems biology.[124] This enthusiasm for nonequilibrium over equilibrium ecology is evident in rewilding in the examples of the Vera cycles of forest–pasture dynamics and the emphasis this model places on the disturbance regimes enacted by large herbivores. This theory offers a polyphonic ecology marked by discordant harmonies and capable of tending toward several contrasting compositions. This model pushes against the pathologization of anthropogenic disturbance in conservation and presents the conservationist as the conductor of ecological dynamics, one informed by the adaptive cycle of resilience theory. Likewise, a nonequilibrium model of human ontogenesis has become prominent in immunology, with a recognition of the haunting power of missing microbes and the hysteresis of dysbiotic gut microbiomes plagued by drug-resistant bacteria. Beneficial microbes are valued for their ability to disturb gut microbiomes and exercise the immune system.

In keeping with this shift away from equilibrium, the ecology of rewilding and biome restoration is more functionalist than compositionalist.

As J. Baird Callicott explains, compositional ecology focuses on units, especially the distribution and dynamics of species and habitats.[125] Compositionalist approaches were prominent in twentieth-century conservation, especially with the creation of the International Union for Conservation of Nature's Red List of Threatened Species and its focus on rarity and extinction.[126] In contrast, rewilding focuses on ecological functions and processes—grazing, predation, decomposition—as they are enabled by species interactions and biotic–abiotic dynamics. Likewise, those mapping the microbiome and dysbiosis in the interests of biome restoration are less concerned with the work done by specific species. Indeed, a taxonomy of species proves hard to apply to bacteria as a result of their interorganism genetic promiscuity. While they might be concerned with pathogenic strains, ecological microbiologists focus much more on understanding the functional ecological configurations that deliver desired bodily processes. As I explore in chapter 7, this functionalist ecology lends itself to the calculation of ecosystem services, in which the ability of an ecology to deliver immunity and metabolism, or to manage flooding and sequester carbon, matters more than the composition of organisms delivering these services.

## Conclusions

This chapter has given an overview of probiotic science and what it means to think like Gaia through an analysis of the science that informs rewilding and biome restoration. To summarize, this work identifies five common sets of concepts that are understood to operate across nested scales. Taken together, this science presents a symbiotic understanding of evolution as well as of ecological function and dynamics. Ecologies are strongly influenced by a small number of keystone species, the absence of which can lead to dramatic regime shifts resulting in situations of ecological dysbiosis and the proliferation of biological miscreants. Aggregated humans are responsible for these shifts, but they also have the knowledge and potential to set them right through proactive practices of enlightened ecosystem engineering to enable resurgence. I identified four axes to specify the type of ecology that informs probiotic science. We should note that this positioning is not universal, nor is it uncontested by other scientists. We should use this figure to highlight the alternative ecological ontologies from which probiotic thinking departs and with which it differs. Indeed, with striking frequency, those involved in developing probiotic science feel obliged to promote their work as shifting paradigms, and in keeping with this model of epistemic change, they often seek to court debate and controversy among their peers.[127] In the chapters that follow, I explore these controversies and the ways in which they encourage wider debates about how ecological science is generated, and how it used, abused, and ignored in its applications to both policy and interdisciplinary research.

# 3

## SYMBIOPOLITICS
### Governing through Keystone Species

> Auto-rewilding is one of the most important processes for
> making our human-disturbed world today. Without auto-
> rewilding, our disturbed landscapes would be thin and bare,
> devoid of organisms except those we put there. But auto-
> rewilding offers ambivalent futures. On the one hand, we owe
> the richness of our feral landscapes to auto-rewilding. On the
> other hand, auto-rewilders often kill the chances of other, less
> aggressive and disturbance-loving species. Auto-rewilders are
> bold. They are weedy. Like us, they do not play well with others.
> They help us make the Anthropocene, the proposed epoch of
> outsized human disturbance.
>
> —Anna Tsing, "The Buck, the Bull, and the Dream of the Stag"

THE PROBIOTIC SCIENCE outlined in chapter 2 presents a dynamic world
of nested social–ecological systems. These are configured by the intensities of their ecological interactions. Passage across thresholds within
the intensities of these ecologies—most notably through the loss of a
keystone species—tip bodies and landscapes into different ecological
configurations. Ecologies thus have multiple states with differing degrees
of stability. Scientists map what Nils Bubandt and Anna Tsing term the
"feral dynamics"[1] of these ecologies, focusing on those haunted by absent
keystone species and plagued by a generalized and anthropogenic state
of dysbiosis. The branches of conservation biology and immunology
from which these ideas emerge are crisis oriented and salvatory: they
aim to diagnose the problems of an antibiotic way of managing life and
to provide new ways of promoting health.

In this chapter I examine the governance practices of rewilding and
biome restoration that are informed by this probiotic science. I identify

the emergence of a distinctive probiotic approach to managing life that is geared toward nurturing and modulating ecological dynamics to secure the delivery of ecological functions and processes. This approach involves the strategic deployment of keystone species within and beyond human bodies, and targeted interventions to create ecological conditions conducive to their flourishing. These mechanisms range across nested scales, from the nature reserves that are the targets of rewilding to the human bodies that are subject to biome restoration. Probiotic governance is targeted at the conduct of individual humans and animals. It also involves interventions into populations of keystone species. But the ultimate aim is to transform the dynamics of the ecologies with which they are entangled. I demonstrate how a probiotic model differs from the command and control of antibiotic approaches to human and environmental health, as well as from approaches to biosecurity associated with previous efforts to institutionalize systems biology as a policy framework.

## Biopolitics and Ontopolitics

To understand this mode of governance, I turn to the writings of Michel Foucault and his acolytes. In tracing his genealogies of modern government in Western Europe, Foucault identifies a diversification in the ways in which power works. He argues that historic modes of sovereign power—the right of the king or another autocratic ruler to "make die and let live"—began to be displaced in the modern period by forms of biopower that "let die and make live."[2] With the rise of the modern state, the focus of power has shifted from the execution of deviants toward the pastoral and productive management of the population. Foucault identifies two techniques of biopower. The first focuses on the disciplining of individual subjects through various governmental regimes (or governmentalities). The second, which Foucault refers to as biopolitics, is targeted at the human population or species. Scholars have since identified a third mode of biopower, sketched by Foucault in the lectures he gave toward the end of his life.[3] This environmental mode describes interventions to modulate the dynamics of the sociomaterial milieu in which social and economic life takes place. Environmental biopower focuses on managing the circulations of bodies and things in order to secure desired systemic properties, like economic growth or national security.

These forms of power—sovereign, disciplinary, biopolitical, and environmental—are not discrete; nor does one replace the other in historical sequence. Instead, they coexist, are interrelated, and are often mutually dependent. All four are at work within the probiotic turn, but in this analysis, I am especially interested in two forms. In the first, the probiotic science introduced in chapter 2 shapes the knowing conduct of human actors in ways that encourage rewilding and biome restoration. I term these holobiont governmentalities. With this concept, I am interested in how the scientific figure of the human as a holobiont— a multispecies chimera that has coevolved with microbes and is dependent on microbial exposure, colonization, and symbiosis—is enrolled to legitimate new projects of subject formation. Holobiont governmentalities draw together genetic, microbial, and ecological governmentalities that have been the focus of existing scholarship.[4]

In the second mode of power, probiotic science informs environmental biopolitical interventions to manage the dynamics of ecologies, within and outside of human bodies. I describe these as forms of symbiopolitics, a term I borrow from Stefan Helmreich.[5] This concept describes deliberate interventions that aim to modulate symbiotic relationships to deliver ecological functions and services. These two forms of power are discursively and materially connected, but they involve different methods and targets for governmental action. The first targets the self-aware behaviors of the human host, and the second focuses on the dynamics of nested ecosystems. Both come to shape the lives of the humans and nonhumans that are governed as populations of keystone species.

Existing scholarship on environmental forms of biopower has tended to focus on practices of biosecurity: the governance of unruly plants, animals, and microbes with the potential to threaten established forms of order.[6] Kezia Barker, via Foucault, explains that this work explores biopolitics as "a matter of organizing circulation, eliminating its dangerous elements, making a division between good and bad circulation, and maximizing the good circulation by diminishing the bad."[7] This approach shifts the analytical focus in scholarship on biopolitics from the management of populations toward the modulation of spatiotemporal dynamics. In geography, this work has centered on the territorial or bordering practices through which the flows of human and nonhuman actors are managed across scales ranging from the nation to the farm to the body.

Steve Hinchliffe and others have examined the different spatialities along which life moves, demonstrating the inadequacies of traditional Cartesian imaginations of space as a discrete and bounded territory. This work draws attention to the diverse topologies within and along which viruses, bacteria, insects, plants, and animals evolve, diffuse, and infect.[8] It picks up on and develops the probiotic science we encountered in chapter 2, flagging network connections and the generative intensities of hot spots that configure pathogenic human–nonhuman interactions.[9]

A further literature has examined the practices of environmental biopower informed by resilience theory. It traces the rise of a "resilience dispositif": a material and discursive apparatus of government seeking environmental security by adapting populations and infrastructure to the new realities of the Anthropocene.[10] David Chandler argues that governance in this model seeks to anticipate and ward off the threat of future apocalyptic change to maintain valued elements of the status quo.[11] Bruce Braun notes how this approach differs from the modernist biopolitics of command and control. Fixed, orderly, and linear systems are replaced by a "decidedly fluid and flexible landscape" of biopower in which "things are permitted to 'flow' and natural processes are allowed to 'function,' such that nature's diversity provides the 'service' of cancelling out the very risks that nature itself presents in an age of anthropogenic climate change. One does not work against nature, as with modernist design concepts; rather, one works with it."[12] Braun suggests that this is a mode of government in which "molding life is replaced by modulating natural processes."[13]

As I noted in the Introduction, some critics have argued that these forms of governance abandon the modern aspiration for control over life. They suggest that biopolitics has been replaced by ontopolitics, in which modernist aspirations for management have been displaced by modes of living with geological and biological disorder and unpredictability. Environmental forms of biopower relinquish so much control, they argue, that they undermine the utility of biopolitics as an analytical framework.[14] I appreciate these concerns, but in what follows, I aim to demonstrate that biopolitics persists in the probiotic turn, undergoing a mutation (a term I use following Rabinow and Rose[15]) with the rise of symbiopolitics. This mutation can usefully be conceived as an ecologization of the biopolitical paradigm, as advocated by Cary Wolfe and others.[16] In the face of doubters, I make the case for the continued relevance of biopolitics as a conceptual framework.

## Governing through Keystone Species

Probiotic science is founded on a functional ecological ontology that is primarily concerned with the processes that drive circulations of matter and energy. It is interested in what creates and what threatens the conditions for the survival of valued forms of life. Symbiopolitics in the probiotic turn therefore aims to secure a desired set of functions that are held to provide and enhance human and/or environmental health. When human and environmental health are separated from either an antibiotic model of nonhuman absence or the certainties of fixed past baselines, the selection of a desired functional archetype becomes a constrained political decision. There are often many possible healthy and functional ecological combinations and dynamics for any given system. Going probiotic through a turn to functionality does not reveal a universal ecological configuration, in the same way that an appeal to nature cannot shortcut (after Latour) politics.[17]

I explore the politics of probiotic transformation in the chapters that follow. Before doing so, however, we need to understand how probiotic science comes to be translated into (and emerges from) the governance practices of biome restoration and rewilding. In the following sections of this chapter, I identify seven sets of practices. The first four involve strategies to change the distribution and character of keystone species themselves. I term these *identify*, *modify*, *distribute*, and *simulate*. Three further sets of practices manipulate the composition and the dynamics of the ecologies to which these species are introduced. I term these *prepare*, *disturb*, and *weed/cull*. This seven-part typology is not exhaustive, but it serves to illustrate some of the most distinct and important dimensions of the symbiopolitics of the probiotic turn.

### Identify

Practical and theoretical knowledge about keystone species has emerged from a wide range of epistemic practices, which I discuss in more detail in chapter 4. These include the controlled manipulation of species composition in experimental settings and the close observation of accidents.

Practical insights have also come from DIY interventions with worms, beavers, and other species by amateur enthusiasts. Scientists have begun to draw together the data and hypotheses generated by this research to develop mathematical models to better identify keystone species for reintroduction. These models try to explain how resilient functional ecologies emerge from the configuration of species interactions. Ecologists commonly describe this configuration as the "wiring" of an ecological network. Keystone species are the most strongly interactive within ecological networks. To continue the metaphor, they are the junction boxes in the wiring. Ecological network models quantify the degrees of interactivity—or topological significance—of different species. They assume that dysbiotic ecologies have problems with their wiring and advocate network rewiring to create desired links and interactions.[18] This work is most advanced in the context of rewilding, but interaction modeling techniques are also starting to inform biome restoration.[19]

These models have been combined with the practical experience of working with different organisms to develop systematic frameworks for identifying potential species for rewilding and biome restoration.[20] These assessments suggest that there are only a small number of species with keystone potential. They also caution that the return of many keystone species would be too politically, ecologically, or clinically transformative for current consideration. For example, although ecologically desirable, wolves have been ruled out in the United Kingdom on public relations grounds,[21] and the immunologically desirable bovine tapeworm is considered too unpalatable, as its motile egg sacs crawl out of the host's anus after dark in search of a new host.[22]

In these situations, rewilders and biome restorers identify substitute keystone species that might be used in situations where a species has gone extinct, or where there are political and/or health risks associated with their return. Examples include the proposals for Pleistocene rewilding with African animals in North America.[23] Much of the discussion of substitution has been speculative, but the most developed application is in the restoration of grazing, browsing, and seed dispersal on island ecosystems through the introduction of giant tortoises.[24] Those involved in biome restoration have also experimented with substitute species to widen the choice of therapeutic helminths. These ventures are largely configured by the availability of clinically acceptable organisms rather than by any detailed knowledge of the ecological functions per-formed by extinct microbial fauna. Therapists have been working with two animal helminths that were a common part of the human diet until recent changes in food hygiene. For example, many of the early clinical trials used the eggs (or ova) of a species of whipworm *(Trichuris suis)* once found in pork.[25] Other helminth users now swallow the larval stage (or cysticercoids) of a rat tapeworm *(Hymenolepis diminuta)* that develops in the bodies of grain beetles that would have been common in bread and other cereal products produced through preindustrial processes.[26]

Identifying keystone species in these ways involves novel forms of biopolitical calculation. The probiotic ecological models we encounter in more detail in chapter 4 perform functions similar to the models cur-rently used to set priorities for species conservation or animal breed-ing.[27] These models are empowered to classify, rank, and thus exclude aggregations of animal life, but they do so for different ends. In this symbiopolitics, species are selected as populations not for their rarity or precarity (as with species conservation), their productive potential (as with agriculture), or their aesthetics and companionship (as with keeping pets).[28] Instead, they are valued for their topological centrality within an ecology of networked relations. As I explore in chapter 7, they are ecological workers prioritized for the leverage they exert over their network and their ability to sustain their modulating agency in a quasi-autonomous fashion. Their authenticity and welfare are secondary to their ability to foment desired circulations. At the same time, the model calculations continue to be tempered by the field experience of the cha-risma of keystone species. The popular affective potential of species like beavers and wolves works in their favor, at least when it comes to re-wilding. They serve as political flagship species that mobilize support for conservation.[29]

## Distribute

Once keystone species have been identified, the next task for conserva-tionists and physicians is to manage their distribution, introducing them into dysbiotic ecologies and developing means of expanding their range and abundance. Creating this new biogeography shifts the focus of sym-biopolitics toward spatial strategies designed to manage the circulation of desired animal bodies. These strategies include securing territories and restoring (or inventing) vectors of movement that have been closed off by antibiotic modes of managing life. Indeed, it is useful to conceive of rewilding and biome restoration as novel modes of biosecurity rather than their antithesis.[30] Many of the spatial strategies I address in this book are akin to those that have been mapped in existing critical work on the governance of infectious diseases and invasive species.[31] What is made to move and what is held back differs, but rewilding and biome restoration maintain the biopolitical emphasis on making live and letting die through interventions that differentiate desired and undesired forms of animal mobility.

For example, the original North American pioneers of rewilding prioritized the three Cs: cores, corridors, and carnivores.[32] Core areas are large territories with low intensities of human habitation and residual populations of keystone species. Those involved in rewilding have sought to secure such spaces through the established practices of conservation geopolitics: the acquisition and/or legal designation of land, or the creation of partnership arrangements with landowners.[33] The concept of corridors for distributing keystone species has been central to the rise of landscape-scale interventions in conservation designed to facilitate assisted migration or colonization.[34] This approach focuses on nonhuman mobilities, connectivity, and the differential permeability of landscapes.[35] Scientists have learned to think like their target keystone species in order to anticipate and facilitate their likely movements. This has led to the creation of linear corridors, stepping-stone arrangements, and the protection of flight ways.[36] This proliferation of ecological networks seeks to enable animals to move to and from core areas.

An additional set of spatial strategies, which have come to prominence since the original 3Cs model, involves the proactive translocation of keystone species from one location to another.[37] Such movements are increasingly planned in light of climate change predictions, such that ecologists talk of future-proofed "prestoration" (rather than restoration): "using species for which a site represents suitable habitat now *and* into the future."[38] Successful translocation may require training animals to be wild: learning how to hunt and be hunted; teaching them to fear people without becoming aggressive to visitors and their dogs.[39] In some pioneering examples, work is also done on the microbiomes and immune systems of captive animals before release: animals are vaccinated, then gradually exposed to new diets and environments as well as their wild kin.[40]

Practices of redistribution have been central to the OVP project. Unlike the autonomous greylag geese that inspired Vera's rewilding experiment, the Heck cattle (and horses and deer) were brought to the OVP by road, on a truck. The OVP was fenced to keep these animals in, but Vera and his colleagues have long imagined this site as a key node (or core area) in the national ecological network that began to be built in the Netherlands from the 1990s.[41] Their ambitious spatial plan for nature development seeks to connect the country's nature reserves with corridors, including ecoducts, or green bridges that span main roads (Figure 13), and to establish transboundary linkages into Belgium and Germany.

A corridor was planned to link the OVP across farmland to the nearby Veluwe forest, giving the cattle (and other herbivores) new vectors to roam and graze.[42] Meanwhile, wolves have made their own way back to the Netherlands, moving out from growing populations in Eastern Germany, and there has been a resurgence in populations of wild boar as a result of the easing of hunting pressure. These animals have

*Figure 13.* An ecoduct in the Netherlands. Source: Apdency, Wikimedia commons.

been observed foraging in the suburbs, sometimes in broad daylight, though they have yet to make an appearance at the OVP.[43] These self-willed, transnational mobilities raise concerns for those charged with biosecurity. While the movement of Heck cattle at OVP has been limited, those involved in Rewilding Europe and the Tauros Programme have raised large herds of hardy cattle for release into a network of rewilding reserves. They have developed the expertise require to breed, quarantine, handle, and transport animals. Securing a network of reserves has required a range of geopolitical work, liaising with a heterogeneous collection of local land managers, politicians, and bureaucrats to fund, permit, and sustain this wild bovine diffusion.[44]

The distribution of keystone species along new networked vectors is also a crucial component of biome restoration. The spatial strategies at this scale aim to modulate the mobilities of both the human host and the microbe to which it is exposed. The early pioneers of helminth therapy engaged in what are arguably the first acts of holobiont governmentality: the reflexive incorporation of another living organism in anticipation of its therapeutic ecological effects. To get worms, they traveled to Southeast Asia, Central Africa, or Latin America to expose their bodies to infective hookworm larvae, then returned home to start their own colonies. Such host mobilities are now rare, as the majority of those using helminths purchase them online from a small number of commercial providers, grow and share worms in small gift networks of fellow users, or are provided them in trial settings. There is now an international network of permissive infection—a network that stands in stark contrast to the antimicrobial geographies that characterize regimes of global microbial security. Worm larvae travel across national borders hidden in the mail. Like returning wolves, these mobilities alarm those, like the U.S. Food and Drug Administration, charged with national biosecurity regulation.[45]

Helminth users facilitate the distribution of their keystone organism through novel modes of what Heather Paxson terms "microbiopolitics": the reorganization of social life through the ordering of microbes and their relations with humans.[46] United on social media across a range of social difference and disease conditions, they share larvae, advice, and experience online. They have developed and express forms of microbial citizenship founded on a postgenomic understanding of their symbiotic individuality and multispecies codependence.[47] Many of those using helminths take a close interest in the science of the microbiome and discuss their use as part of a broader reevaluation of microbial exposure, which we might understand as an emergent "dirty" holobiont governmentality. Here various experts—including several prominent microbiologists—have begun promoting dirt as a popular sobriquet for a broader range of probiotic lifestyle practices designed to redistribute old-friend microbes.[48] Dirty regimes encompass probiotic supplements and hygiene products, alongside specific foodstuffs and diets, and leisure practices designed to secure exposure to desired environmental microbes outdoors, on farms, and with domestic animals. As I explore in more detail in chapter 5, these proactive exposures are coupled with reforms to hygiene practices, leading to less intensive, less frequent, and more targeted regimes of personal and domestic cleaning.

Modify

The success of some early reintroduction projects, as well as the continued absence in some cases of suitable substitutes, has led those involved in rewilding and biome restoration to consider modifying existing keystone species to enhance their performance. The biopolitical target in these interventions shifts here, from ecological relations and spatial circulation to species genetics and reproductive practices. Modification interventions range from modest in vivo breeding programs to (as yet unrealized plans) for in vitro genome editing and de-extinction. These interventions make use of cutting-edge genomic science and biotechnology. But as acts of rewilding and restoration, they are often explicitly counterpoised to, and calibrated against, the long histories of selective animal breeding that led to domestication.[19]

These interventions are evident in the use of Heck cattle. The Heck brothers claimed to have back-bred the aurochs in the 1930s, reversing long histories of productivist selective breeding to restore the wild character of the animal. At the OVP and elsewhere, the descendants of these animals are encouraged to de-domesticate themselves, breeding freely, learning behaviors, and creating social dynamics without an analog in modern agricultural systems. But some rewilders have expressed concern that the Heck cattle do not sufficiently resemble the aurochs. They claim they are much smaller and lack their antecedents' levels of sexual dimorphism. In other words, they look too much like domestic cattle. But we are told that Heck cattle are not being judged purely on these anatomical and aesthetic criteria. Instead, their form is taken to indicate their inability to fully function as aurochs substitutes.[50]

As a consequence, at least three new aurochs back-breeding programs have been proposed. The most established of these, the Tauros Programme, combines archeological and historical information, field knowledge of the behavior of current breeds, and new data generated by the sequencing of the aurochs genome.[51] This sequencing was funded by industry in the hope of finding new genes for agricultural productivity, but the methodology and data have been repurposed first to trace the prehistoric origins of the aurochs, then to differentiate and rank existing cattle to guide future breeding. In creating their Aurochs 2.0, scientists rework cutting-edge genomic technologies to "resurrect" animals that will be "born to be wild."[52] In practice, those involved with the Tauros Programme are not too concerned about authenticity. They favor animals that are not too aggressive; they build public support by producing varieties that resemble different local European cattle breeds. But the promise of resurrection is taken a step further in a rival (but much less coherent) initiative that aims to de-extinct a single ur-ox from frozen DNA.[53] This speculative venture sits alongside a wider range of de-extinction for rewilding projects, and there is a notable enthusiasm for synthetic biology among some prominent rewilding advocates.[54]

In contrast to these acts of de-domestication for rewilding, those involved in biome restoration promote the modification of their human-associated microbial organisms as acts of domestication. "Domestication" is a famously slippery term, not least when it is applied to components of the human multibiome.[55] The idea of the wild is clearly less compelling as an affirmative counterpoint when it comes to the restoration of human symbionts, and wild helminths are considered riskier that de-domesticated cattle. In the microbial context, domestication serves as an index of desired human control. For example, the manufacturer of the rat tapeworm ova *Hymenolepis diminuta* (HDC) seeks to reassure potential customers on their website that their organism is "domesticated, since we control its reproduction for our own benefit."[56]

William Parker is a prominent immunologist who pioneered the laboratory production of HDC. Writing with colleagues, he claims that the "process of domestication with helminths has already begun, with pioneering individuals from diverse backgrounds isolating specific helminths or combinations of helminths and evaluating their effects on disease."[57] He suggests building on the process through which substitute helminths were identified, recommending that "the animals that are the most advantageous with the least drawbacks must be selected and cultivated for human benefit."[58] He goes further in a commentary about the future of helminth therapy:

> Can modern science use naturally occurring helminths as a starting point, and improve them? The use of transgenic helminths and longer-lived helminths are examples that might be considered. Biotechnology to improve production of helminths, including in vitro culture of helminths or cultivation of human specific helminths in genetically modified animals (e.g. humanized or immunosuppressed mice) might also be considered. In addition, irradiation of organisms to achieve sterility and eliminate the possibility of transmission is a possibility. As a specific example, a "designer helminth" might be envisioned which has many of the properties of the bovine tape-worm (long life span, self-limiting colonization) but which is substantially smaller in size, thus eliminating much of the inconvenience when the organism dies and is eliminated from the body.[59]

Parker's proposal for helminth domestication by genome editing is far from realization. As I explore in chapter 7, it is hard to see it being funded in the current funding landscape for drug development. But modified bacterial bugs as drugs are closer to coming to market in translational research on FMT and the developments of ecobiotics. The hope is that these novel pill ecologies can be wired to displace dysbiotic, in situ gut ecologies and found a new colony. Once reset, the new microbiome should resist subsequent infection and establish an immune-modulating relationship with the host gut.

Simulate

This use of bugs as drugs draws attention to the long history of biochemical innovation, in which molecules isolated from living organisms were synthesized and deployed to modulate the circulation of bodily and ecological systems. Often these secretions were turned against the organisms themselves, as through the development of vaccines (e.g., for smallpox) or antibiotic compounds (e.g., penicillin). Such synthetic practices have been common in the history of managing the natural environment, where chemicals derived from plants became the pesticides central to antibiotic modes of managing life. These molecular interventions form part of a broader spectrum of infrastructure modifications in which the agency of keystone species is simulated for the rationalization of ecological systems. Examples include the scarecrow—a rudimentary technology for agricultural security—and the long history of training dogs to hunt, kill, herd, or otherwise control a range of human and animal species.[60]

In the second section of this chapter, I explore how these technologies continue to be used as agents of symbiopolitical control to enable and sustain the reintroduction of keystone species. Here I will focus on situations in which human actors simulate the ecological functions performed by missing keystone species. Often these interventions are cast as intermediate stages toward the eventual self-willed circulation of the target ecology, to avoid chronic dependency. For example, methods for predator simulation have been developed for Heck cattle and the other herbivores at the OVP. These animals were released into the reserve

*Figure 14.* Dead animal carcass at the Oostvaardersplassen. Photograph courtesy of Staatsbosbeheer.

without wolves, their breeding was unconstrained, and their populations quickly expanded. During harsh winters, many animals died of starvation (Figure 14).

Their suffering is highly visible and causes great concern to local citizens and to Dutch animal welfarists. An expert commission was assembled to discuss solutions. In their reports, they ruled out the return of the wolf on the grounds that the reserve was too small and that wolves would use fences to hunt and terrify prey.[61] They discouraged supplementary feeding, as it would compromise the animals' self-regulatory breeding behavior. Their compromise solution is known as the "proactive predator simulation model." Here a ranger, armed with a high-caliber silenced rifle, kills those cattle and horses that the ranger thinks are not in a condition to survive the winter.[62] This assessment is not done according to a fixed quota but by examining them "with the eye of the wolf." Because the ranger cannot observe wolves interacting with cattle and horses in the wild, the human becomes wolf by combining field experience with systems developed to assess cattle and horse welfare in agricultural settings.[63]

OVP conservationists are well aware that this is only a partial solution, as this simulation does not replicate the multispecies affective landscapes—of fear and calm—associated with predator–prey dynamics. A silent and often distant or invisible human wolf provides quite different affective modulation to the anticipation and experience of pack hunting. Without the shaping influence of this landscape of fear, herbivores graze where they like, which takes a visible toll on the vegetation.[64] Conservationists in the Scottish Highlands, concerned about the abundance and grazing patterns of red deer, have been experimenting with the human simulation of the lupine landscape of fear.[65] They have trained volunteers to patrol the landscape as wolves, howling and discharging firecrackers. Meanwhile, a discussion is beginning in conservation circles about developing "hunting-for-fear" approaches to deer management, including training domestic dogs to simulate the hunting behaviors of their lupine ancestors. Advocates acknowledge that these proposals will be controversial.[66]

In the context of biome restoration, many academic immunologists see the molecular simulation of helminths' immunosuppressant abilities as the most likely future for this therapy. There is growing excitement in the pharmaceutical industry about the molecules that might be derived from helminth secretions, and new drugs from bugs are envisaged for standardized pill-based delivery.[67] Such a trajectory would secure the profit margins currently delivered by patients' chronic dependence on immunosuppressants. But critics like William Parker note the difficulties of recapitulating an organism in pill form and dispute their salutary claims: drugs tackle symptoms, but they do not prevent dysbiosis or restore health.[68] Nonetheless, influential public health experts like Peter Hotez see opportunities for helminth-derived molecules to inform new vaccines and anthelminthic drugs that will enable the eventual eradication of hookworms.[69] I explore this debate in more detail in chapters 5 and 7.

## Managing Ecologies for Keystone Species

The four practices we have encountered so far focus on securing the return of keystone species and/or their ecological functions. They are subtended by a further set of symbiopolitical interventions that seek to bring dysbiotic ecologies into a condition such that keystone species can become established and interact appropriately, and to ensure that they are not threatened by the proliferation of other such species. These probiotic acts involve a great deal of care by conservationists, users, and clinicians. However, care can be lethal for those life-forms whose interests jeopardize keystone-enabled circulations; this is biopolitics writ large. Here I identify three further sets of practices for managing the wider composition and dynamics of the ecologies into which keystone species are introduced.

### Prepare

Ecological monitoring of the successional dynamics of early rewilding projects—including proactive experiments at OVP and Knepp, and passive abandonment elsewhere—has revealed how small differences in starting conditions can have big effects on which ecologies subsequently emerge.[70] Similarly, studies of the success of species translocation projects have identified how local disease ecologies, vegetation composition, and competition all configure whether and how keystone species become established.[71] This work highlights the need for good preparation so that rewilding and restoration can be successful. Ideally, rewilders would reset the basic abiotic and biotic conditions of their focal ecosystem before introduction. But in many cases, like the OVP, they have modest political leverage and economic resources for fundamentally reorganizing land management, and they must work with the ecologies they inherit. To illustrate such preparatory processes, I will focus on the case of Knepp, the former intensive dairy farm in Sussex in the United Kingdom that has been transitioned toward an OVP-style naturalistic grazing project with some livestock production.

In her autobiographical account of enabling this transition, Isabella Tree explains how she and her husband, Charlie Burrell, first released the land from intensive spraying with pesticides, herbicides, and fertilizers.[72] But they anticipated that this release would not be sufficient to reset the grassland ecology, which had become dominated by high-yield nonnative species. Some fields were sprayed with weed killer, plowed, and a local wild grass seed mix sown. In other fields, the top horizon of artificially enriched soil was removed to create favorable conditions for the seed mix to take. With these base conditions in place, a further set of preparations was made to welcome the introduced herbivores and to accelerate their ecological engineering. Internal fences were removed to permit the animals' free movement, and stronger external fences were erected to keep them in. Some crops were grown to sustain the cattle during their initial process of de-domestication. At other rewilding projects, conservationists construct windbreaks, nest boxes, or model dams to accommodate new arrivals.

In the terms of probiotic science, Isabella and Charlie were concerned about the potential resistance (or hysteresis) of their agricultural land to resurgence when it was to be given over to high densities of large herbivores grazing in the absence of predators. Observations of the experience at OVP suggested that spatially uniform conditions of former fields and their calm, affective landscape can prevent the cycles of forest regeneration that are so central to the Vera hypothesis. When herbivores can go anywhere, new seedlings are grazed before areas of thorny scrub can become established. Scientists studying this process at OVP and at Knepp have created experimental grazing refuges (Figure 15), small so-called "exclosures" fenced with differing degrees of herbivore permeability, disturbed to simulate the activities of wild boars, and planted with saplings. They suggest that these onetime interventions can help catalyze the regime shift toward forest resurgence. They promote more naturalistic alternatives to fencing, including the use of coarse, woody debris and the creation of small lakes, islands, or floods to restrict herbivore access.[73] This research has informed interventions at Knepp.

Preparatory work is also undertaken by those using hookworms and becoming subject to the holobiont governmentality of helminth therapy. Those taking worms for the first time are encouraged to read the helminth care manual. This wiki collects users' experiences of the effects on their worms of different foodstuffs, medicines, and other everyday chemical exposures. It notes substances that are known to kill worms, as well as those that are believed to favor the establishment and flourishing of both worms and their microbial kin. Users learn to manage their diet and lifestyle in advance of infection, releasing their bodies from some prevalent antimicrobial chemicals. They then change their practices to support their colony. Some even remark on how their dietary preferences

*Figure 15.* Grazing exclosure at Knepp. Photograph by author.

shifted once their worms took up residence, a change they attribute to the preferences of their worms. Some figure this preparatory work as part of a more systematic, dirty model of securing probiotic microbial exposure and colonization.

## Disturb

Probiotic scientists emphasize that a degree of disturbance is normal and necessary for the functional circulation of resilient ecological systems. They note that large-magnitude disturbances can compromise functionality for a period, and that disturbances can tip systems into new states, for better or for worse. Conservationists researching disturbances have identified a wide range of abiotic and biotic processes (including predation, grazing, decomposition, disease, fire, wind, drought, and flood) that shape an ecology.[74] They analyze their spatiotemporal rhythms (extent, intensity, frequency, duration) and examine their ecological effects. This research informs programs of conservation management. They suggest that living with disturbance regimes was central to premodern agriculture and health care, and they trace how antibiotic approaches sought to rationalize disturbance dynamics. While many modern forms of conservation find disturbance anathema, a probiotic symbiopolitics develops new methods for choreographing disturbance, conducting the discordant harmonies of their target ecology for the purposes of rewilding and restoration.

This is exemplified at Knepp. Having prepared the basic biotic conditions of the fields and released them from regular pesticide and fertilizer application, the managers embarked on more systematic modifications to the agricultural disturbance regime. This began with the introduction and management of the large herbivores, and led to the development of a hybrid between rewilding and livestock farming. Animals are selected for their ecosystem engineering potential and for their wild aesthetics. Herd sizes are managed by controlled breeding and/or shooting to generate the desired grazing and browsing intensity.[75] Surplus animals are slaughtered and sold at a premium as wild meat, and the estate runs wildlife safaris in which the herbivores are a star attraction. In this model, de-domesticated agricultural animals are explicitly repurposed as a diversified guild of disturbance agents. This is clear with the reintroduction of Tamworth pigs, a hardy local breed selected over wild boars for their relative docility. These pigs have a formidable ability to turn over the soil in pursuit of worms. In so doing, they break up the sward, aerate the ground, and create ruderal (bare earth) conditions for ecological regeneration. Pigs create scrapes, or shallow, muddy hollows for wallowing, which introduce further complexity into the landscape.

Site managers have also shifted their practices for managing vegetation. They stopped flailing (cutting) the hedgerows to encourage their expansion into the fields. Hedges at Knepp are made up of thorny scrub species (like hawthorn and blackthorn) that surround a small number of mature trees (like oak). They thus resemble established (albeit linear) versions of the grazing exclosures that are central to the Vera cycles. Hedges, which have also become scrubby as a result of lower herbivore densities, have expanded into the field (Figure 16).

This new choreography of disturbance management also extended, more profoundly, to the farm's drainage system. The soil at Knepp is heavy and impermeable clay. As in many parts of Europe, great effort

*Figure 16.* Aerial photograph showing the expanding hedgerows at Knepp. Photograph by Charlie Burrell; reprinted with permission.

had been expended on laying field drains to dry the soil to favor plant species for animal grazing. The managers at Knepp decided to block and break up these drains, and to redirect streams so that more water stayed on and in the soil. New ponds and marshes appeared. This rewetting program was extended to rewilding the river Adur, which flows through the estate. Canalized (straightened) channels were replaced with new sinuous curves, woody debris was left in the channel, and floodwater was allowed to overspill onto the floodplain, restoring water meadows. The river was given more latitude to find its own course. These interventions are understood to have helped prevent flooding downstream by storing water and slowing its flow. Much of this work simulates the ecosystem engineering of beavers, and there are longer-term aspirations toward their reintroduction.[76]

Rewilders at Knepp have also taken a strong interest in the uncharismatic processes of rot and decomposition, which are often sidelined by the tidy aesthetics of European landscape conservation.[77] They try to keep dead trees and animal bodies in the landscape to support populations of saproxylic (deadwood-loving) and carrion (flesh-eating) insects and to encourage the recycling of nutrients. But disease prevention regulations, public sensibilities, and the wild agriculture economy of the estate mean that most carcasses are removed. Feeding stations have been created at other rewilding projects in more remote locations in Europe, where animal bodies are left to support populations of carrion-eating birds, animals, and insects.[78] Stopping the common agricultural practice of deworming cattle and horses at Knepp has also created a rare supply of nontoxic herbivore dung. This is swiftly broken down and returned to the soil by resurgent populations of dung beetles and other insects.

The theory and practice of disturbance management is not as well developed for biome restoration, and hookworm users do not describe their interventions in such terms. Nonetheless, we can stretch this idea to explore the growing awareness among helminth users of the temporalities of the immunosuppression associated with hookworm infection. Experienced users have come to know a cycle in which they enjoy a positive "bounce" in well-being a consistent number of days after infection. Sustaining this bounce requires regular dosing with helminth larvae, a process that is described as exercising or challenging the body's immune system. Sustaining their immunity thus requires choreographing corporeal disturbance in similar ways to those involved in rewilding landscapes at Knepp.

## Weed/Cull

While the modulation of disturbance regimes will inevitably favor some life-forms over others, the final set of symbiopolitical practices are explicitly concerned with managing the presence, abundance, and dynamics of a small number of species that proliferate in dysbiotic ecologies without the top-down regulation performed by keystone species. The presence of these species is understood to prevent the successful return of keystone species and the processes of resurgence that they are intended to initiate. The common focus is on a subset of "invasive" plants and animals and "infectious" microbes. Weeding and culling involve lethal forms of ecological care in which rewilders and restorers exert a sovereign power to kill specific species or errant individuals. Such acts of violence are underpinned by more extensive biopolitical strategies designed to prevent the reproduction and circulation of undesired life-forms. Many of these techniques are borrowed directly from antibiotic modes of managing life. But they also involve novel symbiopolitical modes of control that mobilize trophic relations to manage pest species. Weeding and culling thus describe the regimes of biosecurity that undergird the affirmative practice of distribution.

Interventions to directly kill unwanted organisms use the full armory of antibiotic chemicals and weaponry. At Knepp, fields of agricultural grass were sprayed and plowed before reseeding. The weed-filled margins at the edge of the estate are cut to minimize overspill and to appease local landowners. At Knepp and OVP, especially aggressive animals are preferentially shot by the discriminating human wolf or slaughtered by the stockman and given to the premium butcher to minimize the risk of conflict with visitors and their dogs. Rewilding projects in Highland Scotland dedicate significant resources to culling native ungulate populations that proliferate on local stalking estates. They weed invasive non-native plants like pine trees or rhododendrons, and they trap and shoot animals like gray squirrels and mink. Projects to "retortoise" the Galápagos Islands involve large-scale culling of introduced goats, rats, and other animals through a creative range of lethal biotechnology.[79] Such lethal control remains prevalent in the microbiome: hookworm users take anthelminthic drugs in situations where they are concerned about the size or character of their colony.

These lethal strategies are combined with biosecurity practices that restrict the circulation of undesired invasions and infections. For rewilding, this includes border fencing to prevent the self-willed incursions of deer, wild boars, feral dogs, and other species alongside the internal herbivore exclosures designed to initiate vegetation dynamics. Hookworm users have developed careful biosafety protocols to restrict what travels between bodies when worm larvae are shipped and ingested. As I explore in chapter 5, those concerned with developing dirty governmentalities for probiotic exposure are in the early stages of redesigning antibiotic hygiene protocols to differentiate practices that lead to exposure to either crowd infections or to old-friend microbes. Intimate bodily practices of childbirth and parenting, as well as washing and cleaning, are redesigned to ensure preferential colonization, exposure, and challenge.

Policing invasion while facilitating redistribution requires the careful design of networks for preferential mobility, facilitating the passage of some and restricting the movements of others.[80] But this technological differentiation of desired and undesired mobilities is often impossible. In a U.K. example, invasive mink, rare water voles, and reintroduced beaver can all move along the same watercourses. Aiding or abetting any one of these species' mobilities may affect the others. Likewise, probiotic dirty play in the countryside or the park may grant equal exposure to beneficial commensal bacteria or pathogenic and drug-resistant strains. Here the new modes of circulation imagined for rewilding and biome restoration come into conflict with the established and extensive regimes for antibiotic biosecurity designed to ensure the safe circulation of goods, labor, and services. Mobile wolves in France transgress the securitized territories of livestock production,[81] while worm larvae entering the United States in the mail will be stopped should they become visible to those charged with policing national disease-free status at the border.

A third model of weeding and culling uses top-down trophic relations for symbiopolitical means of ecological control. Biological methods of pest control have a long (and fraught) history in agriculture.[82] Here the otherwise nonproductive predators and diseases are retained or introduced to reduce populations of pests and the agricultural damage they cause. For example, ladybugs are encouraged because they eat aphids, nematodes are used to kills slugs, and the myxomatosis virus is spread to kill rabbits. This thinking has come to inform rewilding. For example, there is much hope in the United Kingdom that the return of the rare pine marten (long persecuted by gamekeepers because it kills game birds) might control populations of invasive gray squirrels (currently shot by the same gamekeepers because they damage commercial trees) and enable the resurgence of the native red squirrel (loved by the public but outcompeted by the gray squirrel) and of native woodland.[83]

Similarly, the experience at Knepp has drawn attention to the co-dependent relations between insects (like butterflies) and the plants they eat. The proliferation of one weed may be checked by a commensurate, but deferred, boom in the population of their consumer. Tree suggests that patience and nonintervention might be rewarded as cycles of boom and bust run their course. Biological competition as a means of pest control is not well developed in the microbiome, though there is a growing interest in the potential of bacteriophages (viruses of bacteria) to tackle the crisis anticipated to result from the declining efficacy of antibiotic drugs.[84] Many of those promoting rewilding and restoration hope that these symbiopolitical models of control might eventually reduce the need for lethal interventions and spatial restrictions. In different ways, they describe how functional ecologies with operational top-down controls have greater "colonization resistance," suggesting that reset, meta-stable ecologies are better able to deal with exposure to a pathogenic microbe or an invasive predator.[85]

## The Controlled Decontrolling of Ecological Controls

This chapter has reviewed a range of governance strategies in rewilding and biome restoration informed by the probiotic science outlined in chapter 2. It has presented these as forms of symbiopolitics involving the strategic use of keystone species to modulate the circulation of ecological systems both inside and outside the human body. Symbiopolitics involves all four of Foucault's forms of power, including sovereign interventions to take animal lives and governmentalities through which both humans and some animals are made subject to probiotic knowledge. In some cases, probiotic helminth users act as self-aware holobionts, changing their practices to manipulate their own microbial composition. The majority of the practices reviewed are biopolitical, targeted at the reproduction and circulation of animal populations and the ecological functions they can deliver.

Taken together, these interventions perform an environmental mode of biopower dedicated to securing the functionality of their target ecology. Existing analyses of environmental biopower as ontopower have tended to focus on governance in situations characterized as unruly and uncontrollable. Here, as Brian Massumi argues, "the figure of the environment shifts: from the harmony of a natural balance to a churning seed-bed of crisis in the perpetual making."[86] But in this analysis I have identified a different figuring of nonequilibrium ecologies, in which an ontology of metastable socioecological systems, subject to top-down ordering by keystone species, enables interventions to modulate ecological dynamics. The aim is to restore ecologies that have already tipped into disaster. In contrast to resilience approaches that seek to head off a change in phase state to preserve homeostasis, probiotic forms of symbiopolitics seek thresholds and try to tip systems across them. These interventions are akin to practices of biosecurity, but they differ in relation to the temporality of the crisis to which they respond. While forms of biosecurity seek to manage circulations to build resilience in the face of imminent but unavoidable disaster, probiotic approaches figure the present as already disastrous. They seek to reverse, restore, or otherwise transform deleterious existing transitions.

Developing this chapter's epigraph from Tsing, this choreography of auto-rewilding seeks to "figure out which kinds of weediness allow landscapes of more-than-human livability."[87] Such interventions frequently work with the proliferating and resurgent dynamics of ecologies while maintaining a biopolitical aspiration toward human control. To achieve this, they mutate existing approaches to biosecurity and focus on modulating the symbiotic intensities of ecological systems to reset their focal ecologies. In using life to manage life, probiotic approaches promote a naturalization—or, better an ecologization (after Cary Wolfe)—of the biopolitical paradigm.[88] The aspirations for control in these approaches go beyond the maintenance of the status quo or the warding off of coming disaster. They share a progressive belief in the future and are transformative in their aspirations. As such, they differ from the ontopolitics of resilience management or more-than-humanist affirmations of precarious survival amid the ruins that I outlined in the Introduction.

As I noted in chapter 2, there is a live conversation underway among the more philosophically inclined advocates of the probiotic turn about the Janus-faced character of resilience: that resilience is not, in itself, a normative good. As the authors of one recent paper put it,

> The fundamental challenge we have in restoration/trophic rewilding is that we want (or do not want) to change the state of the focal ecosystem to follow a certain restoration goal. Depending on this decision, we must actively decrease or increase the resilience, and thus attempt to erode or preserve the ecological memory maintaining the ecosystem in its current (desirable or undesirable) state.[89]

What matters, they argue, is the "resilience of what to what."[90] Dysbiotic systems can be extremely resilient to planned restorative interventions, while ecologies under restoration may require intensive periods of human care before they achieve resurgence, or self-willed, functionality. This might only be sustained through permanent regimes of care and management. In symbiopolitics, resilience itself becomes subject to optimization; it is to be eroded, restored, or engineered for the purposes of ecological transformation. Advocates of rewilding and biome restoration might be lying in the gutter (or amid Tsing's ruins), but they are staring at the stars.

What is perhaps most striking in the transformative and progressivist ethos of these probiotic interventions is their approach to ecological control. It is clear that probiotic approaches involve a recalibration, rather than a rejection, of the modernist command-and-control model and the antibiotic modes of managing life this came to inform. This is not a laissez-faire embrace of natural processes or an unreflexive return to Nature. Rewilding and biome restoration work with novel ecosystems that are inflected by anthropogenic activities and are forever dependent on regimes of human care and conservation. Josef Keulartz describes the governance practices of de-domestication and rewilding at OVP as the "controlled decontrolling of ecological controls."[91] While seemingly oxymoronic, this phrase usefully describes this novel mode of managing life. It resonates with Tsing's exploration of "the Zen arts of managed nonmanagement" involved in *satoyama*: the restoration of peasant agricultural disturbance regimes to encourage the resurgence of matsutake mushrooms in pine forests in Japan.[92]

Keulartz's and Tsing's accounts of what controlled decontrolling involves in the macrobiome, and how it conflicts with the interests of agriculture, biosecurity, and animal welfare, resonate with Heather Paxson's theorization of the microbiopolitics of raw-milk cheese making in the United States. Paxson explains how protagonists in this new craft industry, who value the work done by live and sometimes unruly microorganisms, clash with "a regulatory order bent on taming nature through forceful eradication of microbial contaminants."[93] Paxson terms the latter antibiotic model a Pasteurian microbiopolitics. She contrasts this with the cheese makers' post-Pasteurian alternative. In comparing the two, she suggests, "Whereas a Pasteurian approach treats the natural world as dangerously unruly and in need of human control, a post-Pasteurian view emphasizes the potential for cooperation among agencies of nature and culture, microbes and humans."[94] In further writing, she explains that post-Pasteurians "work hard to distinguish between 'good' and 'bad' microorganisms and to harness the former as allies in vanquishing the latter. Post-Pasteurianism takes after Pasteurianism in taking hygiene seriously. It differs in being more discriminating."[95] In a distinction that will become important as this book proceeds, she also differentiates post-Pasteurianism from anti-Pasteurianism—as expressed, for example, in the antivaccine movement—in that it maintains epistemic and political faith in (some forms) of science and state activity.

We can helpfully expand Paxson's concept of the post-Pasteurian beyond the microbiome. While this is no doubt unfair to Pasteur, her concept nicely captures the controlled decontrolling of ecological controls that lies at the heart of the probiotic symbiopolitics reviewed in this chapter, and the revalorization of the work done by nonhumans, often in collaboration with people, to deliver desired ecological functions. In the following chapter I examine the epistemic practices through which this controlled decontrolling takes place. In subsequent chapters I explore the politics of the transformations envisaged and enabled by probiotic interventions.

# 4

## WILD EXPERIMENTS
### The Controlled Decontrolling of Ecological Controls

IN THE INTRODUCTION I explain how probiotic social theorists like Donna Haraway and Bruno Latour link the parallel emergence of the human as holobiont and the diagnosis of a humanized planet to the demise of the modern figures of Nature and Man that have served as the ontological reference points for antibiotic ordering projects. They suggest that the replacement of Man and Nature by conceptions of nested socioecological systems enables an escape from reductionist modern epistemologies in which life adheres to universal and predictable laws. This figuring confounds the linear models of modern natural resource management that are founded on them. They propose that the resulting uncertainty weakens the modern settlement between science and politics, or science and its publics.[1] Here scientists generate knowledge outside of society and in advance of politics, and policy makers receive scientific knowledge with gratitude and deference, and shape publics in its image. The end of Man and Nature and the loss of certainty have led to the proliferation of knowledge claims and controversies as well as a growing demand for new ways of engaging publics with science and technology.

This chapter describes the diversity of knowledge practices that underpin the probiotic turn and explores how they respond to the challenges of knowing life in the Anthropocene that have been identified by probiotic social theory. Chapter 3 outlined the symbiopolitics of going probiotic, detailing a range of biopolitical mechanisms through which life is governed in rewilding and biome restoration. It identified an environmental mode of biopower geared toward managing the circulation of life to secure desired ecological functions and services. It suggested we understand this as post-Pasteurian, involving the controlled decontrolling of ecological controls, in a recalibration rather than a rejection of modern science and technology. In the examples of Knepp, the OVP, and helminth therapy, we saw how controlled decontrolling is an epistemic art practiced by a range of probiotic experts in novel ecological conditions that are characterized by a high degree of uncertainty and publicity. Cattle, beavers, and helminths are being made public in situations that unsettle the familiar epistemic norms and confines of the nature reserve, clinic, or laboratory.

The first half of this chapter provides a flavor of the diversity of probiotic knowledge practices, tracing a history of the modes of experimentation developed by scientists and other experts. It focuses on the shifting sites of probiotic knowledge and their associated epistemic norms, including laboratories and clinics, computer models, and field locations spanning diverse bodies and landscapes. It explores the strengths, limitations, and interconnections between knowledge produced at these sites, tracing the gradual emergence of forms of real-world experimentation that grapple with the epistemic challenges identified above. The second half engages with the critical evaluation offered by David Chandler and others of two comparable sets of knowledge practices that respond to the challenges of the Anthropocene: the "mapping" activities of advocates of resilience thinking, and the "hacking" practices of those living amid the ruins of capitalism. The chapter aligns and differentiates probiotic knowledge practices from resilience and ruins thinking, suggesting that we understand some forms of probiotic knowledge as wild experiments with transformative sociopolitical aspirations.

Before starting, it is worth emphasizing that the common scientific understanding that was mapped in chapter 2 is the convergent outcome of a great heterogeneity of epistemic activity, even in the subset of policy domains that are the focus of this book. Conservation biology and immunology—in theory as well as in vernacular and scientific practice—are strikingly different in terms of where they are done (nature reserves versus the home, clinic, or laboratory), the visibility and public familiarity of their focal ecology (charismatic animals versus obscure microbes), their necessary technologies (macroscopic monitoring versus microscopic and molecular tracing), and what is at stake in their research (human versus environmental health). In spite of this heterogeneity, it is possible to identify a set of common trends and tensions in the historical development of these fields and in the modes of experimentation on which they have come to depend.

## Modes of Probiotic Experimentation

Sociologists and geographers of science have identified several different modes of scientific experimentation,[2] three of which concern us here, which can be arrayed along an axis measuring the degree of material and political control exerted by the experimenter on the system under study (Figure 17). At one end is the *laboratory experiment*, a highly controlled, private, and artificial intervention for testing hypotheses to reveal universal laws. Some control is relinquished in *experiments in nature*, where the experimenter transposes laboratory practices into the field, manipulating specific variables, with reference controls, and observing the results, sometimes with public oversight. In the final model of *natural experiments*, the scientist is an observer, often in a highly public setting, who monitors and compares natural differences between field sites to identify place-specific patterns and trends.[3] These models offer a useful heuristic for examining how epistemic and political control is achieved in the different modes of experimentation at work in the probiotic turn.

### Natural Experiments

The scientific practices of the probiotic turn emerged from marginal countercurrents in conservation biology and immunology. They are not a result of Big Science, planned and funded by state or private agencies. Instead, they are propelled by the interests of often maverick individuals without great scientific or political resources. Many of the key early scientific insights of rewilding and biome restoration came from close observation of natural experiments—ecological developments in sites

| Laboratory experiment | Experiment in nature | Natural experiment |
|---|---|---|
| Made | ←——————→ | Found |
| Order | ←——————→ | Surprise |
| Secluded | ←——————→ | Wild |

*Figure 17.* Modes of experimentation.

or within bodies that were the outcome of unplanned, accidental, and sometimes disastrous events.

For example, those involved in rewilding make frequent reference to the lessons learned from the large-scale processes of land abandonment that have taken place in marginal areas of Europe and North America. These unplanned changes, coupled with the declining popularity of hunting, have led to reforestation and the unexpected resurgence of wildlife—including top predators like wolves, bears, and lynx. Rewilders have also studied spontaneous ecological resurgence at postindustrial and military locations. In extreme examples, reference is made to postapocalyptic sites, like the exclusion zone around the nuclear reactor at Chernobyl, the demilitarized zone between North and South Korea, or the Siberian aftermath of the collapse of state socialism. The probiotic science of rewilding partly emerges from studies of ruined past futures. The site that is now the OVP was originally earmarked for development as an oil refinery and was abandoned partly as a result of the 1970s oil crisis. The subsequent ecological changes that came to inspire Frans Vera were arguably only permitted because it was an unvalued brownfield wasteland. Even the flagship example of the wolves at Yellowstone National Park was not designed as a rewilding experiment. At the time of their reintroduction, conservationists were interested in the wolf as a threatened species and did not anticipate their landscape-scale impacts.

Although Vera framed the OVP and other early rewilding projects in the Netherlands as experiments to test his alternative hypothesis about the paleoecology of Europe, no clear scientific methodology was established to put this hypothesis to the test. No effort was made to establish control plots within or outside the reserve to enable comparison, and little monitoring was carried out.[4] In part this was due to resource constraints, but it also stemmed from an uncertainty as to what to watch and what to count; processes and functions are harder to surveil than species composition.[5] Meanwhile, open-ended observation fitted with the public celebration of the OVP as self-willed nature development—"the Serengeti behind the dykes"—whose epistemic virtues lay in its ability to generate surprises.[6] Vera's public presentations of what was learned from the OVP make frequent reference to anomalous occurrences, especially those that his critics suggested would never happen. For example, he often celebrates the arrival of the Netherlands' first pair of white-tailed eagles that nest and breed below sea level.

Likewise, the early science on biome depletion and restoration was catalyzed by unanticipated findings on the margins of observational studies designed to explore the beneficial effects of helminth control. Researchers noticed increases in autoimmune disease among their de-wormed cohorts. Subsequent epidemiological research confirmed these effects in populations across a range of tropical countries as well as within populations migrating from these areas to temperate and other regions without helminths.[7] Knowledge about biome restoration also emerged from observations of anthropogenic disasters. In one frequently cited example, neurologist Jorge Correale traced the impact of the early 2000s economic crisis in Argentina on a cohort of patients with multiple sclerosis. The sudden deterioration in public sanitation, livelihoods, and health care provision caused by the peso crash led to an increase in helminth infections among urban residents. Correale persuaded his patients to keep their new worm infections, and for nearly five years he observed improvements in their symptoms. He traced the relapse of those patients who asked to be dewormed.[8]

As their fields of science matured and became more mainstream, prominent conservation biologists and immunologists have reflected on the degree to which rewilding and biome restoration emerged from such "experiment[s] of nature"[9] or "natural experiments."[10] While they celebrate the serendipitous origins of their fields, they often do so with a modicum of dismay and disdain, for observational studies and natural experiments do not have the same epistemic status as those conducted in laboratories, clinics, and field stations. Looking to the future, they consistently advocate more laboratory experiments or experiments in nature involving the controlled manipulation of ecologies and keystone species.

Experiments in Nature

There is in fact a long history of experiments in nature to test theories of top-down ecological regulation. The concepts of the keystone species and the trophic cascade were first proposed by Robert Paine in the 1960s on the back of an experiment in which he forcibly and repeatedly removed starfish from an eight-meter stretch of the Pacific shoreline and compared the results to a control plot.[11] Paine traced how the absence of starfish led to increases in the populations of mussels they would have predated and declines in the diversity of the ecosystem. It is only recently that scientists have begun designing experiments in nature to explore self-declared rewilding projects.

In chapter 3, I introduced a series of experiments conducted by the ecologist Christian Smit and his colleagues in which they constructed grazing exclosures at the OVP with different degrees of herbivore permeability. Working some twenty years after the creation of the site, these scientists planted their enclosed areas with saplings, simulated wild boar furrowing, and observed the effects on vegetation succession in relation to a set of unfenced control sites nearby.[12] These controlled manipulations of disturbance regimes turn the field into a lab to test hypotheses about the ecosystem engineering effects of large herbivores. They aim to prove that excluding animals would catalyze the circular dynamics of forest regeneration that are central to the Vera hypothesis.[13] Through these interventions, the scientists were able to gain control over the local dynamics of the reserve and to generate statistically significant data on the drivers of ecological change.

Although the number of such rewilding experiments in nature is increasing, they remain rare.[14] They are expensive, often as a result of the size of the area that needs surveillance and the length of time a successful experiment must run. It can be politically difficult to gain access to field sites, especially in the case of illegal or self-willed reintroduction, and it is practically challenging to install and maintain scientific instrumentation. Field equipment breaks, is stolen, or vandalized. It gets damaged by animals, who break into or out of their enclosures. The controlled manipulation of keystone species may also be ethically problematic because of the need to exclude or kill species that may be rare and/or charismatic.[15] Finally, and most profoundly, some involved in the practical conduct of rewilding experiments remain skeptical about the ability of scientists to design an experimental apparatus capable of capturing the emergent properties of a resurgent ecology. While they recognize the epistemic power of controlled experiments and the testing of preformulated hypotheses, they place more value on the potential of surprises.

In comparison, immunologists have had more success in transitioning toward controlled experiments in nature with helminths. The small scale of human–microbiome–helminth interactions has made biome restoration more amenable to laboratory and clinical experimentation, and to the types of science conducted in these prestigious "truth spots" of the life sciences.[16] Biome restoration has become subject to preclinical laboratory experiments with animal models and to experiments in nature in the form of clinical trials.[17] Mouse models have been developed by other researchers for many of the human inflammatory diseases associated with the absence of helminths. Technologies for breeding and then colonizing germ-free mice give some control over microbial composition. Immunologists have identified murine helminths that model the human effects of the loss and the return of worms. These developments have enabled laboratory experiments that are beginning to explore interactions between helminths, microbiota, and the host immune system.

The "'ick' factor"[18] of helminth infection has made it difficult to recruit patients for clinical trials. Several phase 1 safety trials involved self-medicating clinicians, like David Pritchard. But the growing awareness of the hygiene hypothesis and the hopes of the hookworm underground enabled the recruitment of patient cohorts. Phase 2 open-label and randomized controlled trials have examined the effects of different numbers of helminth species on a range of conditions. To date, these trials have been inconclusive and disappointing.[19] While they have shown some positive effects, they have yet to demonstrate the efficacy of helminth therapy as experienced by some users. In reflecting on the limited explanatory power and therapeutic success of these efforts to control host–helminth relations in laboratory experiments and clinical trials, immunologists note the difficulty of designing an experimental apparatus that can capture the sheer complexity of the relationships between the microbiome and the host immune system. Like those designing rewilding experiments in nature, they find themselves struggling to control for the multivariate and contingent character of the ecologies under analysis. They note that the experimental apparatus of the mouse model and the clinical trial emerged in an era of reductionist biology. These epistemic tools are designed to test simple cause-and-effect hypotheses on the impact of one gene, one molecule, one lifestyle change, and so on, over short time periods.[20]

In response, immunologists have developed new experimental apparatuses for controlled decontrolling that are equipped to explore longitudinal change, postgenomic interactions, and the effects of ecological diversity.[21] For example, new experimental approaches are emerging with the rise of wild immunology.[22] Immunologists dissatisfied with germ-free mouse models have begun to study the microbiome and immune systems of so-called wild rural, feral urban, and domestic pet shop mice.[23] The microbial diversity of these groups of mice was discovered by accident in research conducted by wild animal biologists. To harness the epistemic potential of these accidents, immunologists have teamed up with ecologists to study these murine natural (or, better, Anthropocenic) experiments in situ. In so doing, their mode of experimentation shifts radically away from the norms of the laboratory toward the field skills, instrumentation, and epistemic norms of ecologists monitoring animals in the wild. Wild immunologists have also begun to bring these mice into the lab, using them in place of inbred and sterile lab animals. This has involved experiments to "rewild" lab strains by rewilding the environments, societies, diets, and lifestyles of their mice to better approximate the real-world conditions of their murine (and by extension human) subjects.[24]

Ecological immunologists like Andrea Graham have been pioneering this controlled decontrolling in helminth research with mouse models. In their experiments, they "dirty up" their mice by manipulating their diets, environments, and microbiomes to simulate some of the probiotic dirty governmentalities reviewed in chapter 3.[25] They have built farm-like spaces that enable the controlled analysis of the variables revealed by their natural experiments with wild mice.[26] In so doing, they invert the push in rewilding to make research in the field more like research in the lab; they try to make their labs more like a set of wild field sites. This work seeks to preserve the epistemic power of mice as model organisms capable of generating generalizable findings to translate to humans. Prominent advocates of wild immunology, like Rick Maizels, promote a continuum of potential model animal ecologies that could stand in for the ecoimmunological realities of humans living in varying degrees of wildness or domesticity. They speculate that the "diseases of development" experienced by animals will approximate those of humans living in comparable conditions.[27] This approach scales the anthropomorphism of laboratory mouse research to the planetary, dissolving any clear boundary between the laboratory and the world it is taken to represent.

Comparable epistemic shifts away from the bounded and controlled practices of the laboratory can also be observed in efforts toward the controlled decontrolling of the helminth clinical trial. Researchers like William Parker and his colleagues have become dissatisfied with the slow pace, high cost, and inefficacies of the standard trial design. They have begun collaborating with and learning from DIY helminth users. In 2014 they began an ongoing program of sociomedical research, in which they interviewed helminth providers and physicians treating helminth users.[28] They worked with providers and support-group gatekeepers to distribute a questionnaire and gather anecdotal reports of individual experiences from social media. This research gathered information about the type of worms used, the methods of application, the conditions treated, and the patients' and physicians' experiences. They propose that the material gathered from these surveys might inform future clinical studies.[29]

This push to engage users in experiments and to learn from their experience and expertise is symptomatic of a wider trend toward citizen- or crowd-sourced science in contemporary efforts to make the microbiome public. For example, companies like uBiome offer consumer microbiome genetic profiling services, similar to those developed by 23andMe for the human genome.[30] They have amassed (at a low cost) significant microbiological and social data that they are using to develop commercial diagnostic platforms. The same is true for the nonprofit OpenBiome, which runs the largest stool bank for FMT in the United States. Donors gift stool and their data to OpenBiome, which are then handed over to Finch, OpenBiome's sister commercial operation, and used for the development of diagnostic and therapeutic products. These developments are indicative of the broader shift in the conduct of clinical trials toward what Melinda Cooper terms distributed experiments, which she defines as "efforts to outsource pharmacological innovation to a distributed public of patients through the use of social networking software. These platforms allow drug developers to escape the limits of the conventional clinical trial by tracking the experimental practices taking place in the distributed clinic of unregulated drug consumption."[31] Cooper notes how this process allows drug companies to outsource the risks and costs of trial recruitment, design, and conduct. It allows them to benefit from the affective corporeal labor of distributed networks of often desperate patients self-experimenting with unlicensed therapies, and to spot and solicit surprising results that would be anathema to the product testing model of the traditional clinical trial.

As Cooper argues, these developments shift the sites of experimentation, demanding a rethinking of the politics of experiment, a theme I take up in chapter 5.[32]

## Real-World Experiments

This short history demonstrates a convergence in the types of experiments that are valued in rewilding and biome restoration. In their controlled decontrolling of ecological controls, probiotic scientists have moved from the close observation of surprising accidents to developing experimental apparatuses capable of disentangling the processes revealed by these surprises. As the experimental apparatuses offered by modern reductionist science were found wanting, so they sought new apparatuses better equipped to grapple with the local specificities, ecological complexities, and public character of the ecologies under analysis. These epistemological developments are exemplary of the rise of what Wolfgang Krohn and Johannes Weyer have termed real-world experiments, in which the site of the experiment moves from the bounded space of the laboratory, clinic, or even the nature reserve. Here the mode of experimentation shifts to encompass collective, public, or distributed forms of knowledge production.[33] As Matthias Gross and others have noted, these experiments often occur in places so inflected by human activities that it is not possible to establish natural controls or to set clear geographic boundaries around the scope of the experiment.[34]

### Mapping and Hacking

Commentators have presented the spread of real-world experiments as an epistemic response to the scientific uncertainties of the Anthropocene. Bruce Braun asks, "Might we be witness today to the *experimentalization* of life as part of a mode of government proper to the Anthropocene?"[35] Examples akin to those detailed earlier in this chapter have been identified across a wide range of policy domains, and critics have begun to identify and specify a diversity of knowledge practices through which scientists, policy makers, and others seek to anticipate, harness, and foreclose on the epistemic and political virtue of surprises.[36] I draw on this work to further specify the approaches to controlled decontrolling at work in the probiotic turn.

I focus in particular on David Chandler's critical review of the modes of knowledge and governance that he sees emerging in the face of the loss of modernist epistemological assumptions associated with the Anthropocene condition.[37] Chandler differentiates three types of response: mapping, sensing, and hacking. He suggests that all three depart from modern command-and-control or solutionist approaches, share a loss in the modernist belief in linear progress, and put the "nature of entangled being at the centre of politics rather than the designs or goals of the human as subject."[38] Chandler offers four criteria to differentiate these modes of governance which in Figure 18 are listed in order of their departure from modernity.

Chandler and others identify sensing as the most established mode of Anthropocene governance.[39] It describes an approach focused on responding to emerging events through the collection and correlation of real-time data. Sensing is not concerned with causation or even prevention but with minimizing impacts or disturbance. Chandler argues that its primary focus is on homeostasis, or the preservation of the status quo, and as a consequence it is the mode of Anthropocene governance that has been most widely challenged by critical social scientists. Sensing is common in the highly database-driven approaches to wildlife conservation that focus on surveilling and counting species, and on calculating and managing extinction risk.[40] It is also common in approaches to biosecurity that survey undesired viral, microbial, and animal movements to anticipate epidemic events.[41] Although the science of the probiotic

| Modernity | Rationality | Linear causality | Culture/ nature divide | Progress |
|---|---|---|---|---|
| Mapping | Autopoiesis | Non-linear causality | Depth/ immanence | Adaptation |
| Sensing | Homeostasis | Correlation | Surface/events | Responsiveness |
| Hacking | Sympoiesis | Experimentation | Entanglement/ becoming with | Radical openness |

*Figure 18.* Chandler's four criteria to differentiate modes of governance: modernity, mapping, sensing, and hacking. Adapted from D. Chandler, *Ontopolitics in the Anthropocene: An Introduction to Mapping, Sensing, and Hacking* (London: Taylor & Francis, 2018).

turn is beginning to borrow approaches, data, and technologies from conservation and biosecurity, sensing is not prominent as a mode of governance within rewilding and biome restoration. Their shared concern with dysbiotic ecologies negates the usefulness of homeostatic approaches, and it demands more transformative models of knowledge and governance. But the real-world experiments of rewilding and biome restoration can be compared and specified through the concepts of mapping and hacking.

## Mapping (and Countermapping)

For Chandler, mapping describes a collection of scientific knowledge practices that model the immanent drivers of complex, emergent events. Mapping approaches do not share sensing's turn away from causation. Instead, they assume that "causality is non-linear and that knowledge is not universal."[42] Mapping does not abandon the possibility of explanation but rather shifts from the pursuit of universalities to focus on how the same external stimulus can produce different responses depending on the internal relations of a particular ecology. In other words, mapping examines the topology—the shape and intensity—of these internal relations. It is interested in how they are wired.[43] Mapping as knowledge making involves tracing nonlinear processes and interactions.[44]

Chandler explicitly links the rise of mapping to the institutionalization of systems biology and resilience thinking in policy making. Although he proposes that mapping adheres to an autopoetic (or self-generating) approach to system change, and is not homeostatic like sensing, his analysis tends to align mapping with the conservative character of what Bruce Braun terms the resilience dispositif: a mode of knowledge and governance informed by resilience thinking that is geared primarily toward warding off the coming apocalypse by maintaining neoliberal capitalism.[45] Chandler suggests that this approach is both "post-epistemological"[46] and "post-political"[47] so that knowledge and governance in this register "seek(s) to adapt or respond to the world rather than seeking to control or direct it."[48]

In developing their probiotic science and designing their symbiopolitical interventions, conservation biologists and immunologists have developed a wide range of methodologies for mapping ecological interactions and nonlinear processes. These include ways of thinking in cycles, flows, and feedbacks that are shared by those interested in both the life cycle of the hookworm and in the grazing dynamics of large herbivores. Similarly, techniques from paleoecology and evolutionary biology have been refined to detect the ecological anachronisms and vacuums left by the absence of coevolved ghost species, whether these be elephants or helminths. Space does not permit a detailed account of all of these knowledge practices, so I will focus on what is arguably the most prominent common set: the use of network analysis for understanding ecological interaction webs, or how ecosystems are wired and rewired by keystone species. These models inform the practices of identification that were outlined in chapter 3.

In simple terms, a network analysis identifies all of the organisms that make up an ecological web and counts the number and frequency of times they interact. This approach tends to focus on trophic (or food web) interactions, with the ultimate aim of measuring the flows of energy within and between different trophic levels. This requires large amounts of observational data. Surveillance has long been a foundational field practice of ecology, and the spatial scope, quantity, and quality of observational data have been significantly enhanced by technological developments. These include new forms of telemetry that reveal movement (like miniature sensors, camera traps, and drones) and molecular methods that identify the plants or animal remains in collected feces.[49] Likewise, interactions in the microbiome have been revealed by new forms of microscopy and by next-generation sequencing of the transcriptome, which makes visible the genetic signature in an organism of past physical contact with another.[50] In practice however, it is not possible to follow every organism, and data tend to only be available for a subset of high-profile interactions between pairs of species— for example, between a wolf and an elk, or a starfish and a mussel.

The subsequent analysis of the gathered data involves statistical modeling tools designed to make visible the topological structure—or wiring—of the food web.[51] These models map links and identify the key nodes in the network. These nodes are the strongly interactive keystone species. The models quantify and rank the influence of different species on the functioning of the web.[52] Working in the absence of comprehensive data sets, advocates propose "probabilistic network models . . . [such as] statistical or rule-based models in which the probability of pairwise species interactions is predicted from a set of parameters, which can often be mapped into one or a small set of species traits."[53] This approach identifies the properties of species that consistently predict how they will interact (like habitat use, abundance, or size). It then builds models to simulate the effects on the food web of removing or adding different species. The aim is to identify which species might best become established to rewire the focal ecology, and to anticipate which species risk becoming invasive and proliferating. These models aim to find thresholds within ecological systems and to test their hysteresis, or their responsiveness to designed interventions.[54]

As experimental sites, these network models are produced by scientists working in offices armed with powerful computers—spaces that offer the privacy and control of the laboratory. They collate and order observational data collected from messy natural experiments, and occasionally experiments in nature like those conducted by Paine. As sociologists of science have frequently observed, models—as a mode of experimental apparatus—can have great epistemic potential and political power.[55] On the one hand, a good model works to generate surprises, revealing relationships that are invisible to field scientists mired in their specific locality and insensitive to the heterogeneous multitude of interacting organisms. Such models generate questions for future investigation and suggest relationships with general application. On the other hand, the findings of a good model can travel to shape understandings of other ecologies elsewhere.[56] A good model allows ecologists to simulate interventions that might be politically and ecologically unfeasible in the field. They help to overcome some of the limitations to practical experiments in nature identified by field scientists. These models have performative power: they work to shape the realities they are taken to represent.[57]

For example, a collection of prominent rewilding scientists have used network models to inform frameworks for adaptive management, "a strategy where management decisions and actions are permanently recalibrated based on *a-priori* defined goals and the knowledge obtained from continuous monitoring of the current response trajectory of the focal ecosystem."[58] Adaptive management emerged from resilience thinking in systems ecology[59] and has been flagged by Chandler as exemplary of mapping as a mode of autopoietic governance that "can not impose or direct outcomes from above but only works indirectly to shape or enable the processes of interactive emergence that are endogenous or internally generated."[60] This framing captures a dimension of the symbiopolitical interventions I detailed in chapter 3, in which network interaction models are used to identify keystone species capable of modulating ecological dynamics by reverting trophic cascades or enabling ecosystem engineering.

Chandler argues that mapping has vestiges of modern regimes of control, but here governance happens from the bottom up, not the top down. He suggests that in mapping, "there can be no external goal of 'progress' to pre-set goals, only the careful management or modulation of interactions to attempt to balance and ease the strains of adaptation as an ongoing process."[61] But in aspiration, if not (yet) in reality, the forms of adaptive management proposed by Svenning and others would seem to differ from those identified and criticized by Chandler. In these probiotic approaches, there is a more proactive intention to think critically about the "resilience of what to what," and to design interventions that will transform dysbiotic ecologies: a priori goals are established.[62]

As I explained at the end of chapter 3, these approaches are not seeking to ward off ecological disaster but rather to recognize that such disasters have already happened. The aim of adaptive management as a probiotic symbiopolitics is less to adapt society to the new realities of the Anthropocene condition and more to find ways of transforming humanized ecologies so that they stay within Holocene boundaries. Although this model of rewilding is open-ended and unmoored to a fixed reference condition, it does have preset goals and thus a trajectory along which progress can be plotted. This transformative model of rewilding involves practices of countermapping,[63] which inform what Wakefield terms "back-loop experiments."[64] These are experiments that explicitly challenge and seek to transform the status quo from a range of different ethical and political economic positions.

The character and politics of this probiotic transformation are the subjects of chapters 5, 6, and 7. To give one (rather outlandish and ecomodernist) example, the idea of transformative countermapping for rewilding has been developed by Bradley Cantrell, Laura Martin, and Erle Ellis in a thought experiment entitled "Designing Autonomy: Opportunities for New Wildness in the Anthropocene." Cantrell, Martin, and Ellis speculate whether an artificial intelligence might be created for the "automated curation of wild places" without "ongoing direct human intervention."[65] Their "speculative design" of a "wildness creator" is inspired by emerging forms of "conservation by algorithm,"[66] as well as by developments at the OVP. The authors imagine an algorithm, developed in part from interaction models, with the transhumanist ability to "take on its own sentience and create a wild ecological space beyond any human control, and even human conception."[67] They argue that this wildness creator would shift the place of technology away from Leo Marx's figure of the "machine in the garden" to the "machine as gardener."[68] As I explore in chapter 6, techno-optimistic visions such as these, as well as the wider ecomodernist movement of which they are a part, try to rescue, resuscitate, and recalibrate Chandler's model of modernist design.

The modeling of ecological interactions and their effects on the functional role of the microbiome and the host immune system is increasingly at the heart of scientific work on biome restoration. Companies like Seres Therapeutics and Finch, which are developing synthetic substitutes for FMT, have sequenced large quantities of healthy human feces to try to establish the minimal standardized microbiome that could be introduced to restore gut functionality. This has involved developing machine learning approaches—based on ecological interaction algorithms—for analyzing the big data generated by high-throughput sequencing. Possible ecobiotic combinations suggested by this modeling work have then been tested in preclinical mouse models and human subjects. Given the relatively early level of development of this science and the complexity of microbial ecologies, there have been few holistic proposals for adaptive management and (counter)mapping at the population scale. Public health is still struggling to work out how (and what) to monitor to sense dysbiosis. Advocates of algorithmic rewilding have yet to explore the potential and limits of their speculative wildness creator, should it be applied to humans and their multibiome.

## Hacking

Although there has been much speculation about the general potential of ecological network models, such mapping approaches are not the most prevalent set of knowledge practices in rewilding and biome restoration. The science that has featured in this chapter so far is subtended by the activities of a range of publics that are invested in the future of their salutary keystone species for different ethical and political reasons, and that come to know their organisms through diverse epistemic practices. It is the individuals that comprise these publics, rewilding in the field and restoring their bodies, that are responsible for most of the symbio-political practices reviewed in chapter 3. In contrast to the calm offices, amenable computers, and rewired utopias that predominate in the worlds of probiotic scientists and policy makers, many of those involved in the everyday practices of rewilding and biome restoration live amid pre-carious, ruined, or painful political ecologies in which the pragmatics of everyday survival predominate. Their approach to dealing with epis-temic uncertainty and to governing ecologies differs in interesting and important ways from the countermapping of probiotic scientists.

To illustrate the modes of experimentation associated with these probiotic knowledge practices, I focus on two small groups of DIY biol-ogists working to introduce keystone species. The first is the small group of hookworm users, who pioneered the craft of breeding, introducing, and caring for worms. The second is a small network of "beaver believ-ers"[69] in Germany and the United Kingdom who have developed the field skills required to catch, transport, release, and protect beavers. I am interested in how they exemplify hacking as a mode of governance. If sensing is homeostatic (maintaining the status quo) and mapping is autopoietic (adaptive management directed toward problem solving), then for Chandler hacking involves "a process of sympoiesis [that] seeks to enable the creativity of the Anthropocene rather than merely resist it or limit its effects."[70] Chandler suggests that hacking involves a distinctly experimental approach to dealing with uncertainty and to governing in the absence of ecological reference conditions. He finds hacking exem-plified in the writings of Tsing on the knowledge and survival strategies of people living precariously in the ruins of capitalism.[71]

Helminth therapy originates in a small group of individuals (mostly, but not exclusively, white men in the WEIRD world) who seek out worms as a means to tackle their autoimmune diseases.[72] They were unsatisfied with the pharmaceuticals and therapy suggested by their physicians and had heard about the scientific research underway on helminths. Some enrolled onto early clinical trials. They then traveled to Africa and Cen-tral America to source their own worms, using published epidemiological data to locate infection hot spots. They paid local people for their shit and self-infected with the larvae they gathered. Returning to the United States or the United Kingdom, they developed methods for cultivating and isolating worms from their own feces. They built incubation facilities in homes and garages from readily available domestic technologies. They learned how to count larvae with basic microscopes, and they learned how to reinfect themselves. They read up about possible pathogens that might pass with the worms and tested themselves to ensure they were clean. They experimented with dosing to work out how many worms they needed and how frequently to reinfect.

With this information, they then began to advertise and sell their worms online, encouraging clients to travel to their clinics and sending regular supplies discreetly through the mail. They now collaborate with their clients to test the efficacy of different combinations of worms for different conditions and users. The collect anecdotal accounts of user experiences and knowledge of what foodstuffs, medicines, and other products harm worms. They have compiled incubation guides for those wanting to try it at home. They share this information online and offer medical advice and emotional support via social media and telephone to a growing community of established users and new inquirers. Some have fled from or have had to circumvent regulatory authorities. In the United States, the hookworm is a classified species, and importing it is a criminal offense. Now they seek to encourage responsible use among users and lobby for new regulations in favor of the legal production and distribution of hookworms and other helminths.

Beavers were reintroduced by the Bavarian local government in the 1960s. By the 1980s, the animals were well established along the Danube and its tributaries, and they were starting to come into conflict with farmers, property owners, and those responsible for managing drainage. The government funded a local NGO to employ two beaver wardens, who in turn coordinate a large network of volunteer beaver enthusiasts who are the first responders if a problem beaver is reported. The beaver wardens have developed a range of technologies to protect trees and other infrastructure, including fencing, tree guards, and beaver deceivers.[73] They have published guides to beaver management and offer training to wildlife managers who travel to Bavaria from all over Europe. In insoluble cases, they trap and relocate beavers; if there is no demand for beavers elsewhere, then they are killed.

Fortunately (for beavers), there is a growing demand for the animals in other parts of Europe. A further network of beaver reintroducers has developed transport methods and quarantine procedures so that animals can be shipped across the continent. The United Kingdom is one of a number of recipient countries. Animals have been reintroduced from Bavaria for many years, initially by private landowners into small zoos and wildlife parks. These beavers bred well and proved hard to enclose. They gnawed through fencing and escaped during flooding events. The private beaver enthusiasts were not overly concerned because they like to see beavers back in the landscape. These self-willed returnees have been assisted by the beaver bomber—that is, a beaver enthusiast who has been proactively introducing beavers into ecologically suitable locations, sometimes with the tacit support of local landowners.

There are at least four common dimensions to the practices of these two groups that evidence a probiotic mode of hacking as knowledge generation. The first is a commitment to inductive, vernacular, and local knowledge. In contrast to those who engage in computer-based algorithms and mapping practices, hookworm users and beaver believers do not generally have much formal scientific training. Their knowledge practices are pragmatic, open-ended, and based on a recursive cycle of trial and error. They learn from experience and seek solutions that would work in this river or with that condition. They develop contextualized interventions for discrete landscapes and bodies.[74] In so doing, they produce local, place-based knowledge of the type most commonly associated with the field sciences.[75] Knowledge of beavers and hookworms has emerged as much from impromptu improvisation and post hoc interpretation as from ab initio deduction. In interviews, they commonly express skepticism about the utility of modeling approaches. Indeed, writing dismissively of an encounter with a conservation biology modeling enthusiast, Isabella Tree notes, "The idea of constructing a computer model to identify the outcomes of self-willed land seemed like trying to predict the lifetime achievements of an unborn child."[76]

Second, those involved in keeping and caring for hookworms or beavers express pride in their craft. This term implies a particular disposition toward bodily knowledge. Craft comes from years of experience through which is developed a means of "learning to be affected" by their organisms.[77] For beavers (and other large mammals), this is knowledge that comes not just from observation but also from listening, smelling, and touching in the field. For hookworms, this embodied attunement requires close attention to bodily disposition: the itch that comes with a successful infection, the "fog" that descends as the worms get established, and the "bounce" that comes when their immunosuppression begins. Users track the presence, absence, and intensities of their disease symptoms, including nausea, lethargy, bowel movements, and streaming eyes and nose. Such "arts of noticing" among amateur ecologists have been the subject of much research and affirmative discussion in more-than-human geography and anthropology.[78] Parallels might be drawn between these beaver believers and the careful knowledge practices of those involved in conserving and reintroducing water voles in the United Kingdom. Steve Hinchliffe and Sarah Whatmore trace how these practitioners treat their animals as epistemic "wild things," marked by their contingent potential to generate surprises. They document their field techniques for "knowing around" these animals, which leave open their potential to become otherwise while still drawing from knowledge gained via experience to guide future practice elsewhere.[79]

The craft of rewilding and biome restoration involves tinkering with technology, with great value in these communities being placed on creativity and handiness. Growing hookworms is archetypal DIY or garage biology.[80] Early pioneers had limited resources or access to laboratory spaces. They hacked the protocols for growing worms published in scientific papers or made available by those providing worms for school experiments. They learned how to repurpose mundane technologies: feces were mixed with garden compost and kept in margarine tubs stored in coolers. The mixture was incubated with heating pads and thermostats commonly used for keeping reptiles. Meanwhile, the Internet remains vital for sourcing materials, sharing information, and enabling anonymity (through e-mail encryption platforms and cryptocurrency). Similarly, beaver lovers repurpose engineering and agricultural technologies. They modify plastic pipes, fencing wire, and other mundane drainage materials to deceive their animals. They refine trap designs borrowed from pest control companies. They express a playfulness when describing their work, and they find joy in recounting past failures.

Third, working with worms and beavers entangles these experts with a wide range of publics. While most of these hold ecological and immunological science in high regard, they are marked by great epistemic diversity. Practitioners are involved in what French sociologist Michel Callon and his collaborators call "research in the wild." Research in the wild requires working across groups with radically different understandings of what counts as natural knowledge.[81] For beavers, such groups include farmers, foresters, hunters, and suburban gardeners, as well as those concerned with animal welfare and animal rights. As we shall see in chapter 6, rewilding in the field entangles wildlife managers with diverse groups of ecospiritualists, primitivists, and others with arm's-length relationships with the epistemic norms of modern science. An even greater epistemic heterogeneity can be found in the hookworm underground. Here the manifest inadequacies of modern immunosuppressant medical regimes have opened up space for alternative frameworks for explaining disease causation and for founding probiotic therapeutics. These tend to share a holistic understanding of health, one often with a comparable primitivist or naturalistic discursive tenor.[82] In some of the forms that we encounter in chapter 6, the decontrolling they prescribe drifts toward an anti-Pasteurian model involving a rejection of science altogether and a laissez-faire embrace of wild encounters.[83]

Fourth, as hacking with beavers and worms is a much more public enterprise than mapping, it implies a different approach to politics. Interventions with beavers and worms have been working around the edges of the law. Some of these activities happen in a legal gray area—for example, regarding the native status of the beaver in the United Kingdom, or the legality of knowingly bringing worms into the United Kingdom or the United States in one's body, then selling them on. In other cases, it is clear that the law is being broken. But the potential public relations damage (of killing beavers) or the envisioned costs and salutary impacts (of prosecuting hookworm users) prevent prosecution. In this way, these hacking interventions have been prefigurative. They have brought novel political ecologies into existence without statutory frameworks and in ways unfamiliar to what Mackenzie Wark terms the "engineering logics" of planning and ecological design.[84] Probiotic hackers have created futures that they felt could not be made through the mechanisms of deliberative democracy.

Stephen Flowers describes helminth therapy as a mode of "outlaw innovation." It is outlaw because the "newly created knowledge obtained . . .

is widely shared with other patients, but mainstream medical research is unable to draw upon it."[85] He contrasts user innovation in helminth therapy with the more collaborative arrangements with industry seen, for example, in the development of drugs for HIV and AIDS.[86] In contrast, those introducing beavers have been more successful in securing the institutionalization of their innovations. As Sarah Crowley documents, by illegally introducing and then publicizing beavers, they choreograph a transformative political event.[87] They playfully pillory the risk-averse culture of the responsible government agency and catalyze widespread public affection and political support. They successfully challenge the preemptive logic of U.K. biosecurity policy, which is configured to act in anticipation by foreclosing unruly ecological surprises. They work with scientists, enabling the government agency to commission retroactive legitimating studies, and they provide the environmental minister with a popular platform as well as a timely opportunity to be seen as proactive, green, and innovative.

Chandler argues that hacking as a mode of knowledge generation and governance shares the post-epistemological paralysis of mapping, and cautions that it decenters humans, reason, and truth to such an extent that it risks losing critical purchase and the ability to think beyond the contemporary juncture.[88] But the practices of DIY helminth users and beaver enthusiasts illustrate a rather different mode of probiotic hacking. Here hackers have transformative aspirations. They are making a living amid the ruined aftermath of antibiotic health care and environmental management, but some seek local solutions, and others have grander aspirations to shift what counts as health and how it might be known. In keeping with the countermapping ethos of their scientific collaborators and the transformative biopolitical aspirations identified in chapter 3, they envision salutary and progressive futures.

## Wild Experiments

This chapter has provided an overview of the probiotic knowledge practices of controlled decontrolling. I have situated these within wider literatures in science studies concerned with the sites and character of different modes of experimentation, and how science-led governance is coming to respond to the ontological and epistemic uncertainties of the Anthropocene and the microbiome. I traced a history of probiotic knowledge, charting its accidental origins in natural experiments and the common efforts to study its dimensions through controlled experiments in nature. I noted some of the successes of this work, alongside a progressive dissatisfaction with the experimental apparatus offered by reductionist modern science. I charted the shift in both rewilding and biome restoration toward modes of real-world experimentation involving distributed public experiments in naturalistic field and laboratory sites less tied to objective controls. I traced how these real-world experiments conjoin knowledge practices for both mapping and hacking political ecological dynamics.

To conclude, I would like to return to the three axes shown in Figure 12, which offer a means by which we might start to differentiate and specify these approaches to controlled decontrolling. To do so, I will expand upon a previous analysis (undertaken with Clemens Driessen) of the nature and politics of experimentation in the knowledge and governance practices of rewilding at the OVP.[89] In that previous study, we explored how scientists and policy makers at the reserve grappled with the three challenges of the Anthropocene—challenges I outline in the Introduction.

The first axis (found–made) examines the contrasting ontological commitments made by scientists to an authentic untouched Nature as the reference point for experimental corroboration. Found sites are most associated with the field settings of natural experiments. They take place without human control and are authentic by virtue of the absence of human influence. They are to be observed, not changed. In contrast, made sites are associated with the laboratory. Here scientists can control all variables to generate objective, universal knowledge. The artificiality of these sites is their primary asset. As I explore in more detail in chapter 6, probiotic scientists shuttle between found and made sites to confer authority on their knowledge claims, but in practice—whether they are working with cows or worms—all the experimental sites involved with probiotic science have some degree of artificiality. The challenge of the Anthropocene is to find ways of acknowledging this artifice and of differentiating which ecologies matter.

The second axis (order–surprise) compares the epistemological and political techniques used by scientists and reserve managers to grapple with uncertainty. Again, there is a distinction between field and laboratory spaces. In the field, the epistemic value of surprises is paramount; local differences in natural experiments give insight into general processes. In contrast, the controlled conditions of the laboratory confer order and enable the rigorous testing of preformulated hypotheses. Surprises are downplayed. In practice, as historian of science Hans-Jörg Rheinberger has made clear, any good experimental apparatus—whether in the lab or the field—must generate surprises.[90] But the expert must be able to detect and then order these events to deduce their epistemic import. Probiotic science involves a practical tension between order and surprise—a tension that is often written out after the event, and that is grappled with differently by those mapping and hacking.

The third axis (secluded–wild) explores how different publics (or epistemic communities) are engaged in political decision making about the conduct of an experiment and the implications of its findings. Secluded experiments—such as those using mouse models—necessarily happen behind closed doors and may have limited public involvement. While such experiments have a place, the normative trend in science studies and Western democracies has been toward experiments in the wild. Such experiments are deliberative. They are designed to happen in the open, involving the early participation of affected publics whose means of representation emerges through their participation. In the conclusion to our article, Clemens Driessen and I promoted the concept and conduct of the wild experiment as a way through which those involved in rewilding might grapple with the epistemic and political realities of the Anthropocene. I will turn to this concept in the three chapters that follow where I explore how thinking with wild experiments helps differentiate probiotic approaches and distinguish them from resilience and ruination. I am especially interested in the geographies, temporalities, and political economies that come to configure who gets to go probiotic and how the benefits of the probiotic turn are distributed.

# 5

# GEOGRAPHIES OF DYSBIOSIS
The Patchiness of the Probiotic Turn

THE VISUAL CULTURE of human–hookworm relations is characterized by two striking archetypes. In one, the hookworm is a nightmare of nature, a monstrous, blood-sucking parasite that is found out there or back then. In the other, the hookworm is a gut buddy or colon comrade, a mutualistic old friend with whom we coevolved and in whose absence our immune system goes awry. These incarnations are captured in Figures 19 and 20. Their coexistence reflects the striking spatial paradox of concurrent programs to deworm and reworm a variegated world. This in turn indicates how hookworms are not inherently pathological. Hookworms are pathobionts, or mutualistic organisms that can become parasitical as a result of shifts in their infection intensities. These shifts give rise to disease situations caused by both excessive presence and ghostly absence. Too many or too few hookworms can cause dysbiosis.

There is a stark and uneven geography to where people encounter hookworms as parasites, ghosts, or mutualists. This chapter examines the geographies of the probiotic turn in the context of a broader exploration of the global patterns and historical drivers of political-ecological dysbiosis. It explores where in the world is it possible for probiotic relations to come into existence, and whether the possibility of going probiotic is conditional on others elsewhere living in pathological relationships with their internal and external environments. Rather than producing a standard map, I develop a relational geography of ecological dysbiosis that places the ecological science outlined in chapter 2 as well as the forms of symbiopolitics outlined in chapter 3 in their political and economic contexts. It explores how the role played by a keystone species is configured by the dynamics and the intensities of its political and ecological

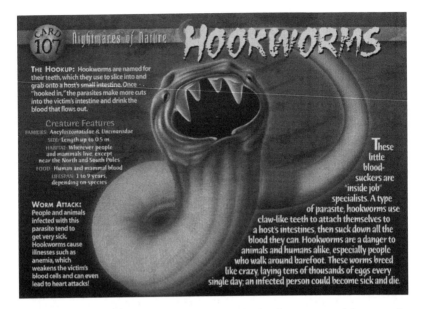

*Figure 19.* A children's trading card showing the hookworm as a nightmare of nature. Author's collection.

*Figure 20.* Hookworms as colon comrades. Screenshot of Colon Comrades website (https://coloncomrades.wordpress.com/).

relations. The chapter focuses primarily on the story of *Necator america-nus*, the most prevalent species of human hookworm. It identifies situations of multispecies mutualism alongside those in which the presence and absence of worms becomes pathological. To illustrate comparable geographies in the macrobiome, I make brief analogies (in italicized sections) to human relations with wolves and dogs.

## Dysbiosis and Its Geographies

The science that informs the probiotic turn presents dysbiosis as the outcome of interactions between different organisms embedded within wider ecological relationships and configured by longer evolutionary legacies. It understands ecologies as nonequilibrium and capable of adhering to multiple stable states. Changes in ecological state are linked to shifts in the intensities of the interactions within any given ecology. A dramatic event, a changing climate, a new competitor, or the loss of a keystone species all have the potential to profoundly shift ecological conditions. Such shifts are described as the outcome of changes in the topology (or spatial configuration) of an ecological interaction web. Dysbiosis is the result of deleterious change in which the functionality of the ecology declines, at least from the perspective of the human observer. Keystone species have a disproportionate ability to configure ecological interactions, and their loss or return has the potential to tip ecologies into new conditions. This science shapes forms of symbiopolitics. An ecological understanding of the human body and of the wider ecologies with which they are entangled legitimates the strategic use of keystone species to manage ecological dynamics and to secure the delivery of desired ecological functions. This involves the identification, distribution, and modification of these organisms, alongside interventions that modulate ecological composition and dynamics to secure their flourishing. These are post-Pasteurian acts of controlled decontrolling.

So far, I have focused on rewilding and biome restoration in Western Europe and North America, but my analysis has not placed these examples in their wider geographical or historic contexts. There are numerous and well-established critical social science frameworks for understanding the unequal geographies of Global Health and environmental risk. These tend to link these patterns to the uneven outcomes of processes of socioeconomic development. This argument is most developed in discussions of the pathologies of natural resource management and the ways in which modern, antibiotic forms of agriculture, forestry, and fishing degrade the resilience and functionality of ecological systems. Critical analyses in different strands of transformative political ecology variously link these degradations to the ecological irrationalities of capitalism, amplified by its associations with colonialism and patriarchy.[1] They focus on the unequal distribution of the benefits of resource extraction and the risks of un-natural disasters. Comparable work in the social sciences of health and development has focused on disease as an outcome of structural vio-lence, with stark inequalities in health outcomes and health care linked to the legacies of colonial exploitation and the operations of capitalist biomedicine.[2]

This work is valuable, important, and more differentiated than this sketch suggests. But it often proceeds by giving limited agency to eco-logical processes and makes little use of the ecological science I address in this book. In this chapter I engage with an emerging body of critical social science that rethinks analysis in light of this ecological ontostory. I aim to ground analysis of dysbiosis in its historical and political eco-nomic contexts while engaging socioecological resilience thinking. As I explained in the Introduction and in chapters 3 and 4, the more estab-lished of such approaches tends toward the technocratic—that is, seeking ways of reforming capitalist resource management to enable sustainable development. But more transformative versions of resilience thinking are emerging that link dysbiosis (or pathogenesis) to the political and economic patterns and drivers of late modern, capitalist modes of man-aging life.

I engage most closely with the work of geographer Steve Hinchliffe and his collaborators that develops elements of probiotic ecological sci-ence in analyses of the geographies of biosecurity in intensive agricul-ture.[3] These researchers take issue with conventional Cartesian mappings of disease as the outcome of universally pathogenic microbes crossing borders and infecting discrete territories. They note the importance of the extensive networks of globalization in enacting and distributing socio-technical diseases,[4] but they suggest that it is more informative to think of disease in terms of topologies of intensive relations. In their reading, topology refers to the complex spatiotemporal relations and intra-actions (after Karen Barad) that configure the intensities of any socioecological situation. Bodies become ecologies entangled and enfolded within a geography of more or less pathogenic landscapes. These are marked by potential disease tipping points.[5] Here infectious disease does not come from an outside but is ecosyndemic[6]—that is, immanent from particular political ecological situations, the shifting microbiologies of which cre-ate conditions of dysbiosis. Pathogenicity, they argue, is best conceived "as a process, rather than a fixed object."[7] Work in this vein by geographers and anthropologists has tended to focus on the topologies of viral hot spots, clouds, and epidemic space.[8] A common concern lies with the intensified socioecological relations of contemporary agriculture, urbanization, and globalization, which generate blowback, or the excessive presence of undesired microbes.[9] Here microbial dysbiosis takes the form of crowd infections and emergent infectious and zoonotic disease. To date there has been little work exploring the topologies of dysbiotic situations generated by microbial absence or the proactive efforts to tip ecologies out of dysbiotic relations.[10]

Hinchliffe and others' relational thinking is inspired by the theories of biopolitics that we encountered in chapter 3, especially by critiques of the binary, or immunitarian, logics of antibiotic modes of managing life. In this chapter as well as in chapters 6 and 7, I link these relational new-materialist understandings of pathogenicity with work on the political economy of colonial and capitalist forms of public health and environmental management. There are important differences between work in this older Marxist tradition and the new relational approaches, and there is scope for a more systematic synthesis than can be offered here.[11] But for the purposes of the analysis that follows, I will identify one common concern and point of conceptual convergence. Hinchliffe and colleagues give a topological twist to established approaches to political economy in their analysis of *Campylobacter* as a just-in-time disease.[12] In working bacterial and viral agencies into theories of capitalism, they link the intensification of agriculture and its endemic dysbiosis to post-Fordist relations of labor precarity, commercial pressure, and regulatory inefficacy. Their argument is that we can only understand the pathological intensification of human–microbial relations by tracing the globalizing logics, practices, and assemblages of contemporary capitalism.

## Worming, Deworming, and Reworming with Hookworms

The heterogeneous character of human relations with *N. americanus* offer rich grounds for developing such a geography of dysbiosis. Here I identify four contrasting human–*N. americanus* relations. I start with the mutualistic relations associated with human–helminth coevolution. I then turn to parasitic relations that reach their extreme expression in the intensities of plantation capitalism. I examine inequalities among those affected by the ghostly absence of a keystone species before turning to the probiotic or mutualistic relations associated with rewilding and restoration, which I term Mutualism 2.0. This typology is not exhaustive, given the large numbers of people and worms involved in these relations, but it offers a useful heuristic for mapping dysbiosis and its inverse, eubiosis.

## Mutualist

The *Oxford English Dictionary* defines mutualism as a "relationship existing between two organisms of different species which contributes mutually to each other's well-being."[13] Helminths have lived in hominids since the Pliocene (5.3–2.6 million years ago), and humans and helminths have a long history of coevolution.[14] Paleoparasitologists who study these relationships differentiate contemporary human helminths into heirloom species, which originated in Africa and have been with us throughout human evolution, and souvenir species, which were picked up from other animals after some humans left Africa approximately sixty thousand years ago.[15] There is uncertainty about the evolutionary origins and about the heirloom or souvenir status of *N. americanus*, but its current form and life cycle likely evolved from relations with small mammals, our hominid ancestors, and then modern humans. *N. americanus* reproduces outside the body when human feces are deposited in warm, moist soil. Once hatched, the larvae eat their way through human skin, then travel through blood vessels, lungs, mouth, and throat to return to the gut. If a larva establishes itself there, then it can live as a worm in the body for many years. By virtue of their need to reproduce outside of their host, resident hookworm populations tend to develop slowly compared to other members of the human microbiome.

Theories of parasite evolution suggest that successful organisms tend to become less pathological over time; those that quickly kill or significantly harm their host run the risk of not completing their life cycle. Initial relations between our ancestors and those of *N. americanus* were likely to have been antagonistic but became more commensal over time. The ecoimmunological research summarized in chapter 2 proposes that during their long coevolutionary history, *N. americanus* learned to train, modulate, or calibrate the human immune system and microbiome to achieve host tolerance. In limited numbers and in situations with modest infection intensity, *N. americanus* became a mutualist or "ancient cloaker."[16] However, every body's immune system reacts to *N. americanus* with differing degrees of severity, so that in some hosts, even a minor infection can cause disease.

Mutualistic relations between humans and *N. americanus* evolved and persist in specific political ecological contexts. The *N. americanus* life cycle (Figure 8) requires situations with warm climate and moist, sandy soil, frequent human defecation, and subsequent opportunities for skin contact. Archaeologists and immunologists suggest that *N. americanus* lived mutualistically in some Paleolithic hunter-gatherer populations, where defecation practices would have ensured repeated reinfection, but infection intensity would have been kept in check by hygiene and medical practices and by frequent movement.[17] No doubt some lost their worms when they migrated to colder climates, or they experienced high rates of infection when confined to a single place, but the general pattern was of a stable worm load.

Such mutualistic relations persist in rural tropical and subtropical parts of the world. Moises Velasquez-Manoff suggests that at least 1.2 billion people, or between a fifth and a sixth of humanity, host some form of helminth.[18] Approximately 439 million people were estimated to be infected by two species of hookworm in 2010.[19] This is a rough estimate. Global hookworm surveillance is patchy and discontinuous, and targeted at situations of high infection intensity in the interests of quantifying the parasitic relations that lead to hookworm disease (or necatoriasis). This focus makes it hard to disaggregate persistent mutualistic relations. Nonetheless, most human infections are likely to be with low levels of *N. americanus* and are asymptomatic. People can tolerate a moderate load of worms without obvious consequence.

There is a shortage of social science research on these mutualistic human–hookworm relations, and the limited epidemiological and ethnographic work has been conducted by anthropologists seeking to enhance the efficacy of deworming programs.[20] Nonetheless, this literature helps theorize hookworm infection intensity as the outcome of a range of social, ecological, and immunological factors that come to configure host exposure and immune system response. This work finds great heterogeneity even at small geographic scales: infections vary between neighboring villages, households, or even family members. *N. americanus* exposure and transmission dynamics are configured by sanitary and occupational practices, alongside access to anthelminthic drugs. Immune response is shaped by age, sex, nutrition, and host genetics.[21]

Reading across these differences, we find that mutualistic relations with *N. americanus* most commonly persist in tropical and subtropical parts of world. Further, these relations occur in impoverished populations, especially those involved in subsistence agriculture, who have only basic sanitation, who have poor footwear, and who have limited access to modern health care. Maintaining mutualistic relationships in these situations is the accidental outcome of daily routines that are often performed in the absence of any biomedical knowledge of the distribution and ecology of *N. americanus* or whether one is infected. To a large degree, these populations continue to inhabit contemporary versions of the evolutionary adapted hunter-gatherer environment eulogized by contemporary ecoimmunologists like Graham Rook. Levels of hookworm infection show strong negative correlations with levels of poverty, and hookworm infection has been proposed to act as a proxy for socioeconomic development.[22] But it is not poverty in itself that tips hookworms from being a mutualist to being a parasite.

*As an apex predator, wolves act as a keystone species shaping the population structure and behaviors of their prey at a landscape scale. Many of the large herbivores we now farm, and the ecologies they graze, coevolved in the presence of wolf predation. These animals inherit behaviors and ecologies shaped by the nonequilibrium dynamics of predator–prey interaction. Humans and wolves also have a long history of coevolution, which culminates in the mutualistic relations associated with human–canine domestication.[23] Antagonistic relations of competition for prey gradually gave way, over long histories of selective breeding, to a diversity of dogs, variously dependent on human care. Canine encounters also shaped human agricultures of hunting and herding, and later practices of recreation, companionship, enclosure, and surveillance. Wolves and then dogs have played fundamental roles in the creation of Anthropocene ecologies and in the evolution of the human microbiome. Anthropologists have recorded numerous examples of mutualistic ecological relationships between indigenous people and wolves, in which the hunting of both wolves and their prey is choreographed in order to maintain wolves in the landscape.[24]*

Parasite

"Parasite" is a polysemous word. In the interests of specificity, I use the rather narrow and negative biological definition provided by the *Oxford English Dictionary*, in which a parasite is "an organism that obtains nutrients at the expense of the host organism, which it may directly or indirectly harm." *N. americanus* becomes parasitic when an increase in infection intensity pushes the human immune system over a disease threshold. The human worm burden grows when the rate of colonization surpasses the rate at which worms are displaced from the body. This growth tends to be incremental and much slower than viral or bacterial epidemics. It is caused by a combination of changes in the body's immune response, increases in exposure to infective larvae, and the availability and efficacy of anthelminthic drugs. Immune response is compromised by age, nutrition, coinfection, and pregnancy, among other factors. Exposure increases when naked skin comes into contact more frequently with warm, shitty soil.

In parasitic situations, mutualistic relations of host tolerance and immunosuppression are displaced by a pathological excess of hookworms that drink so much blood that they cause protein deficiency and anemia. Mortality as a direct result of hookworm disease is rare, but hookworms cause significant morbidity. In 2001 it was estimated that 59.9 million people worldwide had a high-intensity hookworm infection, though only three thousand deaths could be directly attributed to *N. americanus*.[25] Chronic high worm burdens cause lethargy and disrupt cognitive function and development. Parasitic hookworms leave a legacy in the form of growth stunting and cognitive retardation.[26] These effects are often compounded by coinfection with other helminths, and the immunosuppression associated with *N. americanus* can exacerbate infections with malaria and tuberculosis.[27]

The shifts in exposure, nutritional status, and anthelminthic drug availability that make hookworms parasitic are directly linked to changes in political-ecological circumstances. Archaeologists suggest that *N. americanus* infection intensities increased as a consequence of the agricultural revolution. The novel ecological relations associated with what James Scott has termed the "late Neolithic multispecies resettlement camps" created ideal conditions for *N. americanus* and other soil-transmitted helminths.[28] Sedentary populations that live in dense concentrations, that live among domestic animals, and that use human waste as fertilizer and building material experience higher infection intensities.[29] Having made itself at home in the human body, *N. americanus* also came to be at home in the so-called domus of the agricultural village and its immediate environs.[30]

Hookworm infections reach their greatest intensities in the squalor of colonial plantations, early industrial mining operations, and frontier railroad construction, where malnourished migrant bodies are concentrated in conditions of poor health care and sanitation. Critical historians of public health relate how the political-ecological intensities of colonial and industrial capitalism created the perfect conditions for hookworms to become parasitic. As historian Steven Palmer argues, "Hookworm disease was a sickness of ecological, economic, and cultural displacement and recombination; its virulent appearance along these teeming trenchworks of modernity went hand in hand with the increasing density and speed of the globe's road, rail, and ocean connections. A close relative of nineteenth-century capital, the hookworm parasite thrived in heated frontier regions where vulnerable bodies were amassed for hard labor, and it became potent through accumulation."[31] Palmer argues that hookworms were "simmering among rural peoples around the world," but they "boiled over" because of the "intense ecosocial disruptions" of global capitalist expansion.[32] The amplified intensities of the capitalist frontier tipped hookworms over a threshold, further shifting them from a microbial old friend to a parasitic crowd infection. Colonial plantations were ideal hot spots for the genesis of hookworm disease.[33] Necatoriasis became an industrial epidemic,[34] a pre-Fordist precursor to the contemporary just-in-time diseases discussed by Hinchliffe and others.[35] In Anna Tsing's terms, hookworms became an "autowilding weed" immanent from the rationalized and intensified ecologies of the global and colonial Plantationocene.[36]

By the early twentieth century, *N. americanus* had become known as the "germ of laziness" because of the effects of necatoriasis on plantation labor productivity.[37] These were so great that the Rockefeller Foundation found it necessary to fund the first major public health program to try to modulate infection intensities. This began in the U.S. South and was then exported to postcolonial plantation economies.[38] In the United

States, sanitation, footwear, drugs, and education gradually disentangled agricultural workers from transmission vectors. Many people left hookworm hot spots for cities, and gradual improvements in nutrition helped boost host immunity. The Rockefeller campaigns were highly influential and led directly to the creation of the World Health Organization. They pioneered the antibiotic approach to public health that has profoundly shaped subsequent programs for Global Health.[39]

However, the application of such antibiotic approaches is patchy. The biogeography of necatoriasis closely maps onto patterns of uneven global development. Parasitic helminth infections still persist among some largely Black populations living in poverty in rural parts of North America and Europe.[40] But the majority of contemporary necatoriasis hot spots are found amid rural populations involved in subsistence agriculture in low-income countries, especially in sub-Saharan Africa, in South Asia, and on the Oceanic Islands (see Figure 9).[41] Plantation ecologies like the tea estates of South Asia continue to be marked by parasitic hookworm infections.[42] There have been significant reductions in China as a result of recent rounds of socioeconomic development.[43] Some public health experts talk of the "poverty trap" caused by hookworms and other soil-transmitted helminths. They argue that the pathological legacy of early life infection reduces labor productivity and subsequent earning potential in an ongoing vicious cycle.[44]

By the end of the twentieth century, public sanitation projects had become politically and economically unpalatable as strategies for hookworm control. World Health Organization attention, with Gates Foundation funding, is now focused on behavior-change interventions, mass drug administration, and vaccine development.[45] Meanwhile, the widespread but sporadic use of anthelminthic drugs has selected for drug-resistant *N. americanus* phenotypes.[46] Like other microbes, *N. americanus* has evolved to counter antibiotic control methods. With limited demand from the Global North, there has been little investment in drug development, and new products only emerge as a spillover from better-resourced research on deworming pills for agricultural and domestic animals.[47]

This brief analysis of hookworms becoming parasitical suggests that necatoriasis is not necessarily a disease of underdevelopment resulting from the spread of a specific pathogen. Hookworms can be endemic, mutualistic, and asymptomatic. They become pathogenic when infection intensity is pushed over a disease threshold. Archaeologists, immunologists, and historians link this amplification of infection intensities first to the agricultural revolution, then to the exploitative, unequal, or abandoned socioecological relations associated with colonial capitalism, both past and present. But they also caution against reading hookworm disease as a universal outcome of systemic structural violence. The materialities of coinfection, host genetics, soil type, working conditions, defecation practices, and sanitation investments all shape the differential emergence of infection intensities.[48] Nonetheless, decades of public health experience make clear that reducing parasitic relations requires a degree of control over infection intensities that is not provided by drugs or other magic-bullet technological solutions. Necatoriasis has only been controlled through investments in sanitation, nutrition, and wider health care, often coupled with substantial rural-to-urban migration.[49]

*Parasitic relations between people and wolves are less common than with N. americanus and other microbes. However, there are instances in which human relations with wolves, or more recently feral or guard dogs, tip from conditions of mutual benefit to those of one-way depredation. Wolves were once feared as lethal human predators. Alongside other carnivores, they may predate unprotected livestock, especially in situations where wild prey is scarce. Feral dogs that have lost their fear of humans can cause problems. Some wolves and dogs go rogue, transgressing mutualistic norms; others are encouraged to do so in order to exclude or undermine existing land users. Dogs may also be a vector for zoonotic diseases like rabies, fleas, and helminths, which can accumulate in situations of high population density, limited antibiotic veterinary care, and thus elevated levels of infection intensity.*

Ghost

A ghost is a missing keystone species whose absence has a deleterious impact on the functionality of an ecology. Most people in the world do not now host *N. americanus*, either because they live outside of the tropical and subtropical regions in which it can reproduce in the wild, or because they have been disentangled from the cycle of *N. americanus* transmission. Modern barriers to transmission are numerous, but the most significant have been the proliferation of flush toilets, a decrease in the proportion of people working in agriculture, and widespread urbanization. Cities with even basic sanitation provide far fewer sites for *N. americanus* to incubate and reinfect.

Declining relative levels of hookworm infection are indicative of a wider transition that has been claimed by epidemiologists in human–microbial relationships. The rise of modern health care systems (vaccines, antibiotics, and hygiene), sanitation (clean water, sewage treatment), and health education gave modern states some control over infectious diseases like the plague, tuberculosis, smallpox, cholera, and typhus. This antibiotic or Pasteurian (this term used following Heather Paxson) mode of microbiopolitics shifted the balance of causes of mortality toward noncommunicative diseases like cancer or cardiovascular disease.[50]

While access to the benefits of these interventions remains patchy, concerns are emerging that they might go too far: indiscriminately eradicating microbial life may cause pathological dysbiosis (Figure 21). As I explained in chapters 1 and 2, immunologists now suspect that the antibiotic depletion of the human microbiome might help account for the rise in inflammatory conditions—that is, missing old-friend microbes lead to epidemics of absence. In the absence of the calibration and exercise provided by a moderate intensity of infection with some helminths and other old friends, the immune system turns against the body, attacking specific locations and functions.

This trend was captured in a highly influential article published by Jean-François Bach in 2002, which plotted the declining incidence of infectious disease in developed countries against the growth in autoimmune conditions.[51] While subsequent work has rethought the microbiological and immunological connections between these two trends—differentiating old friends and crowd infections—these patterns remain striking.

Data from the U.S. National Institutes for Health on the magnitude and severity of this hypothesized microbial haunting suggest that 23.5 million people in the United States have one of eighty autoimmune diseases.[52] The American Autoimmune Related Disease Association puts this number at 50 million—some 15 percent of the population.[53] Three quarters of those affected are women.[54] Comparable proportions are found in the United Kingdom and in other Western European countries, and the incidence is growing rapidly in newly industrialized nations, especially in urban areas. The global incidence of autoimmune disease is understood to be doubling every twenty years, with some conditions

| Socioecological change | Microbial consequence |
| --- | --- |
| Clean water | Reduced fecal transmission and immune system challenge |
| Widespread antibiotic use | Reduced vaginal transmission, selection for new composition, drug resistant infections |
| Increase in caesarean sections | Reduced vaginal transmission |
| Reduced breastfeeding | Reduced skin-to-skin transmission and a changed gut flora |
| Smaller family size | Reduced early life transmission |
| Changing diet and cooking practices | Alterations in gut flora |
| Diminished contact with soil and farm animals | Reduced early life transmission and immune system challenge |
| Increased bathing, showering and use of antibacterial soaps | Selection for a changing composition |
| Increased exposure to "crowd infections" | Emerging infectious diseases |

*Figure 21.* The modern practices understood to negatively affect the human microbiome. From Martin Blaser and Stanley Falkow, "What Are the Consequences of the Disappearing Human Microbiota?" *Nature Reviews Microbiology* 7, no. 12 (2009): 887–94; Dillan Bono-Lunn, Chantal Villeneuve, Nour J. Abdulhay, Matthew Harker, and William Parker, "Policy and Regulations in Light of the Human Body as a 'Superorganism' Containing Multiple, Intertwined Symbiotic Relationships," *Clinical Research and Regulatory Affairs* 33, no. 2–4 (2016): 39–48; G. A. W. Rook, C. L. Raison, and C. A. Lowry, "Microbial 'Old Friends,' Immunoregulation and Socioeconomic Status," *Clinical and Experimental Immunology* 177, no. 1 (2014): 1–12.

growing at up to 9 percent per year.[55] Moises Velasquez-Manoff helps visualize these numbers:

> If I possessed glasses that afforded me the power to see otherwise non-apparent allergic and autoimmune disease I'd be struck by the sheer abundance of people with these problems. Walking down Broadway in New York City for instance, one in every ten children passing by would have asthma; one in six would have an itchy rash and sometimes blisters—eczema. One in five passersby would have hay fever. . . . One in 250 would suffer from a debilitating pain in his or her intestines,

what's called inflammatory bowel disease. . . . Of every thousand pass-
ersby, I'd note one struggling to move legs or arms. These people have
multiple sclerosis, a progressive autoimmune disease of the central ner-
vous system. . . . I'd note glucose monitors on one of every three hundred children frolicking in Central Park's playgrounds, children afflicted
with autoimmune diabetes.[56]

While the epidemiological connections between missing microbes and
the rise of inflammatory disease are compelling, it is proving hard to
disentangle the significance of different absent old friends. It is also
challenging to explain which of these autoimmune conditions is most
strongly related to microbial shifts rather than other environmental,
epigenetic, and dietary changes. Nonetheless, *N. americanus* has been cast
as one ghostly old friend that has come to figure prominently in clinical
and epidemiological research on global dysbiosis. As such, we can take
the absence of hookworms as a guide to the geographies of autoimmunity and dysbiosis.

Prevalent epidemiological maps showing the decline in old-friend
infections and the increase in autoimmune disease tend to work at the
scale of the nation-state, as this is the principal unit for data collection.
This work plots the rise in autoimmune disease along a north–south
gradient, identifying increased frequency in line with distance from the
equator. Parallel cartography negatively correlates increases in autoimmunity with national indicators of socioeconomic development. Early
accounts suggested that these were Western diseases, or diseases of afflu-
ence, civilization, and modernity.[57] In Europe and North America, allergies like asthma and hay fever were once aspirationally associated with
urbane high society.[58] While such maps are helpful for identifying broad
patterns, they mask smaller-scale subnational variations. By assuming all
those inhabiting these territories have equal experience of the mapped
conditions, they risk ecological fallacy. More nuanced cartography helps
make clear that the absence of hookworms and correlated epidemics
of absence are an especially urban phenomenon: hookworms persist
in some rural areas in the southern states of the United States, while
worms and other old-friend microbes are absent from many cities in the
Global South.[59] But even at this finer resolution, such static cartography
struggles to capture the political-ecological intensities and disparities

that shape the individual human experience of being haunted by hook-worms in urban areas around the world.

To illustrate this point, I use the example of type 1 diabetes, one auto-immune condition whose growing prevalence has been linked to microbial dysbiosis.[60] Type 1 diabetes normally begins in childhood, when the body's immune system attacks the cells in the pancreas that make insulin, the hormone required to take glucose from the blood to fuel the body. Without insulin, the body falls into a coma and dies; such an onset can take just a few weeks. The treatment of type 1 diabetes requires regular injections of artificial insulin, a specific diet, and exercise, coupled with close monitoring of blood glucose levels. Poor blood glucose control leads to complications, including kidney failure, blindness, nerve damage, amputation, heart disease, and stroke. The symptoms, risks, and treatment of type 1 diabetes are fairly well understood compared to some other autoimmune conditions. But the individual experience of becoming and being diabetic is still strongly shaped by socioecological conditions. Two (of many possible) dimensions stand out here that help illustrate what configures the geographies of being haunted by microbial old friends.

The first relates to spatial disparities in the antibiotic processes listed in Figure 21, especially those that provide control over crowd infections. Velasquez-Manoff's children with diabetes in Central Park most likely live in urban situations with clean water, functional sewage and waste collection systems, and pest and vector control mechanisms. The microbial vacuums they inhabit banish both old friends and infectious diseases. This is less likely to be true of populations living in housing projects in poor areas of New York City, and this will certainly not be the case for the growing number of those with type 1 diabetes living in informal settlements in rapidly urbanizing areas of Africa or India.[61] These people may have lost their worms and other old friends, but they are not in control of their exposure to infectious disease like malaria, tuberculosis, typhoid, and cholera. Many of these urban locations are disproportionately exposed to drug-resistant varieties of these pathogens, and type 1 diabetes can amplify the debilitating effects of common infections like tuberculosis.[62] Such people are caught in hot spots of what I call a doubled dysbiosis—a dysbiosis caused by both missing microbes and elevated exposure to established and emerging infectious diseases.

Second, access to artificial insulin and blood glucose monitoring technology is marked by similar spatial disparities. These characterize the availability of the wider collection of immunosuppressant drugs that have come to stand in for absent old friends. The pharmaceutical amulets that ward off and mitigate microbial haunting are developed and distributed by a global and highly capitalist industry with imperfect global reach. Diabetic children in Manhattan may well have sensors in their back pockets and easy access to insulin, but the same is not true for other uninsured Americans. Nor was it true for children in Greece after the 2008 financial crisis.[63] Diabetes specialist Edwin Gale notes that a lack of insulin is the most common cause of death for a child with diabetes, even though the drug is off patent and has been available in the West for over ninety years. He explains,

> Families in the poorest parts of the world must make a choice between insulin for one child or starvation for the rest. The consequence, in parts of India—which has more people with diabetes than any other country in the world—is that girls are missing from the children's clinics. There is no choice at all in some parts of sub-Saharan Africa, where children with diabetes live, and die, as if insulin had never been discovered.[64]

This absence results from the political economy of insulin provision, which is exemplary of the wider and extremely lucrative industry for autoimmune drug development.[65] Companies focus on making minor changes to insulin to secure patent protection rather than producing cheap generic varieties. This economy configures an urban environmental justice disaster that has yet to receive the attention afforded other toxic exposures.[66]

In short, the example of type 1 diabetes illustrates the fine-scale geographies that come to shape how and to what extent different people are haunted by an absence of worms. Haunting can be amplified by the absence of surrogate synthetic immunosuppressants. This can be compounded by the doubled dysbiosis that results from a coincidence between missing old-friend microbes and the hot spots of infectious crowd diseases created by the absence of antibiotic control mechanisms. More broadly, the figure of hookworms (and other old-friend microbes) as ghosts helps extend the topological understanding of Hinchliffe and colleagues of pathogenicity to explore situations in which microbial absences shift ecological intensities to create dysbiosis. Understanding the immune system legacies of absent microbes requires a protracted conception of the microbial geographies of health, where the focus is less on the intensities and immediate spatiotemporalities of outbreak, and more on intergenerational microbial inheritance, the accumulation of microbial exposure and immunomodulation over the life course, and the chronic, cascading effects of dysbiosis.

*As a result of human persecution, wolves now haunt the temperate forest landscapes of Europe and North America where they were once abundant. As Aldo Leopold famously notes, their absence manifests in abundant populations of their herbivore prey, which graze in patterns no longer shaped by canine landscapes of fear. The subsequent proliferation of deer and mesopredators may threaten biological diversity, degrade ecologies, and damage crops. Absent predators also help expand herbivore population densities so that they are capable of harboring zoonotic diseases—like brucellosis or Lyme disease—that affect humans and livestock.[67] Meanwhile, dogs and captive wolves that lose their microbial mutualists as a result of antibiotic veterinary care may develop autoimmune and inflammatory diseases comparable to that of their owners.[68] Children who grow up without exposure to dogs and other domestic animals experience higher incidences of some inflammatory diseases.[69]*

## Mutualism 2.0

The global geography of mutualistic, parasitic, and ghostly relations helps contextualize the probiotic projects to return *N. americanus* and other old friends to us. I term these novel relations Mutualism 2.0, in keeping with the techno-optimistic tenor of the probiotic turn and to differentiate them from the first set of mutualisms that persist in places less affected by the antibiotic practices of modern health care. Biome restoration with helminths occurs either through participation in a clinical trial or by sourcing organisms online and joining the DIY community of self-treaters. One review identified twenty-eight clinical trials with four different species of helminths involving small numbers (fewer than five hundred) of patients living in close proximity to large research hospitals in North America, Western Europe, and Australasia.[70] The geography of the DIY helminth therapy community is harder to establish, but the gatekeepers and most active members on social media are concentrated in similar metropolitan centers. These individuals serve approximately ten thousand helminth users who have a similar Global Northern and urban distribution.[71]

Probiotic users thus make up a tiny proportion of human helminth hosts, united by a common set of characteristics that help illustrate who gets to go probiotic and where in the world such relations are possible. Helminth users tend to be white and WEIRD. They live in urban situations marked by good sanitation and public health, with low levels of infectious disease. They have access to healthy food and nutritious diets. Although they benefit from excellent public or private health care, cutting-edge immunosuppressant drugs, and biomedical expertise, they turn to worms because they are often very unwell. Many have undergone surgery, have limited mobility, and experience social stigma and poor mental health. They feel that modern medicine has failed them. Probiotic users tend to be educated and anglophone, versed in scientific understandings of immunity, and proactive in their pursuit of therapeutic possibilities. They have biosocial capital; they know how to find and interpret online medical advice and to source drugs and materials to support their worm colony. They can get access to effective anthelminthic drugs, should they need them.

Helminth users are also relatively affluent. Although some have learned to grow their own worms, or they receive donations from others, most purchase helminths from private providers. The cost of helminth therapy has come down in recent years as more providers have entered the market, but required outgoings range from $1,000 to $10,000 per year. The cost depends on the species used, the dose required, and the degree of support offered by the commercial provider. Several suppliers offer discounts to those with low incomes, and as a result species like *N. americanus* have become affordable. Providers have demonstrated that it is possible to supply worms at a fraction of the cost associated with the development and provision of immunosuppressant drugs. They are concerned that the current regulatory anxiety about using helminths, coupled with the drive to develop a vaccine and the preference shown toward replacing live helminths with privatized molecular secretions, might threaten the existence and affordability of the therapy. There are tensions within the probiotic turn that point to contrasting political economies of Mutualism 2.0, which I trace further in chapter 7.

Probiotic users tend to take worms for autoimmune conditions marked by intermittent temporalities of relapse and recovery rather than irreversible deterioration, as with type 1 diabetes. Worms are used for modulation (exercising the immune system) with the aim of remission. Advocates promote their use as a preventative measure against the onset of disease conditions. Early pioneers of helminth therapy translocated worms from tropical locations and developed domestic garage ecologies to replicate their life cycles. They established captive populations of helminths in temperate regions, well outside of their tropical home range. Today advocates incubate eggs in containers of feces, then collect and count the larvae. They distribute worms and enable infection through an internationally networked biogeography of quasi-legal exchange. Captive helminth populations in WEIRD bodies are unlikely to interact or interbreed with their wild or free-ranging kin. This latest chapter in the globalization of helminths scrambles their evolved distribution, creating novel spatial patterns and potential speciation events.

We can briefly illustrate these Mutualism 2.0 relations through an example of the use of helminths to treat Crohn disease and ulcerative colitis, two forms of inflammatory bowel disease (IBD). Symptoms of these conditions include cramps and swelling in the gut, recurring and bloody diarrhea, weight loss, and extreme tiredness. Incontinence frequently causes embarrassment, and patients become socially isolated, unemployed, and depressed.[72] State-of-the-art treatment seeks to manage symptoms through immunosuppressant drugs, dietary changes, and surgery to remove chronically inflamed parts of the gut. It is often painful, has adverse effects, and is ineffective. There is no cure. The epidemiology of IBD tracks wider trends in the demise of old-friend microbes and the rise of dysbiosis.[73] These patterns and the inefficacy of current treatment has led immunologists and patients to explore the therapeutic potential of biome restoration with helminths. IBD is currently one of the most common subjects for clinical research and self-treatment with helminths.[74]

Self-treating Crohn disease with helminths involves personalized multispecies experiments to establish and sustain desired infection intensities. Users will frequently host combinations of different species, including human whipworms, human hookworms, and rat tapeworms. These worms take up residence in or stimulate different parts of the gastrointestinal tract. Users synchronize their infection with the worms' life cycles to maintain the diversity and age profile of their colony. They also take immunosuppressant drugs like prednisone to offset their immuno-response on infection and to help their worms establish. They have learned which foodstuffs, mundane modern hygiene products, and medicines boost worm welfare, and which have inadvertent antibiotic effects on their transmission dynamics.[75] They source and share this information on social media, which plays an important role in improving the welfare and sense of community among those with IBD beyond those self-treating with helminths.[76]

The WEIRD configuration of these Mutualism 2.0 relations with helminths helps us map comparable patterns in the broader probiotic turn in health care. Clinicians, policy makers, and citizens are responding to the perceived deleterious microbial shifts outlined in Figure 21 through a range of probiotic experiments, interventions, and regulations designed to recalibrate the intensities of microbial exposure and infection (Figure 22).

These novel probiotic interventions to restore mutualistic microbial relations map onto a specific subset of the situations in which hookworms and other old friends currently figure as ghosts. They are concentrated in the microbial vacuums created by the effective implementation of the practices summarized in Figure 21, which indiscriminately diminish both crowd infection and old friends. The arts of controlled decontrolling outlined in Figure 22 are globally rare, and their occurrence is configured by a highly specific political ecology. Its common characteristics include a temperate climate that minimizes accidental wild infections, the ability to closely manage microbial transmission dynamics through the strategic redeployment of antibiotic infrastructure, the ready availability of diagnostic technologies, the cultivation of vernacular and scientific microbiological expertise, and access to a panoply of affordable consumer products and services to facilitate citizen experimentation.

*Wolves have also been the subject of controlled reintroduction programs in majority WEIRD locations. This symbiopolitics was reported in chapter 3 and involves the self-willed and assisted return of free-ranging populations in North America and Western Europe. The aim here is to use wolves to restore mutualistic relations with large herbivores, as well as to enable the animals to reestablish populations. The pursuit of lupine Mutualism 2.0 relations extends to trials at simulating landscapes of fear through hunting and deterrence by domestic dogs,*

| Socioecological change | Microbial consequence | Probiotic management of microbial intensities |
| --- | --- | --- |
| Clean water | Reduced fecal transmission | Probiotic drinks, like Kombucha |
| Widespread antibiotic use | Reduced vaginal transmission and selection for a changing composition | Ban on some antimicrobial cleaning products<br>Discouraging use of antibiotics<br>Promotion of probiotics |
| Increase in caesarean sections | Reduced vaginal transmission | Discouragement of c-sections<br>Swabbing to replicate the microbial colonization associated with vaginal delivery |
| Reduced breastfeeding | Reduced skin-to-skin transmission and a changed gut flora | Breast feeding promotion<br>New types of probiotic and prebiotic formula |
| Smaller family size | Reduced early life transmission | Pox parties to spread diseases like measles and chicken pox* |
| Changing diet and cooking practices | Alterations in gut flora | Probiotic, raw, and paleo diets<br>Growth in pro- and prebiotic dietary supplements |
| Diminished contact with soil and farm animals | Reduced early life transmission | Children's "messy play" and "Nature reconnection" projects |
| Increased bathing, showering, and use of antibacterial soaps | Selection for a changing composition | Controls on use of antimicrobials (like Triclosan)<br>Growth in probiotic hygiene products and practices |
| Increased exposure to "crowd infections" | Infectious disease | n/a |

Figure 22. Probiotic responses to modern anxieties about microbial dysbiosis. *Some of these practices shade toward the anti-Pasteurian—for example, pox parties organized by participants in antivaccination movements.

*as well as the lethal regimes of care practiced with the eye of the wolf at the OVP. Dogs have also become enrolled as probiotic agents for biome human restoration in the growing popular interest in microbial pet therapy.*[77] *Here parents get dogs to help populate their child's microbiome and to exercise their immune system. Dogs with inflammatory diseases can now be treated through a range of prebiotic and probiotic supplements, though not yet with therapeutic canine helminths.*

## Probiotic Patchiness and Interdependence

This chapter has developed a relational geography of dysbiosis to better specify the spatialities of the probiotic turn. Focusing primarily on human relations with *N. americanus*, it argues that few organisms, big or small, are essentially pathogenic. Pathogenesis is the immanent outcome of political and ecological relations. In the case of helminths, these configure the infection intensities that shape different bodily experiences. *N. americanus* can cause disease as a parasite and as a ghost; it coexists with people in contrasting relations of inadvertent and closely choreographed mutualism. This analysis identifies four types of relation, the properties of which are summarized in Figure 23.

The relational approach I have developed maps the patchy geographies of where in the world these different relations occur. It identifies the persistence of pre-Pasteurian mutualisms, the partial reach and blowback of antibiotic approaches, and autoimmune and inflammatory situations marked by antibiotic excess. It flags the limited incidence of probiotic mutualisms resulting from biome restoration. This way of conceiving relations places the topological understanding of ecological interaction summarized in chapter 2 in political-economic context. It maps the political and ecological drivers that configure the infection intensities that shape patterns of dysbiosis and eubiosis. It reveals a geography of hot spots, vacuums, and inertia. The patterns of these zones of intensities do not conform to simple territorial maps. Parasitic, ghostly, and different sorts of mutualistic relations may occur in close Cartesian proximity, but they are kept apart by the geometries of biopower. For example, helminths were encountered differently in the mansions and the latrines of the colonial plantation, and their contemporary absence is endured differently in the luxury apartments and informal settlements of Mumbai.

|  | Mutualist | Parasite | Ghost | Mutualist 2.0 |
|---|---|---|---|---|
| Global occurrence | (Sub)tropical areas in the rural Global South | (Sub)tropical areas in the rural Global South | Global North and urban Global South | Scattered across urban areas of North America, Western Europe, and Australasia |
| Multispecies composition | Tolerable wormload, polyparasitism, and multiple coinfections. Few crowd infections. | Excessive worms, repeated infection, coinfection with other parasites, poor nutrition, compromised immune system. Crowd infections. | Absent worms. Dysbiotic microbiome. Overactive immune system. Either: microbial vacuum (no crowd infections or old friends) or doubled dysbiosis (crowd infections and no old friends). | Controlled worm population, ready supply of replacement worms, no inadvertent infection risk, access to good nutrition, and managed diet |
| Political ecology | Hunter-gatherer, simple collective, and/or subsistence agriculture | Poor sanitation and health care, intensive plantation agriculture, poverty and inequality, limited access to effective deworming drugs | Common access to sanitation, footwear, deworming drugs, and basic health care. Variable access to immuno-suppressant drugs. | Wealthy. Good health care and sanitation. Worms for sale. Growing private interest in surrogates. |
| Microbiopolitics | Pre-Pasteurian, traditional means of controlling infection intensities | Patchy Pasteurian. "Magic bullet" drugs and vaccines. | Antibiotic. Pasteurian. Involving sanitation, drugs, and urbanization. | Probiotic, post- (and anti-) Pasteurian involving controlled reintroduction by clinicians and expert patients |

*Figure 23.* A summary of the key characteristics of the four types of human–helminth relation.

This approach helps understand the drivers of patterns of dysbiosis. Immunologists suggest that worms made a home in us, developing mutualistic relations through our long history of preagricultural movement, so that hookworms were tailored to the pre-Neolithic domus. These amicable relations were unsettled with the rise of sedentary agricultural systems; they reached extremes of dysbiosis in the high infection intensities of colonial plantations, which pushed hookworms across a political ecological tipping point so that they became a parasitic crowd infection. The patchy efficacy of public health leads to the absence of worms, which pushes the human holobiont over a different autoimmune threshold, resulting in microbial dysbiosis and amplified host inflammation. A dystopic figure emerges of modern humans, bereft of microbial old friends and living itchy, depressed, overweight lives, chronically dependent on expensive and unpleasant regimes of anti-inflammatory drugs. Access to these drugs is distinctly unequal, and the unequal experience of this haunting is compounded by patterns in the persistence and resurgence of crowd infections.

To conclude, I would like to reflect on the connections between these patterns of dysbiosis that are revealed by a focus on their political ecological drivers. In order to understand the import and the potential of the probiotic turn, we need to consider the degree to which the ability of some to go probiotic is conditional on others elsewhere experiencing parasitic and ghostly relations with keystone species, with amplified exposure to the nonhumans that proliferate in their absence. This question applies to both rewilding and to biome restoration. In relation to rewilding—only hinted at in this chapter—it is clear, for example, that the wildlife resurgence that is taking place in parts of Europe and North America has been enabled by the cessation of agriculture and by land abandonment. This process has multiple drivers (including intensification, urbanization, demographic changes, and subsidy reforms), but it is linked in part to the globalization of agriculture and forestry.[78] Global markets for lower food prices make agriculture uneconomic in marginal regions, like the areas of Central Europe to which wolves are now returning, and to which I turn in chapter 6. The sparing of land in temperate regions has been enabled by the conversion and intensification of farming in (sub)tropical regions. It remains to be seen whether rewilding leads to net gains in the space set aside for wildlife, or whether it merely accelerates its poleward redistribution. This shift in the geographies and intensities of agricultural production will have important distributional consequences for the livelihoods of traditional farmers in both the Global North and South.[79]

As I will explain in chapter 6, some forms of rewilding are also haunted by colonial genocide, displacement, and expropriation. In some parts of North America, for example, these practices enabled the historical creation of the national parks, which are now core areas for keystone species reintroduction.[80] Rewilding does little to redress these historic injustices. It has become an emotive, even toxic, label for some campaigning for land rights who argue it remains too strongly wedded to colonial imaginaries of the pristine, never-before-inhabited wilderness. The current resurgence of carnivores in some parts of Europe results from a combination of self-willed and anthropogenic reintroduction, and the cessation of hunting risks enhancing unequal geographies of exposure to the risks of predation and crop raiding. Shepherds in the Alps and Pyrenees experience the damage of returning wolves without the promised benefits of ecotourism.[81] This distanced interdependence between the beneficiaries of newfound mutualisms and novel and historic parasitisms is an important topic in need of further research.

There is also an interdependence between biome restoration and historic and present parasitic and ghostly relations with *N. americanus*. In global historical terms, the rise of the Western metropoles in which people now strive to live well with microbes is fueled by the exploitative relations of colonial and postcolonial plantation capitalism. If we take up Dipesh Chakrabarty's call to provincialize Eurocentric histories of medicine and development, we can understand that *N. americanus* only became a public health concern as a result of its effects on labor productivity; indeed, necatoriasis remains a neglected tropical disease.[82] The profits from plantations flow to cities in the Global North, whose citizens benefit from cheap food, the price of which reflects the abject conditions of its production—production conducive to *N. americanus* proliferation. Current rounds of global industrial urbanization—for example in China—are driving significant declines in hookworm infection. But urban migrants often move to cities that have become hot spots for pollution and infectious disease. They lose their worms but find themselves unable to afford immunosuppressant drugs or other synthetic substitutes for keystone calibration.

Meanwhile, the populations of *N. americanus* taken north for probiotic reinfection were sourced from poor populations in Southeast Asia and Central America in bioprospecting missions that have so far delivered limited benefit to their donors and that promise futures of chronic dependency. The science these worms have come to inform is shaped by established research agendas in Global Health toward both hookworm eradication and development of immunosuppressant drugs. Collected worms are used to develop and test vaccines, while others have their secretions isolated in the hope of identifying promising molecules for pharmaceutical synthesis. This magic-bullet approach to treating necatoriasis through vaccination risks failing to address the socioecological conditions that prevent the management of microbial exposures. If successful, it promises to strip all citizens of their mutualists and drive future epidemics of autoimmune disease. Should the worm go extinct, newly autoimmune citizens would then be required to purchase synthetic versions of hookworms' immunosuppressant molecules. Drugs from bugs are more lucrative than bugs as drugs, but a chronic dependency may not prevent the onset of autoimmune conditions.

Tracing these interconnections poses some hard questions for Global Health and its current antibiotic orthodoxy in which fewer microbes equals better places. Should it prove effective, then the experience of the probiotic turn suggests a profound need to recalibrate the global management of microbial exposure and infection, and the means through which microbial absence is addressed. This would require a new understanding of the unequal and interrelated geographies of infectious and autoimmune disease. It requires the redistribution of technologies that enable people to control their microbial composition. And it will require a more nuanced understanding of the forms of political economy through which the probiotic value of keystone species is calculated and exchanged.

# 6

## FUTURE PASTS

### The Temporalities of the Probiotic Turn

THERE IS COMMON HISTORIOGRAPHY in the stories told by advocates of rewilding and biome restoration that frames the present as a critical inflection point, one occurring at a moment of crisis that is engendered by the pathologies of modern, antibiotic approaches to managing life. Proponents present themselves as seers, visionaries able to look back in time in order to diagnose the problems in the present, to dismiss contemporaneous solutions, and to herald new restorative futures. They offer future pasts. On the one hand, their turns to life are profoundly retrospective, advocating rewilding, biome restoration, back-breeding, and de-domestication. They look to past baselines to identify desired ecological conditions.[1] They tell revisionist histories that challenge the linear and triumphalist narratives of Western agriculture and health care, tracing ecological unraveling through the premodern and then the modern periods. At the same time, such seers are prospecting into the future and anticipating novel ecologies. They seek functional futures that will emerge along unexpected pathways. The quality of these desired future pasts is not directly linked to the authenticity of historical reenactment. Instead it involves a recalibrated post-Pasteurian reentanglement of human ecologies and their keystone species.

This grand temporal vision is perhaps best conveyed in a striking image produced by Rewilding Europe in 2010 (Figure 24).[2] The picture is entitled *The Past, Present, and Future of European Nature*, and it featured in their early promotional material. The implied timeline starts in the top left corner in the deep prehuman past where the strands of the represented ecology are tightly interwoven. We pass through the rise of Paleolithic hunting and then Neolithic pastoralism, where ecological

*Figure 24. The Past, Present, and Future of European Nature.* Illustration by Jeroen Helmer; reprinted with permission.

disentangling starts. The parting of the strands reaches its peak with the advent of intensive, specialized agriculture and the implied antibiotic rationalization of ecological processes. The probiotic turn occurs about two thirds of the way through. It is symbolized by the return of large animals, self-directed rivers, decaying wood, and dead bodies. Walkers and tourists populate a postproductive recreational landscape. Human infrastructure is networked into the ecology, which finishes with the tight recoupling of the parted strands. Agriculture is absent. The iconography is subtle, interweaving the ancient symbol of the cornucopia, the horn of plenty that represents the bounty of the harvest, with the implicitly fragile strands of the DNA double helix.

While I have not been able to find such a compelling equivalent visualization, the temporal imagination presented in this image resonates with the historiography offered by those involved with biome restoration. According to William Parker, Graham Rook, and other immunologists, complete and functional human microbiomes are to be found in the

Paleolithic hunter-gatherer past (or their contemporary equivalents). Microbial unraveling began with the rise of agriculture, with the narrowing of diets and the shift to sedentary and then urban lifestyles. Old friends were lost and crowd infections increased. Dysbiosis peaks with the proliferation of antibiotic chemicals, the excesses of modern sanitation, and the consumption of highly processed food. Restoration begins with the selective return of keystone species as part of the recalibration of lifestyles toward probiotic, or dirty, ways of encountering microbial life.

This chapter explores the political ecologies of the probiotic turn that are enabled by these future pasts. It examines how the invocation of ecological baselines enables those involved in rewilding and biome restoration to naturalize a specific set of desired future socioecological relations. It follows how this historical revisionism is premised on a range of epistemic practices that depart from modernist science without dismissing it. And it explores how these baselines legitimate a set of political relations that mobilize and modify modern modes of biopolitical control. In tracing temporalities, the chapter is ultimately interested in the relationships between the probiotic turn and the progressivist timeline of antibiotic modernity. It develops the analysis that started in chapter 5 to specify and differentiate the alternative or countermodernities that characterize the probiotic turn.

To do so, I engage in this chapter with diverse literature that explores contrasting retrospective or reactive tendencies common to twentieth-century responses to the pathologies of antibiotic modes of managing life. It draws once more on the work of Heather Paxson on artisanal cheese production in the United States, and her argument that these producers are both post-Pasteurian and postpastoral.[3] Idealized imaginations of the rural pastoral have long served urban citizens (in Europe, North America, and other industrial settings) as nostalgic counterpoints to the social, ecological, and economic depredations of urban life. The pastoral is the landscape benchmark for valued recreational experiences on the farm, in the forest, or in the wilderness.[4] Paxson demonstrates how her cheese makers rework these imaginaries, selectively reinventing pastoral practices, knowledges, and forms of value in postindustrial American society. They bring a recalibrated commitment to science, technology, capitalism, and the cosmopolitan values of urban life. Paxson describes these relations as postpastoral.[5] I explore the intersections between the ecological modernities that Paxson reports, and the wider intellectual project of ecomodernism that has come to shape the visions of influential figures in the probiotic turn. In counterpoint to these reformist projects, I identify much less affirmative, or more reactionary, imaginaries of the pastoral in which a return to nature and the disavowal of elements of science, reason, and liberal democracy legitimate darker, fascist, and anti-Enlightenment projects of social and ecological restoration.[6]

To enable this comparison, I turn to some extreme examples from the history of probiotic thinking. I focus in particular on rewilding and the story of Heck cattle, exploring the history of their back-breeding as well as their role in different reactionary and reformist projects for rewilding Europe. I compare the cattle back-breeding and rewilding under national socialism in 1930s Germany with the late twentieth- and early twenty-first-century promotion of Heck cattle and Tauros cattle as tools for rewilding in the cosmopolitan, liberal, market democracies of an integrating Europe. Where relevant, I draw parallels with trends in biome restoration.

## Back-Breeding and Rewilding with Heck Cattle

The Heck brothers began their back-breeding programs in Weimar Germany in the 1920s. Their father, Ludwig Heck, was the famous director of the Berlin zoo, and the boys grew up on the zoo grounds. Lutz succeeded Ludwig in 1932, and Heinz became the director of the Munich zoo in 1928. They used the resources, connections, and networks of their institutions to gather desired bovine specimens from across Europe. Their breeding was informed by a range of archaeological, historical, and mythological materials alongside correspondence with other European scientists, animal breeders, and general enthusiasts. Lutz was a keen hunter and sought out aggressive animals, including cattle bred for bullfighting from Spain. They claimed success after twelve years of experimentation, creating herds of their reconstituted aurochs at their respective zoos.[7]

As Clemens Driessen and I uncovered in our collaborative archival research, by the late 1930s, the brothers had aligned their programs with the diverse constellations of power and the differing political aspirations of the Third Reich. Lutz was a hunting partner of Hermann Göring, whose various positions included Reich hunt master and Reich forest master.[8] With Göring's patronage, Lutz was appointed head of the Nature Protection Authority within the Forest Service in 1938. With Göring's support, Lutz "returned" some of his back-bred aurochs to the occupied Bialowieza Forest, which now straddles the border between Poland and Belarus (Figures 25 and 26). Here they joined reintroduced bison, bears, lynx, and moose, becoming game for Göring's private hunting as part of a wider project to restore the imagined political ecologies of medieval forestry.

In Berlin, Lutz also developed personal and professional relationships with Heinrich Himmler, Konrad Meyer, and others involved in developing and implementing plans for the eastward expansion of the Third Reich. He imagined an expansive role for his back-bred animals in the Generalplan Ost, Himmler's genocidal plan for German

*Figure 25.* Lutz Heck *(left)* and Hermann Göring *(right)* study a relief map with wildlife figurines during a visit to the 1937 International Hunting Exhibition in Berlin. The mounted animal in the background is a wisent; on the table is the horn of an aurochs. Source: Exhibition catalog, *Waidwerk der Welt, Erinnerungswerk an die Internationale Jagdausstellung Berlin 1937 2.–28. November*, Paul Parey, Berlin, 1938. Reprinted by permission of Paul Parey Zeitschriftenverlag GmbH.

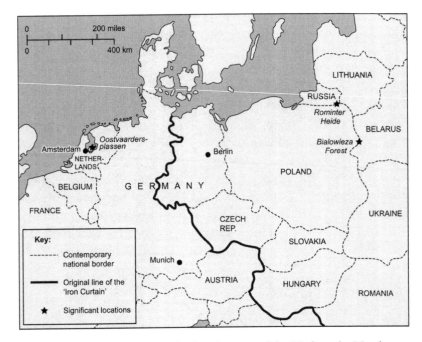

*Figure 26.* Map of key sites involved in the story of the Heck cattle. Map by Ailsa Allen.

*Lebensraum* and the Völkisch resettlement of the territory of Central and Eastern Europe conquered from 1939.[9] Lutz promoted his animals as tools for ecological restoration in publications seeking to address the *Versteppung* (becoming-steppe) of the landscape believed to be caused by the wrong type of racialized (Slavic) land management.[10] Heinz seems to have had a more ambiguous relationship with the National Socialist elite,[11] but in the early 1940s, he was commissioned to write a history of the aurochs for Himmler's Ahnenerbe, a shadowy research organization that sought to use science (archaeology, cultural geography, and even musicology) to legitimate mythological understandings of the Aryan people.[12]

Lutz's cattle at the Berlin zoo and in the forest were killed as World War II came to its bloody end, but some of the animals bred by Heinz in Munich survived. They languished in a few zoos and enclosures as small semidomesticated herds, where they gained a reputation for being especially hardy; they were able to withstand cold winters on poor ground with little human management. It was the descendants of these cattle that Frans Vera brought to the OVP in the 1980s, under very different political circumstances. By the early 2000s, Vera's paleoecology and his experience of working with Heck cattle and other large herbivores came to inspire the continental-scale vision for Rewilding Europe illustrated in Figure 3.

Rewilding Europe involves a network of Western European non-governmental organizations and scientists mobilizing the geopolitical and economic opportunities presented by an expanding European Union to advocate for a dramatic expansion of rewilding through grazing. Their funding comes from private philanthropy alongside grants from the World Wildlife Fund Netherlands and the Dutch and Swedish post-code lotteries, among other sources. Anticipating the continued decline of traditional agriculture and associated land abandonment, they aim to rewild one million hectares of European land by 2022, restoring land-scapes to preagricultural benchmarks. By 2018, they had identified and linked twelve rewilding demonstration projects, working with local part-ners to provide advice, funding, and a supply of keystone species. They promote these as places for ecotourism and as potential sites for invest-ment in emerging markets for ecosystem services.

Meanwhile, doubts about the authenticity of Heck cattle and anxi-eties about their dark provenance led to the Tauros Programme back-breeding initiative. This project involves an interdisciplinary team of diverse scientists, alongside animal breeders, conservationists, and an artist, who combine archaeological and historical evidence with insights gained from DNA sequencing. The project claims to be going back to the future.[13] Herds of Tauros cattle have been bred, and Rewilding Europe is enabling their introduction into areas of Southern and Eastern Europe, along with other large herbivores. In so doing, they promote these cattle as a cosmopolitan European animal, one that can be tailored to reflect local cultural and ecological conditions.

## Contrasting Probiotic Future Pasts

The history of the Heck cattle is an extreme example, one that cannot be made to stand in for the full diversity of rewildings past and present.

But back-bred, de-domesticated, or otherwise wilded mammals feature frequently in the violent histories of modern nations' efforts to naturalize the taming or the wilding of their subject populations. These include Highland and Chillingham cattle in the United Kingdom, Teutonic horses and German shepherd dogs, fascist Italian pigs, Hindu Indian cows, and Polish wisent.[14] Many of these animals continue to be valued in the present day and are becoming central to contemporary rewilding projects. Likewise, there is a dark history of violent, primitivist thought that grounds racialized understandings of biological and cultural superiority in the ecological composition of different human bodies.[15] The hygiene practices and (what we now describe as the) microbiomes of so-called wild, savage, or primitive people have long been coded to evidence the superiority or inferiority of modern Westernized people. As I explore later in this chapter, historical, racialized tropes of "excremental colonialism"[16] and the noble (microbial) savage continue to haunt the biome restoration movement. We need to understand these histories of probiotic thought and the ways in which they invoke the past for a range of political and ecological reasons. Nonetheless, I want to be clear from the outset that I am not accusing any of the contemporary rewilders or biome restorers that feature in this book of being fascist. Nor are they aligned with national socialism, past or present.

## When Was the Wild, and Who Looked after It?

To understand the political and ecological work done by these future pasts, we must first explore their ecological baselines and the types of human activity these normalize. Lutz Heck and Rewilding Europe broadly share the ecological (or symbiopolitical) understanding of rewilding that I sketched in chapter 3. By this I mean that they are primarily interested in cattle and other herbivores as keystone species that reorganize the grazing and other disturbance regimes of the landscapes to which they are introduced. Although they share a concern with questions of anatomy, genetics, and behavior of their animals, they index the character and quality of wildness to how cattle perform as part of a functional ecological system. In spite of this common ontology, they differ markedly in their opinions as to what a functional ecology looks like and what role people played in it. These differences are strongly related to when in the past they set their desired political-ecological baseline.

Clemens and I found that Lutz corresponded with many prominent scientists of his period, but his animal and natural history expertise stemmed less from the systematic observation of field sites and more from his practical experience as a zoo manager, animal breeder, and, most importantly, hunter.[17] Lutz believed that his reconstituted aurochs only displayed their true wildness when they were released in large herds to do battle with human hunters in a specific forested landscape. He shared Göring's enthusiasms for hunting and forestry practices modeled on a contemporary reinvention of the *Nibelungenlied*. This epic medieval tale was prominent in the German nationalist self-understanding cultivated by the Nazis. It tells of Teutonic knights, dwelling in the primordial forest, who hunt deer, boar, wisent, and aurochs according to elaborate codes of aristocratic chivalry.[18] Lutz and Göring went on numerous hunting parties together, during which they sought to relive these myths. Their adventures were documented in a series of intimate photographic portraits (Figure 27).

For Lutz and Göring, reinvented traditional forms of hunting were central to protecting, managing, and restoring the forests of Central Europe. In this *Heimat*[19] cosmology, hunters assume the role of keystone species: apex predators responsible for managing populations of deer and other herbivores. Competing nonhuman predators like wolves are eradicated so that hunters control population size and density. The hunters do so to shape the character of the herd, to manage grazing types and intensities, and to maximize the number of trophy animals. This conservation imaginary sets its ecological (and political) baseline during the medieval period. The earliest manuscripts of the *Nibelungenlied* have been dated to the twelfth century, but the stories they tell are set as far back as the fifth or sixth centuries.

For Göring, this mythology of lost aristocratic hunters in a feudal political system justified the restoration of a heroic ecology of Germanic leadership. Hunting was part of a nostalgic imagination that set medieval forest life as the idealized counterpoint to modern industrial and urban civilization. The forest and its rewilded ecologies were a place for retreat. This vision offered a desired ecological baseline from before the rise of agrarian landscapes. Forests were to be set aside and expanded for elite recreation. Göring's vision of a network of hunting reserves came into conflict with rival modernist plans for the settlement and agricultural

*Figure 27.* Lutz Heck and Hermann Göring on a hunting trip, December 1934. Source: Bundesarchiv, image 102-04224. Photograph by Georg Pahl.

intensification of the occupied territories of Eastern Europe. The Nazi push for agricultural autarky (self-sufficiency), as embodied in Konrad Meyer's Generalplan Ost, focused on the domestication of the wilds, wartime timber production, and the creation of agricultural landscapes— for example, through the reclamation of marshland.[20] Göring's reactionary medievalist fantasies had limited traction on what happened to Eastern Europe, but, as we shall see, they had violent local consequences and played an important role in legitimating a specific mode of postpastoral biopolitics.

In contrast, contemporary enthusiasts for rewilding Europe push their ecological reference conditions further back. They advocate a Paleolithic baseline, set before the premodern hunting and agrarian landscapes of medieval and early modern Europe that have dominated much conservation. Drawing on the writings of Frans Vera, they find their desired past in the preagricultural landscapes of the early Holocene that emerged with the retreat of the ice sheets at the end of the Pleistocene (ca. 11,700 BP).[21] This period is marked by a great diversity and abundance of animal life, including the full guild of predators, grazers, and browsers (many of which are now extinct). As I explained in chapter 3, the focus here is on re-creating functional ecologies by detecting the ghosts that haunt the present. Heck cattle are introduced as ecosystem engineers, aurochs surrogates capable of restoring past grazing regimes, ideally through self-willed processes requiring minimal human management.

In these accounts, Paleolithic humans figure as nomadic hunter-gatherers, living in much lower densities than modern Europeans. These ancestors were blessed with far greater ecological connectedness and awareness, but they were still capable of driving the overkill that led to the extinction of keystone carnivores and megaherbivores.[22] The ecological role afforded to contemporary people differs markedly to that of Lutz and Göring, for whom the beneficent hunter-forester maintains the balance of a *Heimat* cultural ecology. For Rewilding Europe, people figure as urban, postproductivist observers—scientists, tourists, and a few local publics and the employees who will supply their needs.[23] The social and ecological relations imagined here are still pastoral (inasmuch as they reference an idealized rural setting), but they are linked to a prehistoric, not premodern, moment. This is a reinvented paleo, but not one in which people simulate human Paleolithic practices.[24] In this postpaleo, recreational hunting is discouraged. Tourists' trophies will be virtual; they will leave only footprints. When animals need to be killed, this should be done (as at OVP) with "the eye of the wolf": a clinical yet naturalistic death designed to simulate regimes of canine predation rather than those of human hunters (paleo, medieval, modern, or otherwise).

The postpaleo manifests slightly differently in the case of biome restoration. While many take worms and other probiotics as supplements in otherwise unchanged modern, urban, postindustrial lifestyles, there are some for whom an awareness of biome depletion and the idea of evolutionary mismatch necessitates more holistic processes of retrospective recalibration and the reenactment of paleo lifestyles. This approach is best exemplified in the complex character of Jeff Leach. Leach is an anthropologist and a microbiologist who works with the Hadza in Tanzania. He is the author of *Rewild*, which actively promotes the salutary benefits of aligning one's microbiome with those living in the wild.[25] He has undergone an FMT, donated by a Hadza man of his own age, and spends time eating and drinking local foodstuffs to reset his gut ecology. He reports on this as a process of "(re)becoming human."[26] Leach is one a number of microbiologists who have sought out hunter-gatherer populations as repositories of less affected human microbiomes.[27] Less extreme versions of this microbial postpastoral are evidenced in the dirty governmentalities I outlined in chapter 3. These tend to benchmark microbial colonization to a more recent and premodern rural pastoral similar to that described by Paxson. The postpastoral (rather than post-paleo) model of biome restoration is characterized by close encounters with agricultural animals, soil, raw and fermented foods, extended family, and the reinvention of traditional birthing and child-rearing practices.[28]

Returning to the story of Heck cattle, while there are significant differences in the types of human ecological activity naturalized by these postpastoral future pasts, neither model of rewilding advocates a wilderness without humans. Instead, and in different ways, they promote rewilded ecologies as working landscapes that ought to be subject to forms of human stewardship. Paxson identifies the centrality of the working landscape to the "post-pastoral ethos," which she argues "recognizes that culture and nature are not in fundamental opposition to each other; instead, nature, no less than culture, contains and unleashes creative as well as destructive forces—and therefore requires responsible human guidance."[29] In her example, responsible postpastoral artisanal cheese production is a process of "counterindustrial agricultural remediation."[30] For Lutz Heck and Hermann Göring, such remediation required the reinvention of aristocratic hunter-foresters and the feudal socioecological relations from which they emerged. Hunters work the forest to steward its game. As I explore in chapter 7, for some contemporary rewilders, stewardship involves putting nature to work in a novel set of postproductive economic practices in which rewilded ecologies deliver a range of ecosystem services.

Unlike in Paxson's example, both of these postpastoral visions for rewilding struggle with agriculture and the provision of food. While Göring placed mythical value on low-intensity medieval systems, contemporary rewilders castigate these as inefficient and expensive.[31] Neither has much to say about actual productive relations with the land. We are left to assume that food will be provided through the intensification of farming elsewhere. Some of those involved in rewilding now borrow the discourse of ecomodernism to suggest that postpaleo nature tourists will be decoupled from (rather than harmonized with) nature. This decoupling involves models of land sparing, not land sharing, to use the terms of contemporary debate.[32] As I identified in chapter 5, there is a silence in much of the rewilding literature about the spatial implications of this decoupling and whether the passive and active rewilding of Europe and North America is occurring at the expense of tropical biodiversity destroyed by the globalization of agriculture.

To summarize, there is a consistent interest in the history of Heck cattle in using large herbivores to manage ecologies. But these ecologies are linked to very different ecological baselines, and they prescribe very different modes of human management. Both of the formulations presented here are postpastoral in their knowing reinvention of idealized working rural landscapes, but they imagine contrasting roles for people. In the reactionary visions of Göring and Heck, a select few return to the forest as medieval hunter stewards. For Rewilding Europe, people are recreational spectators. These visions differ in the weight they afford the future and the past. Heck and Göring were more nostalgic and reactionary. In contrast, modern rewilding has been inspired by the accidental ecologies emerging in the ruins of past modernist futures (like Chernobyl), whereas activist paleoecologists anticipate the parameters of the new climatic regime of the Anthropocene and its novel ecosystems, and are thus more explicitly future oriented.

## What Political Relations Do These Future Pasts Enable?

These postpastoral imaginaries, with their different socioecological baselines, mobilize a valued past to naturalize specific sets of political relations. These pasts legitimate, anticipate, prefigure, and enact contemporaneous political practices. Here I identify and contrast three dimensions of the political practices that are common to probiotic thinking past and present. The first is a consistent disavowal of the domestic and of domestication that drives a broad antiurbanism in European rewilding. In contrast to many pastoral visions that eulogize livestock, the rewilding postpastoral is based on a denigration of highly bred beef and dairy cattle as well as a celebration of their wild antecedents and contemporary surrogates.[33] More broadly, this antiurbanism shapes the territorial practices through which rewilding is enabled.

For example, although the Heck brothers made their livings by engaging bourgeois urban society in their metropolitan zoos, the idealized locations for their cattle were rural forests. This is partly pragmatic: wild cows threaten urban parks and gardens. But it also territorializes an antimodern conception of the wild that has been prevalent in twentieth-century conservation.[34] Lutz Heck and Hermann Göring's cultural ecology of aristocratic hunting and forestry can be situated within a broader movement in Nazi Germany that elevated Völkisch, rural peasant life as the natural expression of a racialized relationship between blood and soil. This movement informed an antiurban ideology in which the urban figured as a site of individual alienation, moral degeneration, and a racially coded discourse of social decay.[35]

The urban occupies a more ambivalent place in the imagined geographies of Rewilding Europe. On the one hand, it is the processes of urbanization that are creating new opportunities for the return of the wild in abandoned marginal areas: people leave the land and make space for nature. On the other hand, urban life is also understood to cause alienation and a disconnection from nature, causing a host of immunological, psychological, and social problems.[36] This pathologization of the urban is also common among those involved with biome restoration. Far-flung primitive wilds offer reference locations for those like

Jeff Leach involved in microbial projects for paleo or ancestral health, while premodern rural landscapes full of farms, soils, and fermented foods predominate for those advocating a return to a Mediterranean lifestyle.[37] This tension illustrates what Leo Marx describes as a "complex pastoralism": the paradox that nature-saving decoupling drives socially degrading disconnection.[38] So far, the proposed solution is not to rewild the city but to offer some people fleeting tourist encounters with depopulated places. Urban ecologies have played little role in rewilding and its iconography, which has tended toward the untrammeled and the remote.[39]

A second trend in the political mobilization of future pasts is the continental-scale projection of power from Western to Eastern Europe (and sometimes beyond). This has happened through radically different political mechanisms across the period under analysis, but Eastern Europe is commonly conceived as an ecological heartland, a vital biological and cultural resource for Europe and its citizens, and a gateway to the wild territories of Russia.[40] Western European conservation enthusiasts, before and after the Cold War, have placed this territory within their legitimate sphere of influence and have argued that it is in need of their intervention and management.[41]

For example, for Lutz Heck, the conservation of Eastern wilds helped legitimate a desire for territorial control. As Clemens and I have argued on the basis of archival research, as head of the Nature Protection Authority within the German Forest Service, Lutz was well placed to enact his medieval vision of the wild by shaping emerging plans for the eastern Germanic land. Lutz published articles in the official Nazi newspaper (*Völkischer Beobachter*) and in a journal edited by Konrad Meyer aimed at those involved in extending the German population into the east. He argued that the landscape of the *Ostraum* (eastern space) would need to be restored to its Germanic character. Colonizing the east was presented as *Landschaftspflege* (care for the landscape). In this organicist and retroactive geopolitics, the racial and cultural superiority of the German *Volk* in land management legitimated eastern expansion to secure *Lebensraum* for Germanic people and wildlife. With advice from Lutz, Göring ordered a significant expansion of the Bialowieza Forest reserve to create an official hunting ground for his back-bred animals. This led to the displacement and genocidal purging of the indigenous human population. In this way, a particular conjunction of nature conservation, economic requirements, a sense of international geopolitical justice, and a Völkisch racist understanding of what it means to inhabit a landscape were combined into a single murderous logic that promoted ethnic cleansing as a form of landscape restoration.[42]

The Cold War and the Iron Curtain put much of the ecological heartland of Eastern Europe off limits to Western conservationists. When the Iron Curtain fell, conservationists traveling in postsocialist states in the 1990s marveled at a standing reserve of wild animals and landscapes that required both protection and reconnection. German reunification soon made them aware of the opportunities offered by European integration.[43] Conservationists were in the vanguard in establishing ecological conditions for postsocialist countries seeking accession to Europe.[44] Pastoral landscapes were initially secured through legal practices of classification, designation, and action plan implementation, coupled with agroenvironmental subsidies to sustain forms of low-intensity traditional agriculture.

Rewilding enthusiasts—like Frans Vera—subsequently took issue with this protection of agricultural ecologies. They are keenly aware that the European agroenvironmental subsidy regime is unaffordable in an enlarging political bloc, and they eye the opportunities presented by growing rates of land abandonment, though they are publicly cautious in their tacit support for these trends.[45] The current focus on marginal areas of Eastern and Southern Europe is more opportunistic than strategic, but the ultimate aim of this program, as the title of one lobbying event suggests, is toward "Rebuilding the Natural Heart of Europe."[46] Rewilding Europe is a reformist, democratic organization, in stark contrast to the reactionary, fascist political program that enabled Lutz Heck and Göring's interventions. It dedicates a large portion of its time and resources to lobbying for legislative change in Brussels, to communication and marketing, and to developing market and financial mechanisms to fund conservation in marginal areas. It is optimistic about the governmental opportunities opened up by new forms of wild ecotourism and their potential for shaping ecologically aware consumer-citizens. Ethnographic research by other social scientists paints a mixed picture of the local impacts of Rewilding Europe's interventions, but it notes a common commitment to carefully navigating local interests, working with partner organizations to enact largely neoliberal modes of community conservation.[47]

The geopolitics of biome restoration are less developed. However, a discourse is emerging that maps microbial resources to tribes living in premodern conditions. There is nascent enthusiasm for human microbial conservation (in situ and in biobanks) and for exploring methods for commodifying vestigial functional microbiomes through property regimes that would reward their current host. Contemporary microbiologists are self-aware enough to reflect on anthropological tropes they inherit in these proposals and on the political problems of taking the microbiome public.[48] Such risks have been foreshadowed by past projects for bioprospecting as a means for nature conservation in biodiverse but impoverished parts of the world.[49]

The third important political dimension enabled by the future pasts of Heck cattle relates to the geopolitics that is expressed through the different origin stories offered for back-bred cattle. A singular figure of a Teutonic, Germanic aurochs was at the heart of Lutz's vision. In 1939, he claimed that "the extinct Aurochs has arisen again as German wild species in the Third Reich."[50] He flagged the German origins of the word aurochs (*ur-ochse*) and argued that his were the timeless animals of the *Nibelungenlied*. He justified the expansionist, postnational geopolitical aspirations of the Third Reich by appeals for the restoration of a medieval, sylvan, and prenational geography of rooted Teutonic tribes and wild beasts. Lutz and his contemporaries tied the origins of culture, and those understood to possess it, to violent, aristocratic modes of social life rooted in forests. This anti-Enlightenment origin story was mobilized to justify the eradication of Europe's rootless, urban Jews.[51]

The Tauros Programme and Rewilding Europe offer a very different origin story. They trace the cultural significance of the aurochs back to ancient Greece (*tauros* being the Greek word for bull), claiming, "In Greek mythology the god Zeus once took the shape of a bull, when he swam over from Crete to present day Lebanon and snatched away a beautiful princess. Her name was Europa. The aurochs has always been at the very root of the whole idea of a continent called Europe. It is in fact our continent's defining animal."[52] Rooting the animal in an Enlightenment story of Europe's classical origins circumvents chauvinistic and fascistic appeals to medieval folklore.[53] It offers a more pluralistic and cosmopolitan animal identity and geography of belonging. Rewilding Europe and the Tauros Programme promote multiple versions of their back-bred cattle, which will be sourced from a global repository of cattle breeds and will be tailored to local ecologies, climates, and cultures in Europe. They claim the Aurochs 2.0 will be singular in its ability to return ecological processes that are simultaneously differentiated by their local specification. The cattle's provenance matters far less than its functional impacts.

To justify this federal supranational vision of a rewilded Europe, they reference the prehistoric and prenational geographies of early human settlers, who moved alongside herds of animals in a borderless Europe.[54] These nomads were untroubled by petty nationalisms. Aesthetic, epistemological, and political appeals are frequently made in their publications to the famous aurochs cave paintings at Lascaux, and associated text flags the sophistication, longevity, and commonality of human–cattle relations. The continent is made even more permeable and connected by cartographic reference to the lower sea levels during this period and the absence of contemporary barriers created by the North and Black seas.[55]

Contrasting pastoral imaginaries therefore mobilize the past to naturalize specific sets of political relations and forms of contemporaneous symbiopolitics. Neither model of rewilding advocates a direct return to the political organization of medieval aristocracy or Paleolithic hunter-gathering. Instead, they modify these relations to inform novel modes of governance. Here we see how the probiotic turn is premised on post-pastoral political projects that maintain the nostalgic, idealized retrospection of the pastoral, but power up these visions through thoroughly modern governance mechanisms. The clearance of the Bialowieza Forest and the movement of cattle under national socialism were enabled by the vast and violent machinery of the modern German state. Likewise, the translocation of animals and the rise of wild ecotourism in Eastern and Southern Europe involve modern legislation, transport infrastructure, and the finance capital that supports green investment and consumption. Contrary to some criticisms, neither version of rewilding disavows human ecological history or the place of (some) people in the landscape.[56] Neither seeks the impossible political vacuum of wilderness; instead they enforce specific politics of human belonging.

## How Are Future Pasts Known?

These references to past ecological baselines and the forms of politics they naturalize involve knowledge practices, the epistemic status of which is founded on claims of reenacting past ways of knowing animals and the environment. A common argument made by advocates across these examples is that they have rediscovered more authentic ways of knowing and being in nature. The authenticity of these knowledges is expressed through an ambivalence about (or occasionally outright disavowal of) scientific reason as well as about the reductionist epistemologies of modern biological and economic science. This ambivalence creates tensions among prominent advocates that pick up and expand on the differences between the knowledge practices of mapping and hacking identified in chapter 4.

There are three further themes to explore in this regard. The first circles back to the status of the domestic and domesticity as the denigrated spatial and political counterpoint to the wild. The postpastoral antiurbanism in the probiotic turn takes common epistemic form in the equation drawn by advocates for rewilding and biome restoration between relations of domesticity and the alienation caused by modern urban life. In the twenty-first century, this tendency is most clearly expressed in anxieties about a widespread "disconnection" from nature that leads to "nature-deficit disorder."[57] This disconnection is held responsible for a range of social and psychological problems, especially among urban children. For example, in the context of rewilding, George Monbiot bemoans the ecological boredom he feels living in suburban or rural landscapes devoid of large animals, especially those his ancestors would once have hunted and been predated by.[58] He claims this condition to be widespread.

Monbiot finds therapy in immersive, unmediated, and near-death encounters while out fishing in a kayak on the Irish Sea. He values danger, skill, and embodied knowledge; he reflects positively on his past physical experiences with the Hadza and other supposedly primitive hunter-gatherer groups. Andrea Gammon argues that Monbiot's writing exemplifies a reflexive model of "primitivist rewilding," which values risky and sometimes violent encounters with animals and landscapes.[59] Monbiot argues that it is only by immersing oneself in a human ecology of fear that one can throw off the shackles of domestic urban life and rekindle lost ways of being in nature. It is this journey of self-rediscovery that provides the narrative for Monbiot's best-selling book. And it is a journey that, as Anna Tsing argues, is a distinctly masculinist project.[60] Tsing suggests that such violent fantasies of self-making (and of escaping from the wife and kids) have a long cultural history. They are passed down in the biographies of past Romantic male rebels (like Thoreau), who continue to provide cultural and epistemic resources to their contemporary acolytes.[61] With Monbiot as its figurehead, rewilding emerged as a predominantly masculine project. While this is changing, the field is demographically dominated by men. In the narratives told by advocates, its heroes are mavericks who think and work against the grain. They are stubborn, outdoorsy types with no time for the constraints of bureaucracy, the office, or the niceties of domestic life. They are men of action who know nature from their experience in the field. As one such advocate explained to me, they inherit and embody the frontier spirit.

Such fantasies of male self-discovery manifest darkly in the story of Lutz Heck and Hermann Göring and their reenactment of the *Nibelungenlied* mythology of aristocratic hunting. Lutz and Göring sought out violent encounters with large animals like wisent and the reconstituted aurochs. They wore traditional dress and used spears and other primitive technology. Those under their command were encouraged to release their pent-up animal instincts and to use hunting practices in their pursuit of bandits in the forest. Jews and other minorities became game in a violent eugenicist project of racialized personal and social therapy. Heck and his contemporaries gave violent expression to what Boria Sax describes as the epistemic "cult of wildness" among German and Austrian natural historians in the 1930s. The writings of Konrad Lorenz and other prominent figures served to naturalize these violent relations.[62]

But risk and fear are not the only affective logics that come to shape the knowledge practices of rewilding. We can also detect dissatisfactions with the rationality and reductionism of biological and economic science in common enthusiasms for enchantment and intuition. For example, Lutz Heck often boasted of his experience as a hunter and as an animal breeder in accounting for the claimed success of his experiments.[63] This was knowledge gained not from textbooks or in laboratories but out in the field and in the forest, and from growing up in the Berlin zoo. He claimed the earthy vernacular forms of expertise of someone in touch with the folk traditions of hunting and forestry. Lutz was full of wonder at wildlife. He marveled at the power of wild cattle, whereas his brother, Heinz Heck, described the reemergence of the aurochs through his haphazard breeding and reintroduction programs as miraculous,[64] a testament to the vitality of the wild archetype rather than a triumph of mechanistic human control. Lutz's discourse is exemplary of the forms of reenchanted science that Anne Harrington documents in her account of holistic thinking in late nineteenth- and early twentieth-century Germany.[65]

An enchanted holistic epistemology of bovine wildness also underpins the knowledge practices of the Tauros Programme, although with none of its anti-Semitic associations. For example, the steering group of experts assembled to guide this cattle back-breeding program includes field ecologists, breeders, geneticists, an archaeologist, and renowned Dutch cattle painter Marleen Felius.[66] Felius is the author of the definitive illustrated encyclopedia of cattle breeds.[67] She has been given a central role in the breeding program. She applies her aesthetic skill to animate inert archaeological materials through the creation of a series of drawings of cattle. Felius's paintings imagine aurochs' bodies, herds, and ecologies in a range of prehistorical settings. These representations help guide the selection of desirable animals for subsequent breeding. Likewise, Rewilding Europe frequently uses the futurescape paintings of Jeroen Helmer to visualize their future pasts (Figures 3 and 24).

The final epistemological dimension of how the wild is known in this probiotic example relates to the place afforded myth in the practice and the promotion of rewilding. Wild cows and other large herbivores gained some prominence in Nazi propaganda. In 1941, Göring commissioned a film that featured the wisent that had been bred and introduced by Lutz Heck. The film opens with a close-up of an ancient copy of the *Nibelungenlied*. The narrative is set against the backdrop of the ongoing war. The narrator explains, "The swift victory over Poland has brought a welcome return of pure-blooded wisents to Germany . . . giving justified hope for the conservation of the wisent, the strongest wild animal and the last witness of former Germanic primordial forest."[68] In 1938, the Heck brothers joined Himmler's Ahnenerbe and were commissioned to write a popular book on the zoological and cultural history of the aurochs.[69]

In this popular propaganda, cattle and bison are presented as charismatic catalysts for reconnecting elite, alienated, modern, urban German citizens with their true *Heimat* natures. Philippe Lacoue-Labarthe and Jean-Luc Nancy draw attention to the centrality of myth to Nazism.[70] They argue that it is not so much the content of the myth that matters but its performative function, especially the belief in its efficacy in producing a national identity. Lutz Heck's back-breeding project allowed Göring to stalk through the forest carrying a spear, legitimizing his policies and the Nazi war efforts by the purported truth of Germanic mythology, which was taken as seriously as, and fully incorporated within, a form of widely respected natural science.

The power of myth, embodied in the cosmopolitan figure of Europa, is also central to the Tauros Programme, which argues that "the aurochs is now stepping out of the shrouds of myth, but maybe more than ever remains a legend."[71] Shorn of their reactionary associations, charismatic wild cows are afforded political power in the marketing materials of contemporary back-breeders and rewilders. Felius's paintings have been deployed along with professional wildlife photography and film to inspire Western Europe's citizenry in favor of the wild. In their commitment to recognizing local bovine identities, the Tauros Programme deploys cattle as a flagship species, charismatic boundary objects that can bring together different epistemic communities.[72] Although Göring used science to legitimate an exclusive myth, here myth helps communicate and democratize science. Scientific arguments for primeval authenticity, ecosystem services, and future adaptive landscapes are made accessible, alluring, and commercially valuable through the affective logics of wildlife film.

A retrospective, holistic epistemology thus plays a central role in the identification of the wild. This values intuition and field craft over mechanistic science; it celebrates enchantment and myth. At times it manifests in a masculinist, risk-seeking model of reconnection with violent, sometimes fascistic undertones. It also shapes more affirmative propaganda aligned with the ecological aesthetics of contemporary wildlife films and tourism. In both cases, wild cows and their landscapes figure as catalysts for restoring alienated urban publics to nature, and significant importance is attached to the transformative aesthetic potential of wildness to foster a desired citizenry.

## Reactionary and Ecomodernist Ways of Going Probiotic

The story of Heck cattle offers a compelling example of the different geopolitics and biopolitics of rewilding. It is illustrative of some of the most important intellectual, political, and ecological challenges facing this emerging paradigm of nature conservation, and by extension the probiotic turn. At a key historical moment in the first half of the twentieth century, there was a dark paradox to rewilding, when nonhuman flourishing was predicated on the violent, racist, and fascist displacement of large numbers of people. In stark contrast, contemporary rewilders offer a reformist and future-oriented vision of wild cows at the heart of a wild Europe, in which people and nature have been both fundamentally decoupled and (somewhat contradictorily) reconnected. Although I am wary of a genre of critique that tracks all of modernity's paradoxes and abhorrence to the Holocaust, this story demonstrates the analytical benefit of understanding national socialism as an extreme, rather than anomalous, version of modern modes of governance. Engaging with this extremity helps characterize and specify rewildings.

Although it might seem paradoxical, the various rewildings reviewed here are thoroughly modern. They advocate a return to the past, but these are not straightforward primitivist or pastoral retreats. Instead, we should understand rewilding as postpastoral projects, in which nonanalog futures are benchmarked to past ecologies. I have identified two contrasting ecological baselines in this story, among others that organize the future pasts of rewilding. The first refers back to the premodern ecologies of medieval forestry stewarded by human hunters and enabled by managed populations of grazers. The second goes back to the Paleolithic, before the rise of agriculture, but it recasts humans as ecosystem engineers, returning keystone species as tools for naturalistic grazing. Both of these examples share in an ecological understanding of rewilding as the creation and management of functional landscapes.

As exemplified in this chapter, the probiotic turn conjoins retrospection and futurology. Probiotic experts look to a paleo past, tracing ecological unraveling through the premodern and then modern periods. However, they are also prospecting into the future, anticipating the novel or unprecedented political ecologies of the Anthropocene. With this foresight, they are able to redress past changes and to foresee future tipping points, recalibrating or modulating dysbiotic ecologies. While epistemic and political value is found in paleo conditions revealed by archaeology or derived from contemporary, "primitive" surrogates, probiotic alternatives are rarely presented as a straightforward return to the past.

Reading the story of rewilding alongside the account of biome restoration offered in chapter 5, it is possible to identify a common five-part categorization and periodization of political ecological relations within the narratives of the probiotic turn, which I illustrate in Figure 28. Read chronologically, this moves from the paleo, to the pastoral to the urban, to the antibiotic, and after blowback, to the probiotic. But these relations are rarely understood in such a neat chronological fashion. Instead, they offer what historians describe as usable pasts—that is, "elements in history that can be brought fruitfully to bear on current problems."[73] They provide a stark, punctuated temporality for retrospective futurology to guide contemporary governance, with the arrow of time animating narratives of progress and/or degeneration.

Neither of the cases of rewilding reviewed in this chapter involves the pastoral disavowal of science, technology, or bureaucracy. In contrast, they harness cutting-edge science and biological and political technologies to projects to unravel cultures and ecologies of domestication. Laboratory and field knowledges are entangled in scientific experiments to reverse years of agricultural "improvement." Although advocates take issue with the alienating consequences of urban industrial life, they are

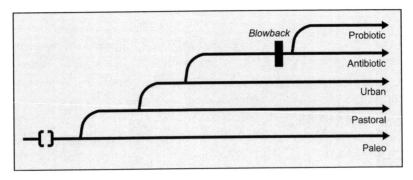

*Figure 28.* The common timeline of political ecological relations in the narratives of the probiotic turn.

not averse to using its tools for governing otherwise. Animals (or their semen) are shipped great distances across the continent for reintroduction, and politicians have sought to leverage the bureaucracy and resources of radically different postnational authorities to fund and regulate diverse forms of social and ecological wildness. These range from the violent military dictatorship of the Third Reich to the local subsidiarity, democratic deliberations, and market mechanisms of the European Union.

This story reveals radically different ways of going probiotic and helps identify two different probiotic modernities. The first is a model that exemplifies the reactionary modernism of national socialism.[74] Trevor Barnes and Claudio Minca explain,

> On the one hand, National Socialism embraced modernity and instrumental rationality; something found, for example, in the Nazi emphasis on engineering, eugenics, experimental physics, and applied mathematics. They were also exemplified in the Nazi technologies of governance around the economy, population, planning, and settlement. . . . On the other hand, cheek-by-jowl was National Socialism's other embrace: a dark anti-modernity, the anti-Enlightenment. Triumphed were tradition, a mythic past, irrational sentiment and emotion, mysticism, and a cultural essentialism that turned easily into dogma, prejudice, and much, much worse.[75]

This Janus-faced modernity is clearly manifest in Lutz Heck's entanglements with national socialism. Lutz valued the modern technologies for animal breeding and transportation offered to him by the zoo, his relationship with Hermann Göring, and the military infrastructure of the German state. He aligned himself with those involved with planning and settling the occupied territories of Eastern Europe. At the same time, he proffered an exclusive and racist understanding of the provenance of his cattle, supported an authoritarian politics modeled on the mythical authority of the charismatic medieval hunter-forester, and was complicit in the genocide of non-Germanic people living in forests that were to be set aside for hunting.

The reactionary extremities of this desire for the wild are largely missing from contemporary practice. In contrast, the twenty-first-century project of Rewilding Europe is exemplary of forms of ecomodernism. This model takes issue with a different reactionary tendency in twentieth-century environmentalism, evidenced in its unwillingness to face the future and its desire to render the premodern eternal.[76] As a form of ecomodernism, rewilding offers a more optimistic belief in the potential of holistic science and technology to decouple people from nature and to secure territories for wildlife. It aligns this science with sophisticated marketing and ecotourism governmentalities that harness the affective power of ecological aesthetics to shape new forms of consumer-citizens. Rewilding is now more explicitly future oriented; some even go so far as to frame rewilding as part of a grander vision for a "good" and "wilder" Anthropocene.[77] But long-standing tensions persist. Rewilding still retains an ambivalence about the future centrality of urban life. The instrumental science and economics of decoupling sit uneasily with the residual romanticism of its champions. Finally, as I traced in chapter 5, important questions remain about the geopolitical economy through which decoupling is proceeding as well as its distribution of political and ecological benefits.

# PROBIOTIC VALUE
## Putting Keystone Species to Work

*Life at Work*

In 2014, the U.S. Public Broadcasting Service released a *Nature* documentary entitled "Leave It to Beavers."[1] The program offers an environmental history of the animal in North America. It tells of its demise as a consequence of the fur trade, before welcoming its return within national parks and other marginal areas. The film updates the allegory of the beaver as a hardworking and monogamous animal for twenty-first-century wildlife conservation.[2] It celebrates the beaver as an ecosystem engineer whose dams clean water, prevent drought and flooding, and create spaces for other wildlife.[3] We learn how industrious beavers are being recruited as natural builders for a range of restoration projects. PBS produced three educational infographics to accompany the film (Figure 29).

*Figure 29.* "Leave It to Beavers" posters. Source: Public Broadcasting Service.

A comparable enthusiasm for keystone animals as workers can be found among some proponents of biome restoration, including those involved with helminth therapy. One blog stands out among those created by the early adopters. It is titled Colon Comrades! (Figure 20), and it documents the experience of Mike, a graduate student in his mid-twenties living in New England, who uses *Necator americanus* and *Trichuris trichiura* to treat his IBD. Reflecting on his first year with worms, Mike writes, "Last but not least, my little comrades, down in my colon and small intestine celebrating their first birthday. Thank you for your hard work in re-balancing my immune system and keeping down my inflammation levels. Gracias mis helmintos lindos. Siempre en mis entrañas y mi corazón."[4] Mike's comic tone is common across these blogs, which nonetheless express a heartfelt gratitude for the "hard work" done by their gut buddies, laboring in the dark against a hostile immune system to restore gut health. In his choice of font and title, Mark expresses a retro solidarity with his comrade worms, articulating the leftist antipathy among some of the early pioneers of the hookworm underground toward the forms of health care provided by the pharmaceutical industry, and the growing appetite within this industry in bringing the work done by beneficial microbes to market.

These two examples illustrate a consistent (though not universal) figuring of keystone animals as workers in the probiotic turn. Some species, like beavers, have long been valued instrumentally as game, or they are afforded intrinsic value as a denizen of the wild. But in their controlled reintroduction, beavers, hookworms, and other keystone species enter into novel economic relations. Their role shifts from products (beaver fur and meat) or threats to human labor productivity (parasitic hookworms) to living labor. They are valued for their capabilities to engineer ecosystems, to construct niches, to exercise immune systems, and to generally manage ecological dynamics. Advocates of rewilding, and to a lesser extent biome restoration, have begun to align their science with the logics and practices of ecosystem services—an increasingly powerful framework for environmental governance. As they become caught up in the networks that enable rewilding and biome restoration, keystone species become lively commodities. They become service providers that are worth much more alive than dead.[5]

This chapter has three aims. The first is to specify the ecological work that is understood to be carried out by keystone species, and to explore how this work creates value within the probiotic turn. The second is to identify the contrasting economic tendencies in the multispecies relations that make this probiotic value circulate—which I term commoning, enterprising, and resisting—and to examine which forms of life they make live and let die. The final aim is to look beyond work to acknowledge the other modes of being that are valued within the probiotic turn. There has long been vibrant two-way traffic in metaphors between ecology and economics. This chapter addresses ideas of how nature works that both deepen and help circumvent the pervasive naturalization of work in emerging and established forms of neoliberal environmentalism and health care.[6]

## Animal Work and Probiotic Value

In trying to understand the value provided by the animals and ecologies caught up in the diverse economic relations of the Anthropocene, anthropologists, geographers, and political theorists have become increasingly dissatisfied with both the instrumental metrics offered by neoclassical economics and the concepts of intrinsic value provided by environmental ethicists. They suggest that the former are too anthropocentric—animals are stock, not subjects—while the latter tend to be apolitical and do not account for the historical material relations that configure human–animal entanglements.[7] Instead, a growing number of theorists have been turning to the work of Karl Marx and his labor theory of value.[8] Marx proposes that the value of a commodity stems not just from the relationship between supply and demand but also from the "socially necessary labour time" required to produce it.[9] However, for Marx, the ability to labor is a defining and exclusive property of human "species being."[10] He famously dismisses the idea of nonhuman labor: "We presuppose labour in a form that stamps it as exclusively human. A spider conducts operations that resemble those of a weaver, and a bee puts to shame many an architect in the construction of her cells. But what distinguishes the worst architect from the best of bees is this, that the architect raises his structure in imagination before he erects it in reality."[11]

Early theorists of nonhuman labor, like Donna Haraway and Tim Ingold, challenge Marx's humanism.[12] They rework Marx's labor theory of value in light of new insights from ethnography and ethnology, and they update it for the age of ecology and animal studies, proposing multispecies theories of labor and of lively capital. In a first step, Haraway offers the concept of "encounter value," which she differentiates from use and exchange value, to capture the work done by nonhuman organisms and their bodily transactions.[13] Encounter value is a concept both generative and vague. It has subsequently been specified by Maan Barua, who defines nonhuman labor as "the productive activity of animals, performed intransitively through a range of carnal and ethological registers, and enacted in the presence of others whose own performances have bearings on the skill agent's activity, be it human or animal."[14]

There is now a rich and variegated body of scholarship that has developed a typology of animal work and labor (the two terms tend to be used synonymously) and examined the critical potential of thinking human–animal relations through Marxist concepts, like exploitation and alienation.[15] Strands of this writing have focused on the metabolic labor performed by agricultural animals, like cattle, sheep, chicken, and pigs—sentient commodities that convert plants to meat, dairy, and other products for human consumption.[16] This interest in animal work has also developed the feminist critique of the preoccupations in Marxist theory with paid industrial work (coded male) and the neglect of reproductive work (coded female).[17] It applies concepts of emotional, body, clinical, and care work to animals, especially those associated with employment in service roles.[18] Research has explored the affective labor performed by flagship species, charismatic zoo animals, and companionable pets whose work gets commodified in various institutions of animal captivity.[19] A third strand within this typology focuses on the ecological work performed by individual animals, or aggregations of nonhuman life, that come to market in the form of ecosystem services.[20] Barua defines ecological work as a "form of ecosocial reproduction necessary for the maintenance of ecosystems."[21]

Ecological Work

The stories of beavers and hookworms allow me to develop this literature. Ecological work comes in many forms, but those involved in rewilding and restoration are most concerned with work carried out by keystone species, the topological centrality of which gives them disproportionate agency to shape ecological dynamics. The scientists we encountered in chapter 4 who are mapping ecological network interactions conceive of keystone species as fulfilling a distinct role within what we might understand to be a nonhuman division of labor, which endows them with an elevated position in the organizational hierarchy of their ecology. One common allegory that emerged from my interviews is of an ecological manager who is employed to supervise the labors of others to maximize organizational productivity.

This figuring emerges mostly clearly in the case of the beaver. "Leave It to Beavers" and the wider literatures advocating their reintroduction tell us that beavers are hardworking and industrious, putting in long hours without holidays. We learn that they are skilled, not just brute manual and dental laborers or lumberjacks; rather, they are endowed with architectural and engineering abilities.[22] By listening to the flow of water, they direct their detailed knowledge of local hydrology and vegetation to engineer landscapes at catchment scale. More importantly, the managerial skills of beavers are valued because they mobilize other nonhumans whose specialist labor helps prevent flooding and drought, enhance water quality, and increase biodiversity.[23] Like all good managers, beavers make spaces in which plants, insects, and microbes might thrive. As Scottish nature writer Jim Crumley claims, "The beaver is the most willing, most accomplished, the most hospitable, and the most tireless of allies that nature conservation can summon to its cause, an architect that designs, re-designs, restores and creates wildness. For nothing. Forever."[24]

A similar ecomanagerialism informs the description of the ecological work performed by *N. americanus* and other therapeutic helminths once they are introduced into the human body.[25] In Mutualism 2.0 conditions, we are told that worms exercise the immune system, keeping it busy, while training it not to overreact to the endogenous elements of the human self. Like beavers, they achieve this modulation by managing their local ecology, in this case the composition of gut microbiome. In dialogue with their host, they modulate the biochemical atmosphere of the intestine to favor bacteria that mask their presence, and in so doing they reset dysbiotic ecologies and deliver immunomodulation. From the perspective of the human host, this immunological work involves forms of multispecies pastoral care that facilitate the metabolic and other labors of hardworking bacteria in a style similar to those celebrated by fermenters and others enrolling live bacteria for economic projects outside of the human body.[26]

## More-than-Human Intellect

A second and connected form of probiotic value emerges when the ecological work of keystone species informs the design of products and practices that simulate or mimic their probiotic agencies, as described in chapter 3. To understand this work, we can draw on the analysis by Elizabeth Johnson and Jesse Goldstein of the rise of biomimicry as an ecological paradigm for industrial and military design. They explain, "Biomimeticists cast nature as a participatory 'mentor' of engineering, an inventive companion capable of generating solutions more sophisticated than our clunky, industrial society can imagine. By working with nature in this way, biomimicry elevates what Connolly (2013) has called pluripotentiality—the generative capacities of living and nonliving processes—as the driving force of technological innovation." Johnson and Goldstein describe these generative capacities as a "more-than-human pluripotent intellect."[27] They trace how this intellect is valued, focusing in particular on how the ecological vision among advocates of biomimicry of sociotechnical transformation has been subsumed to what Marx describes as the general intellect of a capitalist society concerned with industrial growth. Although the discourse of biomimicry has yet to be invoked by advocates of the probiotic turn, there is a comparable valorization of keystone species' intellectual labor.

We see this in the case of the beaver in the United Kingdom, where the animal has been positioned as a guide for the reinvention of naturalistic modes of flood management as part of a wider agenda in the statutory environmental and conservation agencies toward "working with the grain of nature" to deliver "nature-based solutions."[28] Catchment managers have traveled to Bavaria to study hydrological regimes in beaver-engineered landscapes. They are taught to think like a beaver, and their experience has subsequently informed practices that simulate beaver activity in U.K. rivers—akin to those described at Knepp in chapter 3—but directed primarily at flood prevention. These include removing artificial drainage in catchment headwaters, building semipermeable dams to store water by flooding marginal land, and retaining woody debris to slow the flow of the river.[29] The success of these projects laid the foundations for the subsequent return of the animals themselves. Working with this demonstrated ecological intelligence, various advocates promote the wisdom of beavers, proffering the animal as a model for contemporary ecological sustainability.[30]

In a similar fashion, those working with helminths have also come to value their nonhuman intelligence: the insights worms offer on the ecology of the microbiome, the functioning of the human immune system, and our coevolutionary relationships with old-friend microbes. There is an established interest in taking the requirements of the worm and other desired microbes as a guide to naturalistic or paleo practices of biome restoration. But the challenges of living well with a living worm, and the inconclusive nature of clinical trials, have prompted mainstream immunological efforts to identify the worms' molecular secretions. The aim is to recapitulate these in pill form or to incarnate them in the synthetic biology of a modified organism. Future regimes of helminth drug delivery, or of manufactured drugs from bugs, seek to capitalize and improve on the worms' coevolved intelligence. In the emerging field of ecoimmunology, hookworms, other helminths, and the human and mice bodies in which they were raised perform clinical labor.[31] In official trials and the distributed experiments of the hookworm underground, the worms' work informs new and valuable means of conceiving, diagnosing, and treating autoimmune disease.[32]

The probiotic turn is therefore characterized by a novel form of ecological animal work, which is configured within a nonhuman division of labor marked by a specialization of animal activity. Some keystone species, like the beaver, come to play a distinctive and important role in this economized ecology. Beavers emerge as flexible specialists, performing jobs that cut across the spectrum of economic activity described by sociologists. At times they are involved in primary activities, laboring to chop down trees and build dams. On other occasions, they perform skilled tertiary activities: they design, engineer, and manage others. They also entertain and solicit emotion among ecotourist enthusiasts. Finally, as a source of nonhuman intelligence, they figure as quaternary knowledge workers, serving as ecological gurus that inspire landscape-scale programs for rewilding.[33]

## Probiotic Political Ecologies

The multispecies political economic relations through which this ecological work is valued configure who benefits from the probiotic turn.
Differences in the reach and character of these relations have stark distributional consequences. Here I return to the analysis I started in chapter 5 of the partial reach of the probiotic turn to examine the inequalities
between those humans and animals caught up in Mutualism 2.0 relations. I will outline and compare three economic tendencies that characterize the probiotic turn: commoning, enterprising, and resisting.

### Commoning

The early pioneers of helminth therapy and the beaver enthusiasts that
we encountered in chapter 4 were the first to establish proprietorial economic relations with the probiotic value of worms and beavers. These
animals had previously been framed as kin (in various indigenous cos-
mologies), resources (a source of fur and meat by settler colonists), or
pests (parasitic hookworm infections).[34] In building their networks for
biome restoration and rewilding, they converted the probiotic value provided by their keystone species from a situation of open and unregulated
access to a form of multispecies commons: a living resource, commu-
nally managed, and available at cost to those seeking access.[35] They have
established and maintained novel commoning relations that enable the
reproduction of worms and beavers, and that ensure their circulation
into new bodies and landscapes.

To understand the political ecology of these relations, we can draw
on the recent revival of the long-standing interest in the commons as
a precursor and potential alternative to liberal, anthropocentric, and
capitalist models of resource management. Patrick Bresnihan explains
how new thinking about multispecies commoning has emerged from a
conjunction of dissatisfactions among political ecologists with both the
established institutional model of common pool resources management
(famously developed by Elinor Ostrom) and with the autonomist Marxist enthusiasm for the emergent possibilities of the social or immaterial
commons.[36] He suggests that the former preserves the "methodological
individualism, self-interested rationality, rule guiding behavior and maximizing strategies that underpin liberal forms of government,"[37] while the

latter ignores the constraints posed by the material limits and ecological demands of a finite planet, and neglects the work done by nonhumans. As a result, "we end up with one form of the commons that appears to be *asocial* (excluding the socially productive and reproductive labor of humans involved in caring for the 'natural' resources they rely on), and another that appears to be *anatural* (excluding the material limits and properties of more-than-human bodies involved in the re/production of the 'social' commons)."[38] Drawing on his work on fisheries in Ireland, Bresnihan supplements Marxist accounts of the commons with reference to a range of feminist and indigenous scholarship. Alongside others, he proposes a processual and more-than-human account of commoning as the "continuous making and remaking of the commons through shared labours and capacities,"[39] where the commons "is not land or knowledge or rules. It is the way these, and more, are combined, used and cared for by and through a collective that is not only human but also non-human."[40]

We can see a form of this more-than-human commoning at work among DIY helminth users. While some early pioneers sought worms for commercial gain, most providers grow and distribute worms for free or at cost. Some of those involved in online helminth therapy communities are part of a reciprocal economy in which incubators raise and share worms and provide advice and support for free through the helminth therapy wiki. They do so to maintain a safe and accessible stock of organisms, to build a community of experimental users, and as part of a wider mission to bring helminth therapy to the mainstream. This ethos is articulated by those behind Symmbio, an online helminth provider, who explain,

> We are a small group of long-term users/cultivators of symbiotic helminths from different backgrounds, who all believe strongly in the ideal of making effective health-care available and affordable to all those who need it. We see this is a human rights issue and a political one; for far too long human health has been controlled by a select few, bound up in a system that is slow to adapt and all too often gives poor results. As autoimmune issues increase in populations throughout the world we believe it is imperative that safe and affordable alternatives be made available to help people achieve the best possible health for themselves. Our aim is to bolster the growing community of helminth hosts with another option for effective treatment of many immune-related

disorders by providing a cost-effective source of *Necator americanus* hookworm larvae for purchase from almost anywhere in the world.[41]

Symmbio has turned *N. americanus* and other helminths into commodities that can be purchased with cryptocurrency like Bitcoin. They have made the probiotic value of helminths fungible and given it an exchange value to enable circulation.[42] They pay themselves a salary in lieu of their own labor, but they do not take a profit. They offer discounts to those unable to afford treatment, and they suggest that those who can pay more to help secure the supply chain. This commoning ethos is shared by those producing the rat tapeworm ova HDC, like William Parker and the company Biome Restoration. Such advocates have gone to some lengths to make sure the probiotic value of their worms stays in the public domain; they train others how to raise and administer the worms' eggs, and they tailor their published papers to mitigate against predatory private-sector patent activity.

Comparable forms of more-than-human commoning can be seen among the beaver enthusiasts who facilitate the translocation of animals into the United Kingdom. These individuals work for small NGOs or run their own businesses to ease the complex administration required to legally move animals across national borders. They have secured a ready supply of animals from those that cannot be accommodated in the rebeavered landscapes of Bavaria, and briefly take ownership of this open access resource to create a form of common property. They then make this available to any landowner willing and able to host beavers on their land. Working ahead, and sometimes at the margins, of the law, they facilitate the spread of beavers and enable their self-willed transgression into private riparian habits. In so doing, they make a new commons from private property—at least for beavers—by restoring access to enclosed land from which their predecessors had been banished.

These two examples help illustrate some shared characteristics of the type of commoning at work in the probiotic turn. The first is a commitment to public access to ecological work, in which animal managers secure a sustainable population of keystone species that they make available to those willing and able to host them. Key protagonists take on the legal and economic risks and responsibilities associated with these projects, and in so doing, they police who gets access to animals, and on what terms. They build an economically inclusive commons while deciding who gets to join the community of users. As with Anna Tsing's discussion of the commoning practices of those involved in *satoyama* forest revitalization in Japan, this approach is predicated on the entrepreneurial activity of a small number of human gatekeepers.[43] These social entrepreneurs are not in it for profit, but they have gained and deploy significant social capital to make their networks circulate. Worms in Europe and North America are not available through institutional public or private health care systems. While public access to helminth therapy is not limited by affordability, it does require forms of biosocial capital, like a familiarity with social media and online communities, as well as some basic scientific knowledge.[44]

Second, this commoning does not seek a return to past modes of economic organization. Nor does it imagine a world in which animals are not commodified. Instead, and in keeping with the postpastoral future pasts described in chapter 6, the DIY helminth community uses cutting-edge cryptocurrencies and e-mail encryption as means to circumvent the legal and financial responsibilities that come with marketing and retailing medical therapeutics as private or corporate property. Here Bitcoin enables an anonymous and quasi-legal commons. A similar commons is created by those involved in the quasi-legal model of preemptive beaver release, but here the exchange mechanism is cash in an economy based on trust and friendship that is similarly embedded in and enabled through hard-to-trace social media and instant messaging.

Third, these commoning practices with worms and beavers initiate resurgence—Anna Tsing's label for the self-willed restoration of functional ecologies and economies in bodies and landscapes blasted by antibiotic modes of managing life.[45] As Cleo Woelfle-Erskine argues, beaver restoration in North America creates a "latent commons" that shifts the calculus of nonhuman value from dead resources and pests to ecological workers and generative sources of nonhuman intelligence.[46] We might understand Mike's colon comrades as an example of what Battistoni (channelling Haraway) describes as "comrade species," emerging from a relationship premised on "new forms of collective subjectivity shared across species and based not in humans' ethical obligation, aesthetic appreciation, or spiritual experience, but rather, in recognition of mutual dependence, reciprocity, and even solidarity."[47] Perhaps a well-cared-for hookworm colony enters into the type of unalienated relationship with its work that is celebrated by Jocelyne Porcher and other advocates of traditional animal agriculture. Perhaps it even escapes the relationships of exploitation that concern the same agricultures' opponents.[48]

Such affirmative readings of commoning are fascinating, but they need to be tempered by an understanding of the conditions under which revalued keystone species get to work, as well as by a wider appreciation of who and what will be saved in a commons solely comprising hardworking animals. For example, while the return of beavers goes some way to address their historic exclusion, they are not always reintroduced to live in full control of their means of production. Nor do they always enjoy an emancipated or unalienated relationship with their work. Instead, in many cases, their rights to the benefits of their work remain conditional on their economic performance.[49] The areas where land managers want beavers to work are not always the areas where beavers would choose to be. To prevent flooding, beavers need to be introduced into the headwaters of a river catchment. But beavers tend to prefer the floodplain, where water levels are stable, they do not need to build dams, and the living is easy. Beavers are sometimes fenced in to create animal labor camps. On the floodplains, beavers cause problems when their landscaping impinges on the integrity of private property or of public infrastructure. In Bavaria, where people and reintroduced beavers have coexisted for fifty years in modern urban and agricultural landscapes, troublesome beavers are trapped and relocated or killed. Beavers are fenced out of some areas, and technologies have been developed to deceive beavers into building dams where humans want them, or to mask the artificial lowering of water levels.[50]

There are examples, like the self-willed return of wolves and boars into abandoned areas of Europe, of keystone species' resisting human control and gaining access to common lands outside of a biopolitics that elevates human and nonhuman work. But the prevalent practices I have traced here—in which probiotic practitioners enclose probiotic value to create common property—begin a process of naturalizing work and of enterprising nature.

Enterprising

While commoning practices continue to predominate in rewilding and biome restoration, there are moves in both sectors toward more private and capitalistic modes of valuing life. There is a shift, in Jessica Dempsey's terms, toward enterprising probiotic value, thereby creating markets in ecosystem services that reward entrepreneurial private proprietors (individual or corporate), preferentially value enterprising units of nature, and enable capital accumulation.[51] As the probiotic turn goes mainstream, it is increasingly being made to fit models of twenty-first-century environmental management and health care, which critics argue have driven a shared neoliberalization of these domains of public policy.[52] The character and problems with such approaches are well reported in the political ecology literatures, which I draw on.

The primary step in this enterprising process involves the further enclosure of probiotic value. I have explained how pioneering probiotic commoners first converted an open access resource (wild beavers and worms) into a form of multispecies common property, establishing populations of animals and creating a demand for their reintroduction. Probiotic entrepreneurs seek ways to privatize this common resource. This trend is perhaps best expressed in the case of helminth therapy. Some early pioneers, like Jasper Lawrence and his company, Autoimmune Therapies, see helminths as a business opportunity, with worms and follow-up support sold for profit. This venture has had limited success, in part because of the ready availability of cheaper common-property worms. A more significant drive to privatize the probiotic value of helminths is underway among clinical researchers and their backers in the pharmaceutical industry. These researchers have turned away from the ecological work done by bugs as drugs to the more-than-human intelligence that might be synthesized as drugs from bugs.

Researchers seek to identify the molecules secreted by worms in their immunosuppressant interactions with the human body. They hope that decoding this helminth "secretome" will inform the synthesis of the most significant molecules, with an eye toward their eventual industrial production as pills.[53] They would then be able to recapitulate the work done by the worm while circumventing the ecological and affective challenges posed by the therapeutic use of whole living organisms. Synthesized molecular secretions are familiar materials for those involved in drug discovery and clinical trials. Such secretions more easily conform to the legal regimes that enable the enclosure of nonhuman intelligence as intellectual property through the allocation of patents.

Immunosuppressant drugs are a lucrative sector for pharmaceutical companies, and one potentially threatened by biome restoration.[54] Privatizing worms as pills and encouraging chronic use and dependency would secure and enhance this valuable revenue stream. This work is in its early stages, but a handful of helminth-derived molecules have been identified and are currently the subject of clinical trials.[55]

This trend toward the enclosure and privatization of common probiotic property is more starkly expressed in the case of FMT. As with helminths, several organizations (including OpenBiome) have sprung up to make shit a common-property resource, establishing biobanks of safe, donated feces for subsequent transplantation. In parallel, several private companies in the United States (including Seres Therapeutics) are seeking to synthesize the active ingredients of healthy feces to produce ecobiotics.[56] To secure the funding necessary to develop and test these products, they applied to the U.S. Food and Drug Administration for these ingredients' status as investigational new drugs. Such a classification would give the company monopoly rights on their product, thus guaranteeing future revenue earned from the private investment required to support the clinical trials necessary to demonstrate efficacy. This classification would also require all other providers of FMT to cease operations. At the time of writing, this process was being fiercely debated by advocates of commoning and of enterprising the probiotic value of shit, with the decision being deferred until the company can demonstrate the efficacy of their ecobiotic product in a phase 2 trial.[57]

This private enclosure of helminth molecules or active bacteria subsequently enables the presentation of probiotic value as forms of lively biocapital.[58] In a similar fashion to the commodification practices traced by Melinda Cooper and by Elizabeth Johnson and Jesse Goldstein, it is the generative potentials—or pluripotentiality—of keystone species that come to market in the speculative bioeconomy that drives drug discovery and development.[59] This is perhaps the most striking and important difference between commoning and enterprising in the probiotic turn. As capital, animals are no longer perceived as workers engaged across species divides in some common enterprise. Instead, they are monetized as means of (re)production, owned by a select group of people, for the purpose of future capital accumulation. As lively biocapital, animals and their molecular secretions are put to work to generate profit. These processes of monetizing nature are well underway in environmental management and health care, and political ecologists and economic sociologists have traced the elaborate scientific and economic calculation devices they require.[60]

In the context of rewilding, this refiguring frames animals as components of "natural capital,"[61] a concept that has become central to the case for beaver reintroduction in the United Kingdom. The probiotic value of beavers has risen in public prominence and political palatability as a result of their presentation as tools for delivering ecosystem services. The then–secretary of state for the environment, Michael Gove, supported the academic project of calculating natural capital and bringing it into the calculus of public policy.[62] Under conditions of public sector austerity, this department promotes beavers as hardworking and entrepreneurial nonhuman citizens that might be brought to the United Kingdom to do the types of ecological work currently performed by people at public expense, or to mitigate environmental harms that would be covered by the public purse. Prominent beaver enthusiasts, like those involved in writing *The Eurasian Beaver Handbook*, have positioned their animal as an ideal instrument for austerity environmentalism:

> There is now little doubt that many of the habitat maintenance tasks undertaken by human managers of riparian environments to promote biodiversity or to reinstate natural systems of resilience to flood or drought are mimicking the lost activities of beavers, such as willow-coppicing, the cutting/removal of semi emergent plant species, the insertion of brash bundles in rivers to provide refugia for fish fry, canal excavation in reed beds, or open water creation. There is a growing political awareness that, rather that continuing to strive to develop these environments artificially, the reinstatement of sustainable natural processes via nature-based solutions would be more effective.[63]

Beavers perform the kinds of labor-intensive traditional or naturalistic land management that public authorities and NGOs have supported through agroenvironmental subsidies and by donations of volunteer time and money. Leaving it to beavers would help defray the costs of having conservation work done by people while providing a politically acceptable means to continue some forms of agricultural support in the interest of making public money provide public goods.[64] Here beavers emerge as national natural capital rather than strictly private capital associated with the translation of helminth therapy. In Papadopoulos's terms, the probiotic work of beavers has been captured by the state, and beavers will be put to work for the human public interest.[65] More familiar modes of private financial biocapital are anticipated to emerge from any successful synthesis of the immunosuppressant properties of helminths, or from the creation of FMT-simulating ecobiotics. The financial value generated from this capitalization will ultimately accrue to the owners and/or shareholders of the pharmaceutical enterprises, potentially off the backs of the human and nonhuman clinical labor undertaken by those involved in the distributed experiments with FMT or helminth therapy we encountered in chapter 4.

While this shift from commoning to enterprising is still on the horizon and is contested, we can anticipate some of its distributional consequences for the human and nonhuman subjects of the probiotic turn. Existing work by political ecologists on the rise of capitalist modes of managing natural resources, medical technologies, and environmental services traces the various ways in which both people and ecologies are subsumed to the logics of capital accumulation.[66] Neil Smith and others have explored how the lifeworlds, rhythms, and relationships of humans and nonhumans are reconfigured "all the way down."[67] We can anticipate examples of this process in the futures promised by some immunologists and public health experts, in which the mapping of the helminth secretome will enable new drug development—not only to recapitulate the worm but also to produce a new generation of anthelminthic drugs and vaccines. Drugs from bugs will enable hookworm eradication, worms will be outlawed and go extinct, and coevolved relationships will be replaced by a chronic dependence on monetized molecular surrogates.

But as the analysis in chapter 5 made clear, access to current regimes of immunosuppressant drugs is socioeconomically stratified. Furthermore, the history of drug development by private pharmaceutical companies demonstrates how new products circulate in markets that are controlled to elevate prices, such that drugs do not reach all those in need. As with the case of insulin, cycles of subsequent innovation are often designed to deliver new patents preventing the production of accessible generic versions. Securing the future delivery of probiotic value under this model risks disposing significant populations to the future risks of autoimmune disease while making them dependent on health care markets structured to prevent widespread drug delivery.[68] The rise of such capitalistic models of probiotic health care are feared by some leading figures in the DIY helminth community, who highlight several political, technological, and ecological reasons why they believe such futures will not come to pass.[69]

This subsumption of probiotic ecological functions to the logics of capital, as well as the repositioning of probiotic value as an accumulation strategy, will have marked biopolitical consequences.[70] As Morgan Robertson puts it, there are only certain types of natures that "capital can see," and thus a capitalist symbiopolitical model premised on optimizing the circulation and accumulation of natural or biocapital will have stark distributional consequences on which forms of nonhuman life are made to live and which are let die in the probiotic turn.[71] An analysis of the animal work that constitutes probiotic value helps develop the taxonomy of which enterprising units of nature are likely to be produced under this model, and which will be let die as waste.[72] It draws attention to the current and future nonhuman experience of being a valued and a neglected probiotic worker. For example, Dieter Helm, the economist who coined the term "natural capital," chairs the U.K. government's Natural Capital Committee. In a blueprint for rescuing the British countryside, he expresses an unease about (what he understands to be) rewilding, but he makes a case for the return of enterprising units of nature, promising,

> This would not be a wild world, and it certainly would not be a "re-wilded" world. It would be every bit as managed as it is today. Even those areas left aside would be deliberately chosen for intentional neglect. Deer would be culled, hedges would be reinstated and managed, rivers would be built around natural capital deliberately put in place, and city streets would be planted with trees. The prize is not an abandonment of the land to the "forces of nature," but the replacement of a badly managed natural environment with a much better managed one.[73]

Helm offers an instrumental and functionalist view of U.K. nature subsumed to the logics of certain human interests.

In this context, the beaver emerges as the archetype of the desirable nonhuman entrepreneur. "Leave It to Beavers" shifts from the allegorical affirmation of the beaver as a settler colonial frontier laborer to recast the animal for work in postindustrial America. The program's PBS posters show the valorization of the beaver as a flexible specialist, a valued post-Fordist working subject. The beaver is adaptive and resilient, able to start again when shocks happen. They move when food is short, taking their labor to places in need.[74] The flexible specialist beaver is still employed as a laborer, but he (as gendered in the film) is also an engineer, a scientist, and a nature warden. The film depicts the beaver as a charismatic performer whose charming affective work catalyzes political support and enables "spectacular accumulation" through the commodification of beaver imagery.[75] As a source of biomimetic nonhuman intelligence, beavers are taken to naturalize work and a Protestant model of economic citizenship conditional on workplace participation.

We might read these developments as initial steps toward the domestication of beavers—at least when domestication is indexed to degrees of human control. Here beavers would follow the long line of animals that survived by virtue of their ability to work well within the human domus. Valued beavers—like cattle, horses, and dogs—will be those that are subservient, tractable, reproductive, and resilient. While beavers may well flourish in an Anthropocene with biopolitics configured by the value of work done by nonhumans, there are many nonhumans who cannot, or will not, work, and the contemporary of austerity ecologies (like austerity societies) is no place to be unemployed. Even the hookworm, which performs the work so valued among DIY helminth therapists, risks future eradication should its probiotic value prove amenable to molecular simulation and become profitable when delivered in pill form.

In the face of this shift toward enterprising probiotic value, some of the scientists working on rewilding and biome restoration push back by promoting the importance of diversity and functional redundancy. They make a case for life-forms whose work is not yet understood, or who might step in when others get sick or die. They caution that making rewilding fit for policy through frameworks of economic valuation involves processes of abstraction that risk reducing the diversity and complexity of ecosystems.[76] A myopic focus on targeted service providers or molecular surrogates will render many life-forms useless and unnecessary. They will become waste; they will be abandoned, unloved, and left to die. Meanwhile, those animals and microbes that deliver services become a new frontier for accumulation: they offer a probiotic fix for capital.[77] Dysbiosis is not necessarily bad for business; it creates markets for new products and services that put natural capital to work.

## Resisting

A third economic tendency shifts focus from the anthropocentrism of both commoning and enterprising probiotic value to describe the existing (and latent) possibilities that ecologically significant species resist human projects of enclosure and exchange and proliferate as a consequence of probiotic interventions. I take the concept of resistance from Jason Hribal, for whom it describes the various ways in which individual domestic animals circumvent human efforts to make them work—in farms, circuses, zoos, and other centers of animal captivity. Hribal is careful not to tie resistance too closely to intentionality, and he acknowledges a range of animal subjects and their modes of recalcitrance and transgression.[78] "Resistance" is also a common term in health and agricultural research concerned with how insects and microbes evolve and adapt to the lethal effects of antibiotic chemicals to become superbugs or superweeds. While I acknowledge that the proprietorial tendencies of commoning and enterprising are resisted by some human actors, here I draw these disparate meanings together to present resistance as a nonhuman economic practice, with the potential to unmake forms of human property and capital.

As I explained in chapters 2 and 3, some advocates of the probiotic turn promote the strategic use of keystone species as means to tackle the pathologies associated with antibiotic blowback. They argue that modern hospitals, plantations, and factory farms create hot spots that generate novel human, plant, and animal diseases and invasive species. They suggest that keystone species have remedial value in these settings for enabling biological forms of control and for establishing colonization resistance. While this thesis has demonstrated potential, the history of such interventions, as well as explorations within the genre of science fiction, show how such symbiopolitical interventions can or could go awry.[79] For example, concerns have been expressed about what else travels in an FMT, and how a functionally restored gut ecology might nonetheless give rise to novel ecologies with the ability to shape the host along the gut–brain axis. Anecdotal reports among recipients of FMT suggest that donated ecologies can shape moods and dietary preferences, sometimes in common with those of their original donor. Likewise, the experience of biological pest control in Australia and New Zealand—popularized by the cane toad—demonstrates how fast breeding and adaptive animals can transgress their instrumental deployment and make unanticipated futures in novel ecologies.[80]

Even keystone species reintroduced to their native ecologies often return to situations that are radically different from those in which they evolved. In these emergent ecologies, they may be missing competitors as well as the producers on which they depend.[81] Their ecologies may well feature nonnative or recently evolved organisms with which they have never interacted, but with whom they might form new ecological alliances. New species may emerge—like the coywolf[82]—while existing species may take on new roles—like urban cougars, macaques, and leopards.[83] Furthermore, the ecological changes that will accompany future Anthropocene climatic regimes, coupled with the continued acceleration of socioecological mobility and technological change, will only increase uncertainty as to how ecologies will adapt. Future ecologies may have radically different topologies of ecological interaction, which will reconfigure the relational role of a keystone. As we saw with hookworms in chapter 5, coevolved mutualists can swiftly become parasites when their socioecological configuration changes.

Examples of rewilding and biome restoration gone wrong (at least from the perspective of certain groups of humans) are uncommon, but a couple of stories show the potential. For instance, resistance and proliferation characterize the self-willed resurgence of wild boars across Europe. There have been dramatic increases in boars as a result of agricultural land abandonment and diminished overall hunting. Shifting public sensibilities have led to active local releases by both hunters and animal rights activists.[84] In the absence of predators, boars forage, move, and breed at will. They are intelligent animals that are hard to exclude, and they adapt to local situations. They damage crops; because they have lost their fear of people, they scare dog walkers and horse riders. Boars and their ticks have also become vectors for the African swine fever virus. This disease is endemic in warthogs and arrived in Europe in the 1950s. It can cause high mortality in domestic pig herds and has no treatment. It was controlled in Western Europe by the 1990s, but it became endemic in Russia and post-Soviet states. Expanding and mobile boar populations, sometimes assisted by hunters, have returned the disease to Western Europe. Resistant, proliferating boars have become a biosecurity threat.[85]

We might read this feral, boarish proliferation as a process of more-than-human reoccupation and the commoning of once-enclosed agricultural land. Boars are transgressive, like the feral beavers introduced by the beaver bomber in the United Kingdom and the beaver believers in North America.[86] In rural areas, boars move in as people move out, or they trespass on arable land. In so doing, often despite human efforts to control them, they turn over the soil and distribute seeds, making common land for a wide range of species excluded by the past priority given to a narrow diversity of proprietorial grains. Boars have also found a niche in urban areas like Berlin and Barcelona, adapting to forage amid the waste provided by urban excess and aggressively claiming the nocturnal streetscape.[87] Their return has been enabled by their charisma, variously valued as worthy prey (by hunters) or anthropomorphic intelligent renegades (by rural tourists and some urban enthusiasts).

But at the same time, boars return within the globalized disease-scape of pork production.[88] Their bodies could be exposed to an African virus never encountered in their species history, or to diseases emerging from the pathological hot spots of intensive pig rearing.[89] They risk sudden death; they work unwittingly to facilitate an epidemic with the potential to destroy the vast accumulations of animal capital generated on the backs of human and porcine labor.[90] The proliferation of the African swine fever virus, and others like it, have radically disruptive potential. In some apocalyptic predictions, the work done by this virus could lead to massive pig deaths, the cessation of intensive pig production, and a wholesale reorganization of the multispecies relations of livestock production. The potential outcomes for pigs and their work include extinction, emancipation, and yet further intensified enclosure and exclusion, enabled by gene editing for disease resistance. Similar apocalyptic visions abound in current discussions of a postantibiotic future, in which microbial resistance and proliferation lead to the reestablishment of social, medical, and ecological practices and relations once thought to be banished to the past. For some, these coming plagues present an opportunity to rethink status quo political ecological relations.[91]

I finish this section with a different cautionary dystopia of resistance and a probiotic intervention gone wrong. In her Parasitology trilogy, science fiction writer Mira Grant imagines an America in which William Parker's vision of a genetically modified therapeutic helminth has come to pass.[92] A powerful and immensely profitable biomedical corporation sells personalized worms, the secretions of which are tailored to individual bodies' requirements. Worms keep people healthy and productive; they have the ability to shape how we think and feel. Things go awry when a rogue employee splices in genetic material from the brain parasite that causes toxoplasmosis. These so-called toxo worms take control of their host to ensure their reproduction, creating a horde of short-lived flesh-eating zombies. But in a few cases, full integration occurs, and a small number of human–helminth chimeras evolve. One such chimera is the series' protagonist, a young woman tasked with saving her kin and a small number of human carers amid the end of civilization. They find peace in a multispecies commune somewhere in the California hills. The story is rich in Hollywood B-movie cliché, but it enables an ambitious thought experiment as to what might happen should future enterprising worms and their hosts resist their subsumption to the logics of biocapital and make their own multispecies commons.

## Beyond the Naturalization of Work and Enterprise

This chapter has explored the naturalization of work in the probiotic turn. I started with two contrasting examples—beavers as engineers and hookworms as comrades—to specify the ecological work and nonhuman intelligence that are understood to be provided by keystone species in the probiotic turn. I explored how some keystone species are valued as ideal managers and flexible specialists able to orchestrate a nonhuman workforce configured within an ecological division of labor. I identified how the distribution of the value generated by this work is shaped by three economic tendencies in the probiotic turn: early pioneers created a multispecies commons from an open access resource, facilitating distribution and controlling standards. This commoning laid the foundations for a subsequent enclosure of probiotic value as capital that benefits enterprising humans and enterprising units of nature, at the possible expense of those less able or willing to work. But such a subsumption of ecologies to the logics of capitalist enterprise is resisted by some people as well as by some animals and microbes, the unruly agencies of which have the potential to both open up new multispecies commons and to radically disrupt existing accumulations of capital and property.

I want to make two points by way of conclusion. The first is to re-iterate the concerns expressed by many political ecologists about the shift in environmental management and health care toward private property and enterprise over commoning relations. In the case of the probiotic turn, this shift exacerbates the unequal geographies of dysbiosis identi-fied in chapter 5 and risks what Jessica Dempsey terms "an ecological-economic tribunal for life,"[93] in which future survival is conditional on either utility or self-willed circumvention and proliferation, at the expense of both intraspecies and interspecies solidarity. There are many humans and nonhumans not able or willing to work, and the futures sketched by those modeling the Anthropocene are not places in which many would want to be left alone.

Elizabeth Johnson develops the concept of the more-than-human undercommon as a counterpoint to such individualistic, anthropocen-tric, and privatized approaches. She explains that the undercommon "has as its goal the active subversion of the enclosure of knowledge and life's generative processes . . . to imagine enjoining with nonhuman life in a refusal of capitalism's ultimatum to produce or wither away."[94] Empiri-cal manifestations of this undercommon in the probiotic turn are pre-sented by the relations established by beaver bombers and their fellow beaver believers, alongside those involved in developing and sustain-ing the hookworm underground. These scientists and activists changed the status of beavers and worms from open use to common property, not as the first step toward their private enclosure or solely to satisfy wider human interests, but partly out of a desire to build open-ended relations of solidarity with their favored keystone species and to en-able wider human access to the benefits they deliver. The survival of beavers and worms was secured out of a respect for the lives of these animals—and the lives of others that their ecological engineering made possible.

But beavers and hookworms are let live because these organisms lend themselves to Mutualism 2.0 relations in ways conducive to the flourishing of some humans. The proliferation of wild boars illustrates a second undercommon, in which unruly animals unleash their generative agencies in ways that escape and undermine forms of capitalist enclosure. Boars make their own commons in ways not always commensurable with anthropocentric projects. If we replace boars with an emerging infectious disease, like influenza, then we can anticipate uncommoning practices that might threaten human survival.[95] The point I want to make is that there is no single multispecies commons or common wealth. Instead, there is a symbiopolitics of commoning and uncommoning in which some are made to live and others let die. Such an ecologized account—one that attends to parasites alongside mutualists—suggests we should treat with caution some of the more utopian and dematerialized accounts of postcapitalist commoning currently on offer in political ecology. Commoning involves noninnocent acts of symbiopolitics in which, as we saw in chapter 3, others are let die. The admirable undercommons of beavers and hookworms are not generalizable to all human–nonhuman relations, but they do offer a path along which the probiotic turn might proceed.

A second path comes from troubling the status afforded work and labor in conceptual developments in animal studies. While the turn to animal work has reanimated Marxist theories of labor value and forged new alliances between labor and animal activists, it comes at the cost of naturalizing work, and it risks exacerbating the economic conditionality for survival. In developing this agenda, we might want to pause to reflect on the qualitative nature of different types of work to consider other nonwork ways of being that might ground a multispecies political economy. In relation to the former, we can make explicit a difference between animal work and labor—two terms that are largely taken as synonymous in animal studies. Marx condemns capitalism for reducing the worker to a labor with the same status as an animal—what Hannah Arendt calls "animal laborans."[96] The laborer is alienated by the exploitative drudgery of his or her activities, in contrast to the free worker in control of the means of production. This distinction between work and labor is developed more fulsomely by Arendt in her celebration of *Homo faber*.[97] Rather than reducing the animal to the status of a laborer, a more-than-humanist reading of Arendt might offer grounds for a qualitative appraisal of the character of animal work,[98] along the lines suggested in my analysis of translocated beavers and further developed by Matthew

McMullen in his analysis of rewilding in Scotland.[99] Some animals (like humans) can labor, but they can also work. Context matters.

More generally, as the history of twentieth-century authoritarianism and genocide reminds us, a mode of citizenship premised on labor risks productivist totalitarianism and a drift toward the management of bare life (*Arbeit macht frei*—work sets you free[100]). Similarly, as Kathi Weeks and other feminist theorists argue, the promise of work has often been oversold; it is deployed as a political technology to keep agitative publics busy and apolitical while simultaneously radically diminishing what counts as creativity and productivity.[101] We should therefore be wary of the centrality afforded work in the probiotic turn and in certain strands of animal studies. We might attend instead to less productivist modes of being, noting, for example, the status that some scientists and advocates afford to beavers as architects—that is, creative artists who generate and maintain site-specific installations.[102] We might pause to acknowledge the aesthetic dimension of animal activities.[103] We might also note the joy some express when they witness animals at play, noting the affective power of playful juvenile beavers to attract local enthusiasts and traveling ecotourists, and the ways in which some forms of beaver management create spaces (and commodified relationships) that enable such behaviors and interspecies encounters.[104] Creativity and playfulness are not incommensurable with work; nor do they exhaust the myriad ways of animal being. But they do offer glimpses of new grounds for a more-than-work multispecies ethics.

# Conclusions

# A SPECTRUM OF PROBIOTICS

> It matters which stories tell stories, which concepts think
> concepts. Mathematically, visually, and narratively, it matters
> which figures figure figures, which systems systematize
> systems. . . . We need stories (and theories) that are just big
> enough to gather up the complexities and keep the edges open
> and greedy for surprising new and old connections.
>
> —Donna Haraway, *Anthropocene, Capitalocene,*
> *Plantationocene, Chthulucene*

THIS BOOK has woven together a collection of stories. It tells of worms
and wolves, beavers and shit, boars and cattle, and of the worlds these
animals and their humans bring into being. I have ranged across spatial
and temporal scales, guided not by character or historical development
but by a wager that a probiotic turn is underway in the management
of life in WEIRD parts of the world. I offer the concept of going probiotic as a counterpoint to antibiotic modes of conceiving and managing
life, and I develop a figure of Gaia as a way to comprehend planetary
entanglements, identifying a common model of systemic thinking across
seemingly disparate domains of knowledge and praxis. In keeping with
Haraway's appeal in the epigraph that opens this conclusion, I have told
these stories across complexities in the hope of making new connections.
Of course, my argument is also incomplete and contestable. Here I will
recap what my wager makes apparent before reflecting on what the probiotic turn offers to those of us concerned with multispecies flourishing
amid and beyond the Anthropocene. I will sketch a spectrum of probiotics, presenting a typology for classifying ways of going probiotic and for
differentiating modes of probiotic social theory.

First to recap. In the Introduction, I identified six common elements to the probiotic turn, which I then elaborated in the chapters that

followed. I argued that anxieties about antibiotic blowback—the pathologies caused by the excesses of antibiotic modes of biopolitics—are driving experiments with probiotic alternatives. These alternatives recalibrate, rather than reject, antibiotic models. Shifting from a warlike biopolitics of division, simplification, and eradication, they use life to manage life, manipulating symbiotic relationships to secure desired circulations and processes. Probiotic approaches reconfigure the composition of socioecological systems by using significant nonhuman actors to modulate their dynamics and intensities. Keystone species, like hookworms and beavers, feature prominently as nonhuman agents of change. This probiotic turn is founded on a common scientific ontology of a world made up of connected spheres, linked across scales, marked by nonlinear cycles and feedback loops, and capable of multiple stable states. This world is dysbiotic, haunted by missing parts, and tipped by human activities into degraded conditions. Going probiotic therefore involves thinking like Gaia, where Gaia is neither a holistic totality nor a beneficent mother but rather a system that is ticklish yet amenable to some degree of human understanding and control.

This Gaian thinking emerges from a range of wild experiments in which scientists and other experts, in both the lab and the field, work with keystone species toward the controlled decontrolling of ecological controls. These experiments involve open-ended knowledge practices calibrated to generate and learn from surprises. Different protagonists map and hack ecological interactions, working with heterogeneous publics in whose bodies, or upon whose land, the experiments occur. Gaian thinking leads to a common set of governmental activities in which citizens are shaped by an awareness of their ecological entanglements as holobionts and as hyperkeystones. These forms of symbiopolitics modulate the circulation of life such that some ecological processes are enabled and some are constrained, with implications for which humans and nonhumans are made to live and which are made or left to die.

I then mapped the unequal and interdependent geography of the probiotic turn, arguing that only some people, in some places, currently get to go probiotic, and the possibilities of them doing so are often conditional on others elsewhere experiencing the blowback from which probiotics seek to escape. I mapped these geographies of dysbiosis by attending to the socioecological configurations that shape where a keystone species is absent or present, and how this presence or absence intensifies or ameliorates differing pathologies. Through the example of hookworms, we saw how the same keystone species can deliver both health and disease. Shifting to cattle, I then traced the temporalities of the probiotic turn, identifying the central role of future pasts, those anticipatory histories that mobilize ecological baselines to legitimate very different future-oriented restoration projects. We saw a consistent periodization that indicated a fall to a prehistoric socioecological shift, in which immunological and ecological relations departed from evolutionarily adapted baselines. I traced how the different ideas of the wild—when it was and how it might be known—that inform these future pasts are entangled in contrasting attitudes about science, technology, and modernity. Chapter 7 examined elements of the political economy of the probiotic turn. It developed the geographical analysis of dysbiosis by specifying the types of ecological work that gain economic value. It traced which humans and nonhumans benefit from the different ways in which this value is realized and distributed. Focusing on beavers and hookworms, it identified a creeping, but not hegemonic, naturalization of work and entrepreneurship in how the probiotic turn is configured.

Taken together, this preliminary analysis identified some of the key common dimensions of the probiotic turn, which were manifest in examples that ranged across scales and between domains of knowledge and praxis. But these commonalities also mask significant differences. A mutual interest in thinking like Gaia, controlled decontrolling, and wild experiments does not generate a consensus on how life is conceived and managed; nor does it offer a naturalistic solution to the political questions of who decides and who benefits. There was dissensus among my informants and their opponents, within and across the case studies I examined, as to what the probiotic turn was and how it ought to proceed; there are many ways to go probiotic. We can begin to parse this discord by looking at three themes that emerge from my analysis and that cut across the concerns of chapters 5, 6, and 7. These relate to hoary questions of humanism, progress, and justice. Different conceptions of the place of people, the possibilities of progress, and the notion of social good allow me to specify a typology of modes of going probiotic. This spectrum is color coded, using established schemes for differentiating political and ecological positions by the hue, saturation, and brightness of their symbolic color.

## Humanism

The first tension that runs through the probiotic turn relates to the primacy that is afforded human interests. As I argued in chapter 2, the predominant model of going probiotic, as exemplified in rewilding and biome restoration, is characterized by what Josef Keulartz describes as a mode of enlightened anthropocentrism.[1] Here new forms of ecological science enable new models of ecological modernization that can sustain high levels of economic growth and development through the decoupling of economic systems from their ecological footprints. Humans assume their Enlightenment destiny by rationally managing ecological and immunological systems and processes to ward off tipping points and keeping (or reverting) planets, environments, and bodies at desired stable states. This ethos is best conveyed by the ecomodernist movement, in which life is permitted to survive on the condition that it conjoins its labors in the primary interest of human survival, bringing its ecological value to market through the delivery of heterogeneous ecosystem services, including the labor of beavers and the spectacle of wolves and wild cattle. Awkward, resistant, or redundant life-forms are killed or abandoned. And, as we saw with the imagined futures for hookworms, even such useful species might be eradicated should their ecological agencies be amenable to recapitulation in molecular form. At an extreme, we are presented with a functional probiotic planet with its life-forms subsumed to the logics of probiotic capitalisms as enabled through state control. This position is plagued by its internal contradictions, manifest, for example, in discussions about how best to decouple people from nature without disconnecting them from the immunological, psychological, and cultural benefits associated with nature-based activities.[2]

A more tempered version of this light green anthropocentric environmentalism maintains the primacy afforded people as the agents of environmental change and their responsibility for tackling it. But this version challenges the argument that the earth should be subsumed to human interests. Advocates think that although humans have the power to know and to govern Gaia, it does not mean that they should do so guided only by human interests. This position is akin to the "radical anthropocentrism" that Clive Hamilton advocates in his 2017 book *Defiant Earth*. We find advocates among prominent figures in the rewilding movement who take issue with the conditionality of nonhuman survival implicit in the shift to ecosystem services, and who argue that space should be made for the functional redundancy and transgressive complexity of wildlife.[3] Likewise, some of those hosting worms resist the proposal to create molecular surrogates partly on the grounds that the worms themselves should have a place in an ecologically modernized world.[4]

These light green ways of going probiotic are contested by two sets of deeper green opponents. The first set is concerned for the rights and/or welfare of individual animals. They take issue with how surplus keystone species are managed in rewilding, or with the fate of those subject to their predation. For example, the projects that use cattle, horses, and beavers for naturalistic grazing in the absence of predators often kill animals or let them die of disease or starvation. Beavers in Bavaria and in Scotland may be killed when they conflict with human property. Cattle and horses at the OVP are killed when there is insufficient food to see them through the winter, while the deer are let die of disease or starvation. This practice raises challenging ethical questions that I and others have explored elsewhere.[5] In the Netherlands, it led to a political opposition to rewilding so powerful that in 2018, the Dutch government brought an end to the OVP wild grazing experiment, enforcing stricter population management and the supplementary feeding of the animals. Instead, animal advocates suggest that reintroduced keystone species should be subject to the forms of care afforded laboratory or farm animals in situations where their welfare is compromised by anthropogenic restrictions to their movement, reproduction, and feeding.

A different set of opponents advocates deeper green (or more ecocentric) approaches to going probiotic that prioritize the integrity, abundance, and functionality of ecological systems. Channelling Aldo Leopold's famous land ethic, they argue that rewilding and biome restoration will require a reduction in some humans' claims on ecological systems as well as a reduction in the need to compromise with nonhuman interests. This position is most strongly advocated by the early North American rewilders—like Dave Foreman—who founded the radical environmentalist organization Earth First! We see it expressed in this book in the actions of the beaver bomber and other vigilante rewilders across Europe, who surreptitiously move and introduce species, including wolves, pine martens, and boars, thereby facilitating their self-willed uncommoning of human property and ecological management regimes. Some of these actors relish the transgressive potential of these animals and the novel ecologies they bring into being. They value the autonomous, self-willed character of such changes over designed acts of ecological restoration.

Thinking beyond the human, we see how resistant animals and microbes—like boars or drug-resistant superbugs—pull at the centrality of human interests even in the more ecocentric versions of rewilding. Such ecocentrism is riskier when it comes to hookworms, but more permissive microbe-centered realms of ethics can be identified among those advocating for the paleo dirty governmentalities we encountered in chapter 3. For figures like Jeff Leach and George Monbiot, reconnecting with the earth through traditional eating, hunting, and other lifestyle practices provides grounds for a radical ecological ethics counterpoised to the anthropocentrism of modern antibiotic health and hygiene. In extreme form, this generates a laissez-faire anti-Pasteurianism that folds the human into a microbial cosmos.

## Progress

These tensions about the primacy of humans in the probiotic turn are connected to a second set of debates about the character, possibility, and desirability of progress. The most powerful protagonists we encounter going probiotic in this book share a bright green optimistic idea of progress and the possibilities of symbiopolitical control. Scientists and policy makers see great potential in Gaian thinking to recalibrate modern governmental regimes. They maintain faith in science and reason; they hope that by (counter)mapping ecological interaction networks across a range of scales, they will be able to know, anticipate, and govern earth system dynamics. They are confident of devising interventions that will revert, ameliorate, or circumvent undesirable tipping points. They redesign and repurpose cutting-edge biotechnologies, like those involved in gene sequencing and synthetic biology, to redirect long histories of animal domestication, seeking to back-breed and genetically engineer cattle or worms. They aim to harness the power of artificial intelligence, developing automated algorithmic tools for designing ecobiotic drugs and for managing rewilded ecologies. They foresee a green dividend accruing from the biotechnologies that enabled the rise of antibiotic health care and agriculture. In extreme cases, this techno-optimism tends toward a hyperbolic accelerationist celebration of green innovation in which rewilding and biome restoration figure as the latest in a series of disruptive innovations with the autologous power to transform human environmental relations.

This bright green techno-optimism is more tempered by those practitioners we encountered in chapter 4 who hack, rather than map, ecological interactions in their experiments with worms and beavers. The difference is subtle and relates as much to their relative power and the scale of their ambitions as it does to any profound alternative disposition toward modernity. Those involved in hacking—like the DIY helminth users or those using beavers for local restoration projects—are concerned more with the development of appropriate technology, crafting and altering mundane practices and materials to refine workable solutions. They work amid the ruined aftermath of antibiotic interventions; they are informed by site-specific vernacular epistemologies at some remove from the models and experiments of their scientific counterparts. They are less interested in general laws, predictive capacities, or even definitive explanations. They use life to manage life to get by, to cope with autoimmune disease, or to address degraded ecologies with little long-term capacity or ambition toward restoration or adaptation. In chapters 3 and 6, I presented these as post-Pasteurian and postpastoral or postpaleo approaches that recalibrate and rework (rather than reject) modern science, technology, and market capitalism. They share a modernist belief in control, but this is delinked from linear narratives of human triumph.

In contrast, the history of the Heck cattle and the brief encounters with primitivist rewilders evidences a much darker green model of probiotic thinking. This comes in different forms, which are premised on some combination of a retreat from modern, urban life; an explicit disavowal of some technology and forms of science and reason; and a celebration of vernacular knowledge that is variously coded as folk, indigenous, or alternative. Here there is a common antihumanist desire to return to the wild, where the quality of the wild is linked to the absence of alienating modern human activity. The aim is to reconnect by re-creating lifestyles that approximate those of Paleolithic ancestors, generally in rural, off-grid, and small-community settings. Some of these dark greens are disaffected environmentalists who have lost faith in the power of science and reason to shape policy. Others appeal to myths and beliefs that are set over and above science, dismissing scientific expertise in conspiratorial accusations of fabrication, vested interests, and state control. They see themselves guided by feeling, instinct, or intuition as they rediscover their humanity through reconnecting with their wild-animal or microbial selves. There are overlaps here with the anti-Pasteurians, exemplified now in elements of the antivaccination movement.

## Justice

A third tension relates to the politics of the probiotic turn as it manifests in differing opinions as to how it should be managed and who it should benefit. Different facets of this politics were outlined in chapters 5, 6, and 7. Chapter 5 described the unequal geographies of dysbiosis and of going probiotic that have emerged from long histories of colonial capitalism. These geographies are exacerbated by the persistent and stark inequalities in access to health care, sanitation, and food caused by the global political economy of capitalist health care, agriculture, and infrastructure development. Common trajectories toward the amplification of antibiotic methods of health care and of agriculture in some parts of the Global South will raise standards of living, but they risk compelling subject populations to the types of blowback experienced in the Global North. There is a racialized, structural violence to this dysbiosis that acts both quickly and slowly, exacerbating epidemics of infectious and inflammatory diseases.

However, those involved in deworming and reworming occupy very different worlds—worlds with surprisingly little overlap. Public health researchers and immunologists working on vaccines rarely rub shoulders with immunologists trying to understand the inflammatory response to absent worms and experimenting with reintroduction. They go to different conferences and publish in different journals as they respond to what they perceive to be different crises. As a result, the political ecology of global microbial dysbiosis is not always clear to those working with hookworms. Likewise, those currently engaged in rewilding in Europe are not always aware of how their projects are enabled by the globalization of agricultural production and deleterious land use changes in tropical regions. While many share a political commitment to social justice and to improving global or planetary health, the narrow focus of their praxis means that they often miss the paradoxes and interdependencies of their activities. As a result, and albeit inadvertently, the prevalent model of going probiotic performs a lifeboat model of politics, in which only a select public gets access to its benefits. At a global scale, it risks performing the type of abandonment of solidarity by the globalist ruling class that Latour detects in his analysis of Trumpian responses to climate change.[6] A few secure their own futures, and some may even get rich in a fast-unraveling world.

Chapter 6 identified a reactionary manifestation of this exclusive model, one in which going probiotic was entangled in the violent, racist, and nationalist biopolitics of national socialism in early twentieth-century Germany. In the discourse of Nazi propogandists, dysbiosis was the result of racially degenerate forms of knowledge and land management associated with people who did not naturally belong within desired premodern ecologies. This discourse was mobilized by Lutz Heck and Hermann Göring in their plans to restore the forest and steppe of Eastern Europe. It legitimated the genocide of thousands of Jewish and Slavic people, killed to make way for an ecological vision of a racially purified, Germanic landscape. Only a select group of native Aryan people were to benefit from this intervention. This was an extreme example of the abuse of ecological knowledge to naturalize geopolitical interventions, but it is certainly not unique. The twentieth-century history of fascism is replete with dark green xenophobia in which ideas of nature, the native, and the nation informed future pasts with unjust outcomes. One does not have to look far online to find contemporary manifestations in far-right solutions to the problem of immigration. In short, there is a powerful ecofascist—or brown-green model—of going probiotic. Elements of this model overlap with the antimodernism of the dark greens, but as Jeff Herf argues, it also took a reactionary modernist form that mobilized science and technology in the pursuit of ecological control.[7]

This exclusive vision of the wild and of ecological integrity remains marginal at present. Instead, private, public, and NGO advocates for rewilding and biome restoration promote a more cosmopolitan understanding of whose ecologies matter, and they advocate a more democratic model of ecological management. We are told that there are many different desired functional ecologies, and that these are not exclusively tied to bounded locations or racialized typologies. For example, Rewilding Europe and the Tauros Programme are willing to tailor their backbred cattle—Aurochs 2.0—to accommodate the aesthetic preferences for local breeds. Those close to the project of Scottish independence involved in promoting the return of beavers do so as part of a cosmopolitan nationalism that welcomes hardworking European immigrants who might make themselves at home to build a greener economy. They advocate a vision of a green Scotland set, at the time, against the hostile immigration policy of the Westminster (i.e., English) government and their seeming unwillingness to accommodate beavers.[8] While rewilding has often been pioneered by private landowners, the economic fortunes of their visions are conditional on public subsidy, the market appeal of ecotourism, and the democratic preferences of consumer-citizens.

Some of those involved in mapping the human microbiome and developing microbial therapies do tend to link microbial quality to nonmodern indigeneity, seeking the ur-stool in the guts of the most noble of savages. There is an earthy indigeneity to some of dirty governmentalities that are promoted for reconnecting with nature, and some of these err toward the celebration of violent paleo masculinities, like Monbiot's search for fear as the antidote to ecological boredom.[9] But most are wary of any simplistic ranking of microbiomes by provenance or primitivity. Instead, they note that functional microbiomes come in myriad forms that relate more to host genetics, diet, and local environment than to fidelity to a single origin. They willingly exchange worms and feces in democratic social networks that cut across exclusive or racialized categorizations of social difference, noting common evolutionary origins and immunological composition. Managing microbiomes becomes a story of making kin across social and species difference. In contrast to the brown-green ecofascist ways of going probiotic, these examples illustrate a rainbow model calibrated to twenty-first-century identity politics.

Finally, chapter 7 identified a different tension to the politics of going probiotic that relates to contrasting ideas about how the value of the ecological work provided by keystone species should be captured and distributed. This made clear a political economic spectrum marked by differing opinions on property relations, the state, and mechanisms of market exchange. One early version advocates for common-property regimes in which probiotic pioneers bring an open source resource under control, securing a standardized and safe stock of keystone species—like worms, human shit, or beavers—that are made available at cost or on an ability-to-pay principle. A red model of this version of going pro-biotic seeks to make keystone species available to all by securing their public provision either through state-sponsored agroenvironmental sub-sidies (for rewilding) or by licensing them for national health care provi-sion (for biome restoration). But in the case of much of the hookworm underground, the DIY FMT community, and the beaver believers, this approach works outside of, or at least ahead of, state organization through outlaw forms of innovation. This presents an anarchist, or black, version of going probiotic that distrusts the public institutions of conservation and health care, seeing them as too closely associated with the interests of either the pharmaceutical industry or big agriculture.

In contrast, as the efficacy of probiotic approaches is demonstrated, and as they become established, they become more palatable for delivery through regimes of private property, market exchange, and speculative financial investment. Worms or their molecular surrogates become com-modified and are put to work in the interests of capital accumulation. They become biocapital, patented and owned in corporate property regimes, secured by the law and enforced by agencies like the U.S. Food and Drug Administration. The character, scope, and future of probi-otic therapies are to be determined by the pursuit of profit. This blue (as the political-economic inverse of red) model of going probiotic can also be detected in the maturation of rewilding and its alignment with the policy frameworks of ecosystem services, ecotourism, and other financial mechanisms like species banking. As we saw in chapter 7, this shift from black to red to blue is contested, even as it follows the well-documented transition of the mainstreaming of environmentalism and alternative health care. These three themes of humanism, progress, and justice help identify a spectrum of modes of probiotic thought and prac-tice (Figure 30).

This list of common elements and axes of difference is not exhaustive; nor can I claim to have done full justice to the empirical and conceptual

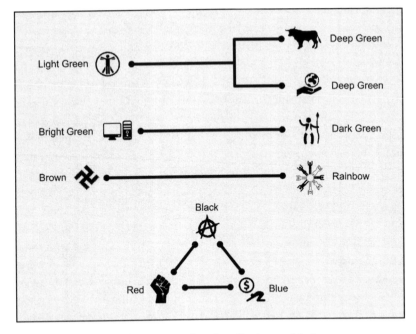

*Figure 30.* Color-coded spectrum of modes of going probiotic.

diversity of the probiotic turn. More work remains to be done in at least four areas. The first would be to explore the other domains of knowledge and praxis identified in Figure 10. For example, I have had little to say about going probiotic in relation to agriculture and food—key domains that link ecological change across scales and that have long histories of probiotic experimentation.[10] Second, there is a need to expand the geographic scope of this analysis, which has largely focused on white WEIRD contexts. It has had little to say about non-Western, provincial, or indigenous ways of going probiotic, and the ways in which these develop and confound the themes of humanism, progress, and justice.[11] Connectedly, the book has only begun to explore questions of race and the ways in which biologized categories of difference justify the partial application of antibiotic and probiotic modes of managing life.[12]

Third would be to attend to further themes that cut across the probiotic turn. There is more to be said about the nature of politics at work across these domains, as well as the character of the experimental knowledge practices sketched in chapter 4. But if I were to write one further chapter, it would explore the aesthetics of going probiotic, examining the contested valorization of mess in debates about rewilding, biome restoration, and probiotic agriculture. Here a holistic shift to processes shapes a neo-Gothic or baroque aesthetic that stands in sharp contrast to the high modernism of valued forms produced by antibiotic modes of managing life. Finally, science studies scholars will want to know more about the origins of probiotic thinking, more about the traffic in ideas between the different domains reviewed, and how they are coproduced by popular anxieties associated with the age of ecology.

## Symbiosis Now?

After describing and specifying the probiotic turn, the third aim of this book has been to examine what it can tell us about the current enthusiasms for Gaian thinking in influential strands of social theory. In the Introduction, I identified the turn to Gaia and symbiosis by figures like Donna Haraway, Bruno Latour, and Isabelle Stengers, which is manifest in their use of Gaian science and their collaborations with Gaian scientists. I suggested that we understand this as a probiotic turn in social theory that results from a triangular coproduction among public anxieties about ecological dysbiosis, crisis science that responds to ecological dysbiosis, and a desire among social theorists for approaches that overcome the inadequacies of twentieth-century social constructivism. I suggested that the intellectual zeitgeist of the Anthropocene and the microbiome has promoted a rapprochement between social theory and natural science, working with publics toward new forms of ecological explanation and management.

Some influential philosophers have suggested that this is evidence of the rise of a general ecology in Western thought, in which theory responds to a new age of ecology that is marked by the proliferation of ecological and environmental thinking across a wide range of domains of knowledge and practical action.[13] This intellectual history is convincing at a general level, but the history of ecology makes it clear that there is no such thing as a general ecology. As Donald Worster and others demonstrate, ecology is a preeminently social science, one that has long been shaped by its social context and riven by debates over its implications for human behavior.[14] This remains true for the probiotic turn in social theory; as the title of one recent workshop suggests, there are "A 1000 Names for Gaia."[15] These multiple Gaias could ground many different ways of ecologizing social theory, and they could be used to legitimate a spectrum of ways of going probiotic. As Stengers puts it, Gaia "has no unifying power other than to authorize sounding the alarm."[16] With this caveat in mind, I would like to return to the three concerns with symbiotic social theory that I outlined in the Introduction—scientism, quietism, and hubris—and reflect on how they might be informed by an understanding of the spectrum of ways of going probiotic.

## Scientism

Critics of scientism are concerned with the tendency they detect in some strands of new materialist theory to take scientific knowledge at face value, and to ignore its contingent emergence, internal heterogeneity, and contested dynamics. They caution against making naive epistemic alliances with the biological sciences, now that they offer more palatable ontostories to ground liberal, antiracist, and more-than-humanist projects. The salience of this warning is affirmed by my analysis of the epistemic practices and politics of the probiotic turn. As chapter 4 makes clear, probiotic thinking has serendipitous origins. It comes from close observations of accidents and their surprises as much as from the testing of any clear hypotheses. It required social institutions and scientific truth spots, and it was pioneered by charismatic, sometimes maverick, individuals self-consciously working in crisis disciplines bent on changing paradigms. While probiotic science should not be reduced to its social context, this context matters. Furthermore, although my story has tended to seek conceptual commonalities across the disciplines studied, it is clear that probiotic thinking is neither universally applicable nor uncontested. Wild experiments in the field, lab, and clinic push against the reductionist norms of modern science. They struggle to capture ecological complexity, may be hard to replicate, and rarely generate definitive laws. They seek to represent a holism, the complexity of which has long thwarted ecologists trying to make their science more like physics, caught by the epistemic allure of mathematical clarity.

As I traced in chapter 6, there are also occasions in which probiotic thinking has been legitimated with reference to epistemic criteria that

tend toward antiexpert, antirationalist, and antiscience appeals to intuition, myth, emotion, and tradition. Such tendencies have long characterized the reception of Gaia among New Age environmentalists and
some alternative health care practitioners. These sentiments inform some
contemporary enthusiasts for dirt and legitimate their dirty governmentalities for reconnecting with nature. They orient some of those who see
a return to earthy, paleo indigeneity as the best route to microbial and
spiritual rewilding. Darker versions of this anti-Enlightenment thinking
characterize reactionary modernist Nazis like Lutz Heck and Hermann
Göring. Taken in all their differences, these antirationalisms illustrate
the types of anti-Pasteurian tendencies that Heather Paxson identifies in
her work on microbiopolitics.[17] Their existence partly drives her argument (with Stefan Helmreich) for new materialists to attend to both the
epistemic and the ontological in making alliances with the biological sciences.[18] Such dark histories and contemporary practices of folkish coproduction need careful navigation as social theory turns to face Gaia.[19]

I sought to navigate these challenges in chapter 5, in which I borrowed and refined the conceptual vocabulary of socioecological systems
and ecoimmunology to develop a political ontology for understanding
the geographies of dysbiosis. Thinking with others—especially Steve
Hinchliffe and Anna Tsing—this chapter combined new materialist Gaian
thinking with political ecology to develop a critical framework for understanding the human and nonhuman drivers of dysbiosis. The coherence
and credibility of my argument partly rests on the claims made in immunology, microbiology, and epidemiology about the causes of inflammatory and infectious diseases. For a critical political ecologist, there is
a beguiling appeal to the links these sciences make to context, history,
and multispecies relations. As others have demonstrated at length, this
approach continues a long history of work in disease ecology that permits political analysis of contingency relations and structural violence
that undermines deterministic accounts of disease causation and environmental degradation. I make an alliance with this science aware of this
siren call, acknowledging that the claims of immunologists and conservation biologists are provisional and contested.[20] The microbial origins
of diseases like diabetes or IBD may prove to be unfounded, and the
claims of advocates may be overplayed. Nonetheless, this skeptical affin-
ity with Gaian science allowed me to develop the figures of the parasite,

ghost, and mutualist in ways that would not have been possible without a sincere engagement across disciplines and a strong commitment to the objectivity of hookworms and their diverse relations across plantation and postindustrial ecologies.[21] In so doing, this approach inherits the modern pursuit of progress that drove the antibiotic Anthropocene. It makes an alliance with the brighter green elements of the probiotic turn (and of probiotic social theory), maintaining the faith that tempered science and reason offer epistemic grounds for recalibrating modes of managing life.

## Quietism

Accusations of political quietism have been made most strongly against two contrasting social science engagements with Gaian thinking. The first is the fifty-year history of using socioecological systems science, and latterly earth systems science, to inform governance. This endeavor is critiqued for using resilience thinking to naturalize the status quo and to ward off changes that might threaten established state or private interests. Resilience thinking has been framed as part and parcel of the rise of neoliberal models of both health care and environmentalism. Accusations of a different form of quietism have been made against the more recent efforts by anthropologists—exemplified by the writings of Anna Tsing—to document the survival strategies of those living amid the ruins of capitalism. Tsing's writings draw on ecology to understand the process of resurgence and proliferation through which multispecies communities come to live and die in the Anthropocene. While this work is critical of mainstream resilience thinking, commentators like David Chandler suggest that it too readily abandons commitments to the ideals of progress, truth, and improvement, and forsakes modernist aspirations for socioecological transformation.[22]

The political compass that guides this book was calibrated to these critiques. I have identified situations in which probiotic thinking tends toward the conservatism of mainstream resilience thinking—for example, in the subsumption of keystone species and their ecologies to the logics of ecosystem service delivery. I examined ruined situations akin to those documented by Anna Tsing, making use of her helpful framework for theorizing and differentiating the socioecological processes that drive dysbiosis. But in so doing, I engaged a persistent strand of critical thinking that offers a revisionist reading of resilience theory as a means to imagine and promote more transformative ways of thinking with Gaia. Drawing on the writings of Bruce Braun, Sara Nelson, Stephanie Wakefield, and others, this suggests that some of the wild experiments associated with the probiotic turn—like those performed by DIY helminths users or the beaver believers—avert political quietism. By hacking and countermapping socioecological dynamics and securing multispecies commons, they offer deliberative political models of thinking and composing with Gaia. This model of going probiotic contrasts with mainstream state and private interventions toward biosecurity; this approach operates in situations marked by disaster in which ecologies have already tipped across thresholds. It looks for means to restore ecologies rather than to ward off future shifts. It makes resilience itself the subject of symbiopolitical intervention rather than assuming it to be a transcendent property.

This approach seeks to evidence Nelson's claim that the historic intersection between resilience thinking and neoliberalism was contingent rather than necessary, and that resilience—or, better, resurgence—can be enabled under different political ecological models.[23] In Braun's and Wakefield's terms, the examples of ecological transformation and multispecies commoning that I explore in this book demonstrate that the value provided by probiotic ecologies should not be taken as testament to the inherent creativity of capital. Instead, as we saw in chapter 7, ecological value comes in part from the ecological work and the nonhuman intelligence performed by keystone species. This reading helps ground Braun's appeal to both new materialists and critical theorists that they free notions of nondeterministic nature—or what Johnson and Goldstein refer to as nature's generative pluripotentiality (after William Connelly)—from the hold of capitalism, and imagine and describe examples where these properties are put to use for other more-than-capitalist projects.[24] To place it on my color spectrum, this approach develops red models of going probiotic.

Drawing this interest in transformation into conversation with Foucauldian concerns with biopolitics also helps address a further political problem that has been associated with the rise of neovitalism in strands of symbiotic social theory. In the work of Roberto Esposito, for example, a Deleuzian commitment to becoming shapes a somewhat totalizing opposition to the immunitarian impulse of antibiotic modes of managing life and a libertarian commitment to differentiation.[25] For Esposito, the modern martial and individualistic models of immunity associated with the antibiotic Anthropocene must be replaced with communitarian alternatives premised on continued encounters with difference. We are encouraged to live and let live, testing and blurring the bubbles of social identity in a rainbow model of going probiotic. But as Bruce Clark and Cary Wolfe caution, when applied to ecological relations and the lives of nonhumans, this laissez-faire or affirmative model can give rise to a thanatopolitics that naturalizes interspecies violence, dissolves the animal subject, and abnegates human responsibilities in the novel ecosystems of the Anthropocene.[26] The analysis in this book helps foreground an alternative model of multispecies hospitality. In chapters 3 and 7 in particular, I traced how the symbiopolitics of the probiotic turn requires a controlled decontrolling of ecological controls in which human actors position themselves as hyperkeystones to modulate evolutionary, ecological and immunological dynamics. Going probiotic involves an affirmative biopolitics of multispecies hospitality, but one that stays with the trouble by recognizing the noninnocent character of ecological relationships and the ways in which the lives of some are always conditional on the pain and death of others.[27] Tracing going probiotic as symbiopolitics offers an ecoimmunological account of the unavoidable politics of responsible multispecies living.

Hubris

The final criticism of probiotic thinking is of hubris, or an excessive anthropocentric confidence in the power and potential of human ingenuity to grapple with the challenges of ecological health and to avert catastrophic change across scales that lead inexorably to the planetary. Critics suggest that a persistent commitment to biopolitics that creates planned projects for managing life, however affirmative, ecological, or symbiopolitical, is doomed both to fail and to generate further blowback. They caution that the world remains fundamentally unknowable and intractable. Further, the world is now sufficiently intensified—most prominently by the fossil-fueled Great Acceleration—that there is no realistic prospect of staying within the Holocene-like conditions that made human life possible. They argue that we need to focus on geopower and ontopolitics. We need to use these tools to find ways of living with uncertainty and the radical asymmetry of a ticklish Gaia, making do in the current and anticipated sites and relations of everyday ruination. They offer a deep green caution to bright green ecomodernist optimism.

This is the most profound challenge facing the probiotic turn, and one that troubles many of my informants as they come to terms with dysbiosis. Working in crisis disciplines, they are well aware of the magnitude of the environmental and immunological pathology—and of the short time frame available for action. In their troubles, they illustrate the different hopes offered by symbiotic theory and action for life within and beyond the Anthropocene. On one level, and in different ways, they have hope for the future. Some, like those conservation biologists and immunologists involved in mapping ecological interaction networks and building models for restoration, cling to what David Chandler and Jonathan Pugh describe as modern hope— "a hope based on human capacities, imaginaries and the telos of progress."[28] As ecomodernists, they remain hubristic; they hope that more and better science, an enlightened citizenry, strong democratic leadership, and technological innovation will secure ecological stability. In Latour's terms, they see us "coming down to earth," recognizing our entanglements with the nonhuman world as a catalyst for redistributive projects for planetary and personal stability and sustainability.[29] Their visions are beguiling, and will no doubt be shared by many of the readers of this book, but they run counter to prevalent and amplifying trends in the Trumpian present.

Others caught up on the front line of the probiotic turn express more modest hopes. They find hope in the local here and now. For example, those seeking biome restoration with hookworms find hope in personal bodily relief—a temporary remission of allergy and inflammation, a suspension of ruination unsecured to any final solution or reversion. Their hope is not of a return of global order but of a livable future. For those hacking with beavers and cattle, hope comes out of grief.[30] Keystone species making new landscapes help rewilders work through the lived memories of extinction and the loss of biological abundance. They offer practical therapy for the failures of past endeavors. As one prominent commentator puts it, rewilding offers a redemption narrative.[31] These hopes help ground the types of transformative commoning politics imagined by Nelson, Wakefield, and others, but they begin (and often end) in situ, working out of modest experiments.

The probiotic turn also evidences a third form of hope that is an explicit counterpoint to the hubris of antibiotic modes of managing life. Chandler and Pugh describe this as "beyond hope"—a hope freed of the hubristic illusions of human control, ascendency, and perpetuation. This is a hope that although the world and the realities of the Anthropocene are always withdrawn from us and insoluble, life will continue. A minimalist ethics finds solace merely in the knowledge that the world will outlast us. These are profoundly nonhumanist, or even antihumanist, hopes, found only on the margins of the stories in this book. They manifest, for example, in the respect articulated for the self-willed proliferation of some species, like boars, and their ability to make life amid the agricultural ruins, even in ways that threaten human property and their bodily security. There are others who, in marked contrast to the transhumanist dreams of silicon immortality, find hope in death, in the life that their decomposing carboniferous bodies will make possible, and in the hope of their future molecular reincorporation into earth system cycles. In Roy Scranton's terms, they are "learning how to die in the Anthropocene."[32]

# ACKNOWLEDGMENTS

THIS BOOK draws on nearly a decade of conversation, collaboration, and disagreement with a rich and varied group of students, scholars, practitioners, and activists. They are too numerous, and my memory is too poor, to list. So apologies if I forget anyone, and thank you all the same.

Much of this book emerges from research and writing with a small number of colleagues. I would like to thank three in particular, starting with Clemens Driessen. Clemens came to work with me in London in 2011, putting his life in the Netherlands on hold. He willingly dug into the archives to track down the dark story of the Heck cattle and the wider entanglements of the Heck brothers with national socialism. Working across several languages, he gathered copious and diverse data on rewilding past and present. He proved adept at asking difficult questions and exploring sensitive issues with charm and diplomacy, even successfully navigating a sensational TV program titled *Hitler's Jurassic Monsters*. Clemens's research and some of our joint writing feature in this book. Clemens has become a good friend, a trusty plumber, a guide to Dutch culture, and a gifted conversationalist. Thank you. I am also grateful to Tim Hodgetts, who started working with me as a PhD student and morphed as a postdoctoral researcher into a geographer microbiologist. Tim led our participatory project on the domestic microbiome. Thank you for your friendship, efficiency, and intelligence. You proved a wonderful foil to my exuberant interest in all things microbiome. Finally, thanks to David Overend, who has become an amiable and enthusiastic collaborator in an ongoing series of art–science and public engagement events exploring the cultural resonances of ideas of rewilding, ghost species, and animal landscapes in contemporary Britain.

I would also like to thank my colleagues in the School of Geography at Hertford College and elsewhere in the University of Oxford with whom I have developed and taught these ideas. Special mention to Andrew Barry, Maan Barua, Jamie Castell, Nathan Clay, Thomas Cousins, Marion Ernwein, Tara Garnett, Charles Godfray, Beth Greenhough, Richard Grenyer, Paul Jepson, Ann Kelly, Keith Kirby, Claas Kirchelle, Ian Klinke, Javier Lezaun, Marc Macias-Fauria, Yadvinder Mahli, Derek McCormack, Carmen McLeod, Jasper Montana, Michelle Pentecost, Gillian Rose, Alex Sexton, and Sarah Whatmore. Further afield, this book has benefited enormously from the writings and insights of a wide network of fellow thinkers, including Irus Braverman, Nik Brown, Henry Buller, Bram Buscher, Clare Chandler, David Chandler, Nigel Clark, Sarah Crowley, Gail Davies, David Demeritt, Caitlin DeSilvey, Rob Dunn, Steve Hinchliffe, George Holmes, Ann Kelly, Eben Kirksey, Hannah Landecker, Alex Nading, Yamini Narayanan, Laura Ogden, Heather Paxson, Jonathan Prior, Tobias Rees, Harriet Ritvo, Krithika Srinivasan, Heather Swanson, Anna Tsing, Thom van Dooren, Kim Ward, and Kathryn Yusoff. Thanks in particular to Stefan Helmreich, who read a full draft of the book while he was working with waves. Many of you organized seminars and conference sessions at which I have presented and developed these ideas. Thank you for your time and hospitality.

I have been fortunate to work with a wonderful collection of bright and motivated graduate students, whose suggestions and analysis have come to shape this work. I am especially grateful to Filipa Soares, who read chapters of the book. Thanks also to Myung-Ae Choi, Kelsi Nagy, Cyrus Nayeri, Jenny Dodsworth, Josh Evans, Dong-Li Hong, Hibba Mazhary, Gregory Anderson, Mark Bomford, and Oscar Hartman-Davies, as well as to the cohorts of students in the Oxford Nature, Society, and Environmental Governance MSc program, for which I developed and taught some of these materials. Special mention to Sarah Vaughn, Diane Borden, Adam Searle, and Jonny Turnbull, all of whom are bright prospects for the environmental humanities.

Thanks also to those who participated in this research, especially those who willingly gave their time and opinions in interviews, at work, and through a collection of collaborative exercises. Special thanks to John Scott and William Parker, from whom I learned a great deal about helminths and their users. Thanks also to Donna Beales, Judy Chinitz, Don Donahue, Peter Hotez, Jasper Lawrence, P'ng Loke, and Rik Maziels for advice on helminths and immunology. Thanks to all those households who participated in the Good Germs, Bad Germs project, and to the scientists who joined the Oxford Interdisciplinary Microbiome Project and came to speak at our seminars—especially Andrew Singer and Lindsay Hall. Thanks to Sally Bloomfield for her expertise on hygiene and for introducing me to the International Scientific Forum on Home Hygiene. Thanks to Elizabeth Bik, David Coil, Tom Domen, Glenn Gibson, Majdi Osman, Barry Smith, Glenn Taylor, and Larry Weiss for information on the microbiome and probiotics. Thanks to Derek Gow, Charlie Burrell, Isabella Tree, Penny Green, George Monbiot, Paul and Louise Ramsey, Jens-Christian Svenning, Mark Elliot, Hugh Graham, and Gerhard Schwab for their guidance on rewilding and on beavers.

I am grateful to Alice Roques, the librarian at Hertford College, and to Ailsa Allen, who helped prepare the illustrations.

This research was supported by two grants from the U.K. Economic and Social Research Council, as well as funding from the Wellcome Trust and the John Fell Fund. The writing of this book would not have been possible without a British Academy midcareer fellowship, which gave me a year of teaching relief in 2018. Thank you to Beth Greenhough, Marion Ernwein, Janet Banfield, and Bharath Ganesh for covering during my absence.

A big thank-you to Jason Weidemann and the team at the University of Minnesota Press, alongside the two readers of my proposal and the book's early chapters. It has been a great pleasure to publish again with the Press. It is also an honor to be featured in the Posthumanities series alongside so many of my academic luminaries and touchstones.

# GLOSSARY

**Antibiotic** (n.)—a substance produced by an organism and designed to inhibit the growth of another organism; (adj.)—the management of life through the eradication, suppression, and rationalization of ecological processes.

**Dysbiosis**—an ecological imbalance that causes a degraded ecology or disease. Generally used in association with the microbial flora of the intestinal tract or other part of the body.

**Ecobiotic**—a functional microbial ecology assembled in a laboratory to restore the intestinal microbiome.

**Ghost**—a missing, sometimes extinct, species, the absence of which changes the functional dynamic of an ecology, sometimes leading to dysbiosis.

**Gnotobiotic**—an organism (usually a higher animal) or its environment artificially rendered devoid of bacteria and other organisms which would normally be present as parasites, commensals, symbionts, etc.

**Holobiont**—the eukaryotic organism or host plus its persistent symbionts.

**Homeostasis**—the maintenance of a dynamically stable state within an ecosystem by internal regulatory processes that tend to counteract any disturbance of the stability by external forces or influences.

**Hysteresis**—the dependence of an ecology on its history, including the lag between an input and a change in ecological state.

**Microbiome**—(*a*) an ecology of organisms invisible to the naked eye comprising bacteria, fungi, protozoa, and viruses; (*b*) the collective genome of this ecology.

**Multibiome**—the microbiome plus a small number of eukaryotic animals, like helminths.

**Mutualist**—one of two organisms of different species that contribute to each other's well-being.

**Parasite**—an organism that obtains nutrients at the expense of the host organism, which it may directly or indirectly harm.

**Pathobiont**—organisms that can switch between mutualistic and pathogenic relationships depending on their ecological context.

**Prebiotic**—A nondigestible food that selectively promotes the growth of beneficial intestinal microbes.

**Probiotic** (n.)—an organism used to modify the composition of the microbiome of the body; (adj.)—the management of life through the targeted encouragement, simulation, and enhancement of ecological processes.

**Proliferation**—the spread of species with ecological effects that impoverish the functionality, diversity, and livability of a landscape or body.

**Resurgence**—(after Anna Tsing) the remaking of livable landscapes through the actions of many organisms.

**Symbiosis**—an association of two different organisms (usually two plants, or an animal and a plant) that live attached to each other, or one as a tenant of the other, and contribute to each other's support. Sometimes used more widely to describe any intimate association of two or more different organisms, whether mutually beneficial or not.

**Tipping point** or **threshold**—the point at which a series of small changes within an ecology become significant enough to cause a larger, more important change.

**Topology**—the mathematical properties of figures and surfaces that are independent of size and shape, and that are unchanged by any deformation that is continuous.

# NOTES

## Introduction

1. I take the acronym WEIRD from Henrich, Heine, and Norenzayan, "Weirdest People," who coined the term to criticize sample bias in psychology research.

2. I take the phrase "use life to manage life" from Jim Crumley and his description of beaver reintroduction in Scotland. See Crumley, *Nature's Architect*.

3. Grassberger et al., *Biotherapy*; Sonnenburg and Sonnenburg, *Good Gut*.

4. Finlay and Arrieta, *Let Them Eat Dirt*.

5. Hajek, *Natural Enemies*.

6. Crawford and Crawford, *Bioremediation*.

7. Esteves, *Managed Realignment*.

8. Chien, Hong, and Lin, "Ideological and Volume Politics."

9. Morton, *Planet Remade*.

10. Wallace and Wallace, "Blowback."

11. Holling and Meffe, "Command and Control."

12. Hinchliffe et al., *Pathological Lives*.

13. Logan, Jacka, and Prescott, "Immune–Microbiota Interactions."

14. Blaser, *Missing Microbes*.

15. Davies, Grant, and Catchpole, *Drugs Don't Work*.

16. Rockström et al., "Planetary Boundaries."

17. Lynas, *God Species*.

18. Asafu-Adjaye et al., *Ecomodernist Manifesto*.

19. Latour, *Down to Earth*.

20. Bordenstein and Theis, "Host Biology."

21. The figure of *Homo microbis* is introduced (with irony) in Helmreich, *Sounding the Limits of Life*.

22. Chandler, *Ontopolitics*; Tsing, *Mushroom*.

23. Latour, "Why Has Critique Run Out of Steam?"; Chandler, *Ontopolitics*.

24. This case has been made by Latour, *Politics of Nature*.

25. These shifts in the character of expertise and the relationships between science and its publics are explored in Callon, Lascoumes, and Barthe, *Acting in an Uncertain World*; Chilvers and Kearnes, *Remaking Participation*.

26. Clarke, "Rethinking Gaia"; Stengers, "Autonomy"; Latour, *Facing Gaia*; Haraway, *Staying with the Trouble*.

27. Worm and Paine, "Humans as a Hyperkeystone Species."

28. Foucault, *Birth of Biopolitics*.

29. Helmreich, *Alien Ocean*.

30. For a discussion, see Grove, *Resilience*; Hinchliffe et al., *Pathological Lives*.

31. Paxson, *Life of Cheese*.

32. I take this phrase from Josef Keulartz's description of rewilding in the Netherlands. See Keulartz, "Emergence of Enlightened Anthropocentrism."

33. Lorimer and Driessen, "Wild Experiments."

34. I take the idea of mapping from Chandler, *Ontopolitics*.

35. The idea of hacking is introduced in Chandler, *Ontopolitics*.

36. Despret, "Body We Care For."

37. Estes, Terborgh, and Brashares, "Trophic Downgrading."

38. Velasquez-Manoff, *Epidemic of Absence*.

39. Rook, "Hygiene Hypothesis."

40. DeSilvey and Bartolini, "Where Horses Run Free?"

41. Collard and Dempsey, "Life for Sale?"

42. Dempsey, *Enterprising Nature*, 10.

43. Stengers, "Autonomy."

44. See Lenton and Latour, "Gaia 2.0"; Arènes, Latour, and Gaillardet, "Giving Depth." Latour's meeting with Lovelock is reported in Latour, "Bruno Latour Tracks Down Gaia."

45. Latour, *Facing Gaia*; Latour, *Down to Earth*; Latour, "Why Gaia Is Not a God of Totality."

46. Haraway, *Staying with the Trouble*. Haraway is acknowledged in an important article that summarizes the science of thinking with holobionts; see McFall-Ngai et al., "Animals in a Bacterial World." She is also cited in Gilbert's introductory textbook, *Ecological Developmental Biology*. Further, she was included in a roundtable discussion; see Haraway et al., "Anthropologists Are Talking." Haraway taught Scott Gilbert when he was a graduate student.

47. Tsing and Bubandt, eds., *Arts of Living*.

48. The highest profile of these was the Anthropocene Project at the Haus der Kulturen der Welt in Berlin (2013–14). For an overview of the Anthropocene zeitgeist, see Lorimer, "Anthropo-scene."

49. I take the concept of an ontostory from Jane Bennett, for whom it describes a mode of naive realism through which idealist social theory might reengage with the vital materiality of the world by (selectively) engaging with natural science. Bennett, "Force of Things." Anna Tsing prefers the term "ontics" over "ontology" in her work on the lifeworlds of animals and objects involved in ecological restoration. For Tsing, ontics describes "the practices in which modes of being are enacted," as is made apparent to her and her collaborators through various forms of expertise, scientific or otherwise. Tsing, "The Buck, the Bull," 15.

50. Latour, "Why Gaia Is Not a God of Totality"; Haraway, *Simians, Cyborgs, and Women*; Martin, *Flexible Bodies*.

51. For an introduction to postmodern synthesis, see McFall-Ngai, "Noticing Microbial Worlds"; Gilbert, *Ecological Developmental Biology*.

52. Ingold and Palsson, *Biosocial Becomings*.

53. Haraway, *Staying with the Trouble*. See also Hird, *Origins of Sociable Life*.

54. Latour, *Down to Earth*.

55. Lenton and Latour, "Gaia 2.0," 1066. This piece can be read alongside a growing number of Anthropocene agendas for research and global governance. See, e.g., Biermann et al., "Navigating the Anthropocene"; Gibson, Bird Rose, and Fincher, *Manifesto for Living in the Anthropocene*; Haff, "Humans and Technology"; Lovbrand, Stripple, and Wiman, "Earth System Governmentality"; Ogden et al., "Global Assemblages."

56. Hörl and Burton, eds., *General Ecology*.

57. Latour, "Why Has Critique Run Out of Steam?"; Latour, *Down to Earth*.

58. Latour, "Attempt."

59. Tsing and Bubandt, eds., *Arts of Living*.

60. Bubandt and Tsing, "Ethnoecology for the Anthropocene"; Tsing, *Mushroom*; Gan, Tsing, and Sullivan, "Using Natural History."

61. In her 2015 Association of Social Anthropologists Firth lecture on symbiotic anthropology (https://www.theasa.org/downloads/publications/firth/firth15.pdf), Tsing argues, "Concerns about the Anthropocene make new conversations possible between natural scientists and humanists, and these can interrupt a previous era of closed doors between the sciences and humanities. I understand the concerns that closed those doors; I was schooled in that era and participated in the critique of science. But now, I think, something else is possible: a new mutuality based on common interests in liveability."

62. Tsai et al., "Golden Snail Opera"; Kirksey, *Multispecies Salon*.

63. Swanson, Bubandt, and Tsing, "Less than One"; Tsing and Bubandt, eds., *Arts of Living*.

64. Paxson and Helmreich, "Perils and Promises," 186.

65. Paxson and Helmreich, "Perils and Promises," 165.

66. Paxson and Helmreich, "Perils and Promises," 169.

67. O'Malley qtd. in Paxson and Helmreich, "Perils and Promises," 170.

68. Braun, "New Materialisms," 3–5.

69. In his writings on Gaia 2.0, Latour is careful to note that "the examples of social Darwinism, sociobiology, and dialectical materialism suggest that drawing political lessons from nature is problematic." Lenton and Latour, "Gaia 2.0," 1066. In other writings, Latour makes clear the political indeterminacy of Gaia. Latour, "Why Gaia Is Not a God of Totality."

70. Chandler, *Ontopolitics*; Massumi, "National Enterprise Emergency"; Massumi, *Ontopower*; Wakefield, "Infrastructures of Liberal Life."

71. Braun, "New Urban Dispositif?"

72. Cooper, *Life as Surplus*; Grove, *Resilience*; Walker and Cooper, "Genealogies of Resilience."

73. Wakefield, "Infrastructures of Liberal Life"; Chandler, *Ontopolitics*.

74. Tsing, *Mushroom*.

75. Tsing, "The Buck, the Bull"; Tsing, "Threat to Holocene Resurgence"; Tsing, *Mushroom*.

76. Nelson and Braun, "Autonomia in the Anthropocene"; Nelson, "Resilience"; Wakefield, "Inhabiting the Anthropocene Back Loop"; Johnson, "At the Limits"; Braun, "From Critique to Experiment?"

77. Stengers, "Autonomy"; Latour, *Down to Earth*.

78. See, e.g., Battistoni, "Bringing in the Work of Nature"; Bresnihan, *Transforming the Fisheries*; Moore, *Capitalism in the Web of Life*; Nelson and Braun, "Autonomia in the Anthropocene."

79. Clark, *Inhuman Nature*; Colebrook, *Death of the Posthuman*; Povinelli, *Geontologies*; Yusoff, *Anthropogenesis*; Yusoff et al., "Geopower"; Hird, "Indifferent Globality."

80. Nigel Clark develops the concept of radical asymmetry in *Inhuman Nature*.

81. I take the concept of learning to die in the Anthropocene from Scranton, *Learning to Die in the Anthropocene*.

## 1. The Probiotic Turn

1. Jørgensen, "Rethinking Rewilding"; Prior and Ward, "Rethinking Rewilding"; Gammon, "Many Meanings of Rewilding"; Keulartz, "Rewilding."

2. Svenning et al., "Science for a Wilder Anthropocene."

3. Fraser, *Rewilding the World*; Lorimer et al., "Rewilding"; Pettorelli, Durant, and du Toit, *Rewilding*.

4. Vera, "Large-Scale Nature Development"; Marris, "Conservation Biology."

5. Cornelissen, "Large Herbivores."

6. Vera, *Grazing Ecology*.

7. van Vuure, *Retracing the Aurochs*; Daszkiewiez and Aikhenbaum, *Aurochs*.

8. For more information about the Tauros Programme and the breeding program, see Goderie et al., *Aurochs*.

9. Baerselman and Vera, *Nature Development*. More information on this process is provided by van den Belt, "Networking Nature"; Jepson, Schepers, and Helmer, "Governing with Nature."

10. Rewilding Europe, *Vision for a Wilder Europe*.

11. Pereira and Navarro, *Rewilding European Landscapes*.

12. For more information, see Mike Wade, "Meet the Knepp Group: The Billionaires Dedicated to 'Rewilding' the Environment," *Times*, December 8, 2018, https://www.thetimes.co.uk/.

13. More information about the Endangered Landscapes Programme, supported by Rausing's Arcadia Fund, is available at their website (https://www.endangeredlandscapes.org/).

14. The figure of $170 million comes from Jonathan Franklin, "Can the World's Most Ambitious Rewilding Project Restore Patagonia's Beauty?," *Guardian*, May 30, 2018, https://www.theguardian.com/.

15. Rewilding Britain was founded by George Monbiot and is run by his partner, Rebecca Wrigley. It seeks to capitalize on the popular interest in rewilding catalyzed by his best-selling book, which makes a case for rewilding in the United Kingdom. Monbiot, *Feral*. Burrell is well connected with local landowners in Sussex and with key figures in the upper echelons of the Conservative government. More information on Rewilding Britain is available at their website (https://www.rewildingbritain.org.uk/).

16. Crumley, *Nature's Architect*.

17. Campbell-Palmer, Gow, and Needham, *Eurasian Beaver*; Puttock et al., "Eurasian Beaver Activity."

18. Law et al., "Using Ecosystem Engineers."

19. The story of the reintroduction of beavers in England is explored in Crowley, Hinchliffe, and McDonald, "Nonhuman Citizens on Trial." Illegal reintroductions through such so-called van jobs have become a common but poorly understood strategy in European rewilding, subject to much rumor and hearsay. See Hodgetts, "Wildlife Conservation."

20. Campbell-Palmer, *Eurasian Beaver Handbook*.

21. Chapron et al., "Recovery of Large Carnivores."

22. Tack, *Wild Boar* (Sus scrofa) *Populations in Europe*.

23. Drenthen, "Return of the Wild."

24. Gavier-Widén et al., "African Swine Fever."

25. Levi et al., "Deer, Predators."

26. Ripple and Beschta, "Wolves and the Ecology of Fear."

27. Laundré, Hernández, and Altendorf, "Wolves, Elk, and Bison."

28. Foreman, *Rewilding North America*.

29. Donlan et al., "Pleistocene Rewilding."

30. Griffiths et al., "Assessing the Potential"; Griffiths et al., "Use of Extant Non-indigenous Tortoises."

31. Zimov et al., "Mammoth Steppe."

32. Nikita Zimov spoke of the carbon bomb at a lecture he gave in Oxford in 2018. The phrase "the proper tool to fight global warming" features on a promotional image for the idea of mammoth deextinction, which serves as the Twitter handle for the Pleistocene Park (https://twitter.com/pleistocenepark).

33. This narrative is best set out in Goderie et al., *Aurochs*. See also Vera, *Grazing Ecology*.

34. Orland, "Turbo-Cows."

35. Mitchell, "How Open Were European Primeval Forests?"; Birks, "Mind the Gap."

36. Caro and Sherman, "Rewilding"; Hodder and Bullock, "Really Wild?"

37. Drenthen, "Rewilding in Layered Landscapes."

38. Korthals et al., "Battles of Nature"; Lorimer and Driessen, "Bovine Biopolitics."

39. For an introduction, see Yong, *I Contain Multitudes*.

40. Filyk and Osborne, "Multibiome."

41. See Blaser, *Missing Microbes*.

42. In so doing, Pritchard joins a long and illustrious list of self-infecting helminthologists. See Lukeš et al., "(Self-)Infections with Parasites."

43. There are two species of hookworm that colonize humans. *Necator americanus* is generally known as the New World hookworm and predominates in the Americas, sub-Saharan Africa, and Southeast Asia. *Ancylostoma duodenale*, the Old World hookworm, predominates in the Middle East, North Africa, India, and (formerly) in southern Europe.

44. Loukas et al., "Hookworm Infection."

45. Overviews and reviews of these trials are provided by Elliott and Weinstock, "Nematodes and Human Therapeutic Trials"; Fleming and Weinstock, "Clinical Trials of Helminth Therapy"; Wammes et al., "Helminth Therapy or Elimination."

46. Babayan et al., "Wild Immunology."

47. Maizels and Nussey, "Into the Wild."

48. Rook, "Hygiene Hypothesis"; Cheng et al., "Overcoming Evolutionary Mismatch."

49. Perry, "Parasites and Human Evolution."

50. Parker and Ollerton, "Evolutionary Biology and Anthropology."

51. Allen and Maizels, "Diversity and Dialogue"; Wammes et al., "Helminth Therapy or Elimination." For an overview of the rise of ecoimmunology, see Tauber, *Immunity*.

52. Zaiss et al., "Intestinal Microbiota."

53. Parker et al., "Prescription for Clinical Immunology."

54. Rook, "Review Series on Helminths."

55. Logan, Jacka, and Prescott, "Immune–Microbiota Interactions."

56. Strachan, "Hay Fever."

57. Parker, "Hygiene Hypothesis"; Hanski et al., "Environmental Biodiversity"; Parker and Ollerton, "Evolutionary Biology and Anthropology."

58. The case of AIDS activism is discussed in Epstein, *Impure Science*.

59. Cheng et al., "Overcoming Evolutionary Mismatch."

60. Helminthic Therapy Wiki, https://helminthictherapywiki.org/.

61. Biome Restoration (https://biomerestoration.com/) is the name of one commercial provider of helminths.

62. "Gut Buddies" and "Colon Comrades" were the titles of two blogs run by early helminth users.

63. Briggs et al., "Hygiene Hypothesis."

64. The term "hookworm underground" was coined by Moises Velasquez-Manoff in his autobiographical account of becoming a helminth user. See Velasquez-Manoff, *Epidemic of Absence*.

65. Parker et al., "Prescription for Clinical Immunology."

66. Navarro, Ferreira, and Loukas, "Hookworm Pharmacopoeia"; Briggs et al., "Hygiene Hypothesis."

67. ClinicalTrials.gov identifier NCT01279577.

68. Loukas et al., "Hookworm Infection"; McKenna et al., "Human Intestinal Parasite Burden."

69. van Nood et al., "Duodenal Infusion"; ClinicalTrials.gov identifier NCT02148601.

70. Market research conducted by Lumina Intelligence suggests that the global probiotics market was worth $35.5 billion in 2016, and it is predicted that its value will exceed $65 billion by 2024. See Lumina Intelligence, "Global Probiotics Market: Trends, Consumer Behaviour and Growth Opportunities," https://www.lumina-intelligence.com/. The advent of Probiotic 2.0 is discussed in O'Toole, Marchesi, and Hill, "Next-Generation Probiotics"; Maxmen, "Living Therapeutics."

71. For a discussion, see Young, "Of Poops and Parasites"; de Vrieze, "Promise of Poop"; Sachs and Edelstein, "Ensuring the Safe and Effective FDA Regulation."

72. Katz and Morell, *Wild Fermentation*; Sonnenburg and Sonnenburg, *Good Gut*; Spector, *Diet Myth*.

73. See, e.g., Finlay and Arrieta, *Let Them Eat Dirt*; Gilbert and Knight, *Dirt Is Good*.

74. Louv, *Last Child*.

75. Monbiot, *Feral*.

76. Leach runs the American Gut Project in collaboration with Tim Spector at Kings College London. He reports on his personal microbiome project in *Rewild*. See Gammon, "Many Meanings of Rewilding," for some wider reflections on these primitivist models of rewilding.

77. See, e.g., the range of work funded by the Alfred P. Sloan Foundation on the Microbiology of the Built Environment (https://sloan.org/), including projects on homes, hospitals, schools, and other institutions, as well as on transport systems and other public infrastructure. The ambitions of this probiotic approach to managing life are articulated in Jacob et al., "Urban Habitat Restoration"; Flies et al., "Cities, Biodiversity and Health."

78. Blaser, *Missing Microbes*; Velasquez-Manoff, *Epidemic of Absence*.

79. Hajek, *Natural Enemies*.

80. Münster, "Performing Alternative Agriculture."

81. Puig de la Bellacasa, "Making Time for Soil"; Granjou and Phillips, "Living and Labouring Soils"; Krzywoszynska, "Caring for Soil Life."

82. Di Gioia and Biavati, *Probiotics and Prebiotics*; Hinchliffe and Ward, "Geographies of Folded Life."

83. Paxson, *Life of Cheese*; Enticott, "Lay Immunology"; Ingram, "Fermentation, Rot"; Spackman, "Formulating Citizenship."

84. The cultural and gastronomic value of fermentation has been promoted by Pollan, *Cooked*. Pollan popularizes the work of Katz, *Art of Fermentation*. The New Nordic food movement is most associated with the Danish restaurant Noma, which has pioneered new fermentation practices. See Redzepi and Zilber, *Noma Guide*. I am indebted to my PhD student Josh Evans for this resource.

85. Crawford and Crawford, *Bioremediation*; Head, Jones, and Röling, "Marine Microorganisms." In 2017, there was excitement in the press about the potential of a recently discovered plastic-eating fungus to tackle the proliferation of oceanic waste. See Khan et al., "Biodegradation."

86. There is an established and growing market for domestic and commercial probiotic hygiene products led by the Belgian company Chrisal. The Boston-based AOBiome manufacturer produces Mother Dirt, a personal hygiene spray

containing ammonia-oxidizing bacteria that they claim reduces the need to wash. There is also a growing demand among high-end consumers for probiotic beauty products. See Vicci Bentley, "The Probiotic Beauty Products Revitalising Skincare Regimes," How to Spend It, *Financial Times*, January 9, 2018, https://howtospendit.ft.com/. This growth has been fueled in part by concerns about the proliferation of antibacterial chemicals and the rise of what one of my informants terms the "anti-anti-bac movement."

87. Warner, van Buuren, and Edelenbos, *Making Space*. Some of the social and cultural dimensions of this shift are explored in Drenthen, "Developing Nature."

88. "Slowing the flow" is a term used in a British context by flood managers who seek to address upland drainage systems to try to prevent downstream floods. This approach was pioneered at Pickering in North Yorkshire.

89. In a North American context, this has also involved the removal of artificial dams to restore the movement of fish species. See Hawley, *Recovering a Lost River*.

90. Campbell-Palmer, *Eurasian Beaver Handbook*.

91. Esteves, *Managed Realignment*. Bruce Braun explores the character and politics of comparable forms of naturalistic flood defense in New York in "New Urban Dispositif?"

92. Miller, "Contribution of Natural Fire Management." For a wider discussion of natural fire regimes, see Bowman et al., "Human Dimension of Fire Regimes." There has been some discussion of how changes to both fire and grazing regimes might facilitate rewilding. See Fuhlendorf et al., "Pyric Herbivory."

93. I take this definition from the Partnership for Ecosystem-based Disaster Risk Reduction (https://pedrr.org/).

94. For overviews, see Renaud, Sudmeier-Rieux, and Estrella, *The Role of Ecosystems*; Moos et al., "Ecosystem-Based Disaster Risk Reduction." I am grateful to my former PhD student Cyrus Nayeri for bringing this work to my attention.

95. This approach is discussed in Wakefield and Braun, "Oystertecture."

96. Zimov, "Pleistocene Park"; Kintisch, "Born to Rewild."

97. Seddon et al., "Grounding Nature-Based Climate Solutions." See also University of Oxford, "Nature-Based Solutions Policy Platform," https://www.nbspolicyplatform.org/.

98. Shepherd, *Geoengineering the Climate*; Morton, *Planet Remade*; Stilgoe, *Experiment Earth*.

99. Diprose, *Corporeal Generosity*.

100. Hörl and Burton, eds., *General Ecology*.

## 2. Thinking Like Gaia

1. The idea of a ticklish, intrusive Gaia is proposed in Stengers, "Autonomy." Haraway advocates the concept of sympoiesis (the process of collective creation) in place of autopoiesis (self-creation) in *Staying with the Trouble*.

2. Gilbert, Bosch, and Ledon-Rettig, "Eco-Evo-Devo," 612.

3. Lovelock and Margulis, "Atmospheric Homeostasis"; Lovelock, *Gaia*.

4. Margulis and Sagan, *Acquiring Genomes*.

5. Gilbert uses the term "domestication" in his textbook on eco-evo-devo. See Gilbert, *Ecological Developmental Biology*, 442.

6. For an overview, see Yong, *I Contain Multitudes*.

7. Gilbert, Bosch, and Ledon-Rettig, "Eco-Evo-Devo," 612. See also Bordenstein and Theis, "Host Biology."

8. Gilbert, Sapp, and Tauber, "Symbiotic View of Life."

9. William Parker, interview with author, February 2018.

10. Rook, *Hygiene Hypothesis*, 15.

11. Rook, *Hygiene Hypothesis*.

12. Maizels and Wiedermann, "Immunoregulation," 48.

13. Gilbert, *Ecological Developmental Biology*, 466.

14. Laland, Matthews, and Feldman, "Introduction to Niche Construction Theory."

15. Laland, Matthews, and Feldman, "Introduction to Niche Construction Theory."

16. Estes, Terborgh, and Brashares, "Trophic Downgrading"; Terborgh and Estes, *Trophic Cascades*.

17. Filyk and Osborne, "Multibiome." They caution that much less is known about the ecological dynamics of the microbiome.

18. Filyk and Osborne, "Multibiome." 46.

19. Filyk and Osborne, "Multibiome." 48.

20. McFall-Ngai et al., "Animals in a Bacterial World."

21. Gunderson and Holling, *Panarchy*; Keulartz, "Emergence of Enlightened Anthropocentrism," 58.

22. A focus on strongly interactive species was first encouraged by Soulé et al., "Strongly Interacting Species."

23. Paine, "Note on Trophic Complexity"; Paine, "Conversation."

24. Caro and Girling, *Conservation by Proxy*. Some have expressed concern that this proliferation of the use of the keystone metaphor leads to a loss of terminological specificity. They call for more systematic ways of differentiating keystone roles and for quantifying the topological significance (or interactional magnitude) of different species. See Jordán, "Keystone Species."

25. Estes et al., "Trophic Downgrading," 311.

26. Eisenberg, *Wolf's Tooth*.

27. I have developed the concept of animals' affective atmospheres in collaborative writing with Tim Hodgetts and Maan Barua. See Lorimer, Hodgetts, and Barua, "Animals' Atmospheres."

28. Brown, Laundré, and Gurung, "Ecology of Fear"; Laundré, Hernández, and Altendorf, "Wolves, Elk, and Bison."

29. Ripple and Beschta, "Wolves and the Ecology of Fear."

30. The concept of ecosystem engineers is introduced and developed in Jones, Lawton, and Shachak, "Organisms as Ecosystem Engineers."

31. Byers et al., "Using Ecosystem Engineers," 493.

32. Caro and Girling, *Conservation by Proxy*, 143.

33. Darwin, *Formation of Vegetable Mould*.

34. Gilbert, *Ecological Developmental Biology*.

35. Historians of microbiology tend to link the concept (if not the term) of probiotics to Russian immunologist Elie Metchnikoff, who developed a theory that aging was caused by toxic gut bacteria. See Sangodeyi, "Making of the Microbial Body." Metchnikoff suggests that the lactic acid produced by strains of *Lactobacillus* would prolong life. He drank fermented or sour milk every day to seed his gut. Subsequently, scientists and commercial enterprises have promoted the health-giving properties of specific stains of *Lactobacillus* (e.g., in products like Yakult) and *Bifidobacterium*. The probiotic industry has since grown, often without a strong understanding of the mechanisms through which prebiotics and probiotics might work, or evidence of their efficacy.

36. Maxmen, "Living Therapeutics."

37. For example, Harry Sokol and Kenya Honda have identified the anti-inflammatory effects of clusters of clostridial bacteria, especially strains of *Faecalibacterium prausnitzii*. *F. prausnitzii* has been promoted as one keystone species with probiotic potential in its ability to regulate gut ecologies and prevent the onset of IBD. See Velasquez-Manoff, "Among Trillions."

38. Work exploring the possible role of keystone species in structuring these gut ecologies suggests that analysis should focus instead on keystone genes. These are genes shared between bacterial species that code for specific ecological functions. It is notoriously difficult to apply the species concept to many strains of bacteria. Song et al., "Microbiota Dynamics."

39. See Marchesi et al., "Gut Microbiota."

40. Parker and Ollerton, "Evolutionary Biology and Anthropology."

41. Allen and Maizels, "Diversity and Dialogue"; Wammes et al., "Helminth Therapy or Elimination."

42. Giacomin et al., "Suppression of Inflammation by Helminths"; Zaiss and Harris, "Interactions."

43. Brosschot and Reynolds, "Impact of a Helminth-Modified Microbiome."

44. Filyk and Osborne, "Multibiome."

45. Brosschot and Reynolds, "Impact of a Helminth-Modified Microbiome."

46. Hooks and O'Malley, "Dysbiosis."

47. Hinchliffe et al., *Pathological Lives*.

48. Botkin, *Discordant Harmonies*.

49. Anna Tsing develops the metaphor of polyphonic ecologies in *Mushroom*.

50. Vera, *Grazing Ecology*.

51. Keulartz, "Emergence of Enlightened Anthropocentrism," 57.

52. Hooks and O'Malley, "Dysbiosis," 8.

53. This term is used by Isabella Tree in *Wilding* to describe the changes at Knepp that resulted from agricultural disturbance regimes.

54. Prehistoric megafauna are commonly defined as having a body mass of greater than a thousand kilograms and being of such size that they were immune (as adults) to predation pressure.

55. Malhi et al., "Megafauna"; Galetti et al., "Ecological and Evolutionary Legacy."

56. Estes, Terborgh, and Brashares, "Trophic Downgrading."

57. Terborgh and Estes, *Trophic Cascades*, 367.

58. Blaser, *Missing Microbes*; Velasquez-Manoff, *Epidemic of Absence*.

59. Rook, "Review Series on Helminths."

60. Brosschot and Reynolds, "Impact of a Helminth-Modified Microbiome."

61. Hanski et al., "Environmental Biodiversity."

62. von Hertzen, Hanski, and Haahtela, "Natural Immunity."

63. von Mutius and Vercelli, "Farm Living"; Böbel et al., "Less Immune Activation"; Jacob et al., "Urban Habitat Restoration."

64. Parker et al., "Prescription for Clinical Immunology," 1195.

65. Schweiger et al., "Importance of Ecological Memory"; Lake, "Resistance."

66. Carpenter et al., "From Metaphor to Measurement."

67. Hooks and O'Malley, "Dysbiosis."

68. Lloyd-Price, Abu-Ali, and Huttenhower, "Healthy Human Microbiome"; Lloyd-Price et al., "Strains, Functions and Dynamics."

69. For an introduction, see Schweiger et al., "Importance of Ecological Memory."

70. Terborgh and Estes, *Trophic Cascades*.

71. Hastings et al., "Ecosystem Engineering"; Malhi et al., "Megafauna."

72. Laland, Matthews, and Feldman, "Introduction to Niche Construction Theory."

73. Galetti et al., "Ecological and Evolutionary Legacy"; Janzen and Martin, "Neotropical Anachronisms."

74. Monbiot, *Feral*, 89.

75. Soulé et al., "Strongly Interacting Species," 168.

76. Parker and Ollerton, "Evolutionary Biology and Anthropology"; Rook, *Hygiene Hypothesis*.

77. Cheng et al., "Overcoming Evolutionary Mismatch," 1.

78. These arguments are made in publications like the journals of *Evolutionary Medicine* and *Evolution, Medicine, and Public Health*, and in popular science books including Taylor, *Body by Darwin*; Giphart and van Vugt, *Mismatch*; Lieberman, *Story of the Human Body*; Gluckman and Hanson, *Mismatch*.

79. Rook, *Hygiene Hypothesis*.

80. Schweiger et al., "Importance of Ecological Memory."

81. Tsing, "The Buck, the Bull"; Tsing, "Threat to Holocene Resurgence."

82. Rook et al., "Evolution."

83. Sassone-Corsi and Raffatellu, "No Vacancy."

84. Sears and Pardoll, "Perspective: Alpha-Bugs," 307.

85. Terborgh and Estes, *Trophic Cascades*, 365.

86. For a more extensive discussion of deer as a weedy species, see Tsing, "The Buck, the Bull."

87. Ripple and Beschta, "Wolves and the Ecology of Fear."

88. Richardson et al., "Naturalization and Invasion."

89. Marris, *Rambunctious Garden*; Pearce, *New Wild*; Thomas, *Inheritors of the Earth*.

90. Hinchliffe et al., *Pathological Lives*; Tsing, "Threat to Holocene Resurgence."

91. Tsing, "Threat to Holocene Resurgence," 51. Tsing focuses on the resurgence that results from low-intensity peasant agriculture, which she contrasts with the proliferation associated with plantation agriculture systems.

92. "Self-willed" is a common affirmative adjective in discussions of the ecological changes associated with rewilding. See Fisher, "Self-Willed Land." For a discussion of the importance afforded nonhuman autonomy in rewilding, see DeSilvey and Bartolini, "Where Horses Run Free?"

93. Tree, *Wilding*, makes reference to emergent properties in her description of her experience of rewilding at Knepp as it was explained to her by Frans Vera.

94. Svenning et al., "Science for a Wilder Anthropocene."

95. Hobbs, Higgs, and Hall, *Novel Ecosystems*.

96. Lorimer, "Anthropo-scene."

97. Malhi et al., "Megafauna"; Ruddiman, "Early Anthropogenic Hypothesis."

98. Strathern, *After Nature*.

99. Paxson, *Life of Cheese*, 18.

100. Keulartz, "Emergence of Enlightened Anthropocentrism."

101. Jepson, "Rewilding Agenda."

102. The idea of humans as hyperkeystones was coined by the ecologist Robert Paine, the originator of the concept of the keystone species, in the last article he published before his death in 2018. Worm and Paine, "Humans as a Hyperkeystone Species." See also Albuquerque et al., "Humans as Niche Constructors."

103. Ed Yong, "Humans: The Hyperkeystone Species," *Atlantic*, June 21, 2016, https://www.theatlantic.com/.

104. The clearest example is provided by this report from the Breakthrough Institute on the future of nature conservation. Blomqvist, Nordhaus, and Shellenberger, *Nature Unbound*. This thinking is echoed in future-oriented appeals for rewilding, such as that of Svenning et al., "Science for a Wilder Anthropocene"; Svenning, "Proactive Conservation.

105. Logan, Jacka, and Prescott, "Immune–Microbiota Interactions."

106. Blaser, *Missing Microbes*; Yong, *I Contain Multitudes*.

107. Bested, Logan, and Selhub, "Intestinal Microbiota"; Podolsky, "Metchnikoff and the Microbiome"; Ackert, *Sergei Vinogradskii*. In the foment of early twentieth-century Russian science, Vinogradsky was a colleague of Vladimir Vernadsky, whom some claim was the originator of the concept of the Anthropocene and a forefather of earth systems science. See Grinevald and Rispoli, "Vladimir Vernadsky."

108. Myelnikov, "Alternative Cure."

109. Anderson, "Natural Histories of Infectious Disease"; Podolsky, *Antibiotic Era*.

110. Sangodeyi, "Making of the Microbial Body."

111. Worster, *Nature's Economy*.

112. Evans, *History of Nature Conservation*; Cocker, *Our Place*.

113. Grove, *Green Imperialism*; Rangarajan, *Fencing the Forest*; Beinart, *Rise of Conservation*; Adams, *Against Extinction*.

114. The idea of a crisis discipline was proposed by Michael Soulé, one of the original founders of conservation biology. In a much-cited 1985 article, Soulé argues, "Conservation biology differs from most other biological sciences in one important way: it is often a crisis discipline. Its relation to biology, particularly ecology, is analogous to that of surgery to physiology and war to political science. In crisis disciplines, one must act before knowing all the facts; crisis disciplines are thus a mixture of science and art, and their pursuit requires intuition as well as information. A conservation biologist may have to make decisions or recommendations about design and management before he or she is

completely comfortable with the theoretical and empirical bases of the analysis . . . tolerating uncertainty is often necessary." Soulé, "What Is Conservation Biology?," 727.

115. Worster, *Nature's Economy*; Oelschlaeger, *Idea of Wilderness*.

116. Warde, Robin, and Sörlin, *Environment*.

117. Tauber, *Immunity*.

118. Tauber, *Immunity*. Tauber suggests, "The ether of environmentalism has enveloped all of us, and immunology is finding its own expression in response. The environmental movement has heightened awareness of human dependence on intricate ecological balance, so, as opposed to only a generation ago, the immune system is now being firmly placed in an environmental context in which immunology and ecology have formed a new disciplinary amalgam." Tauber, *Immunity*, 218.

119. The idea of helminths as ecological engineers has been more comprehensively discussed by wildlife biologists, perhaps because they are more acquainted than immunologists with this branch of ecological theory. See Rynkiewicz, Pedersen, and Fenton, "Ecosystem Approach."

120. Worster, *Nature's Economy*.

121. McFall-Ngai, "Noticing Microbial Worlds."

122. Worster, *Nature's Economy*, 2.

123. Gilbert, Sapp, and Tauber, "Symbiotic View of Life"; Brown, *Immunitary Life*; Anderson and Mackay, *Intolerant Bodies*; Tauber, "Immune System."

124. Worster, "Ecology of Order and Chaos."

125. For an explanation of the philosophical differences between a compositionalist and a functionalist approach, see Callicott, *In Defense of the Land Ethic*; Whittaker et al., "Conservation Biogeography."

126. Takacs, *Idea of Biodiversity*; Lorimer, *Wildlife in the Anthropocene*.

127. For example, all of the following claim their respective science to constitute a paradigm shift: Vera, *Grazing Ecology*; Jepson, "Rewilding Agenda"; Estes, Terborgh, and Brashares, "Trophic Downgrading"; Guarner et al., "Mechanisms of Disease"; Blaser et al., "Toward a Predictive Understanding"; Parker and Ollerton, "Evolutionary Biology and Anthropology."

## 3. Symbiopolitics

1. Bubandt and Tsing, "Ethnoecology for the Anthropocene."

2. Foucault, *Society Must Be Defended* and *Birth of Biopolitics*; Foucault et al., *Security, Territory, Population*.

3. Anderson, "Affect and Biopower"; Massumi, "National Enterprise Emergency"; Braun, "New Urban Dispositif?"; Gabrys, "Programming Environments."

4. For example, Rose, *Politics of Life Itself*; Braun, "Biopolitics and the Molecularization of Life"; Mansfield and Guthman, "Epigenetic Life"; Lock, "Comprehending the Body"; Meloni, "Postgenomic Body"; Landecker, "Food as Exposure"; Anderson, "Excremental Colonialism."

5. Helmreich introduces symbiopolitics in Helmreich, *Alien Ocean*, but does not really expand on its implications for conceptualizations of biopolitics. The version I develop is shaped by Heather Paxson's concept of microbiopolitics and by Valerie Olson's concept of ecopolitics. See Paxson, *Life of Cheese*; Olson, *Into the Extreme*.

6. For an overview, see Dobson, Barker, and Tayor, *Biosecurity*.

7. Foucault qtd. in Barker, "Biosecurity," 358.

8. Hinchliffe et al., *Pathological Lives*. See also Ali and Keil, *Networked Disease*; Braun, "Biopolitics and the Molecularization of Life"; Clark, "Mobile Life"; Blanchette, "Herding Species"; Lowe, "Viral Clouds."

9. Brown and Kelly, "Material Proximities."

10. Braun, "New Urban Dispositif?"

11. Chandler, *Ontopolitics*. See also Wakefield, "Inhabiting the Anthropocene Back Loop"; Grove, "Agency, Affect"; Grove, "Security beyond Resilience"; Massumi, "National Enterprise Emergency."

12. Braun, "New Urban Dispositif?," 58–59.

13. Braun, "New Urban Dispositif?," 60.

14. Chandler, *Ontopolitics*; Massumi, *Ontopower*. See also Colebrook, *Death of the Posthuman*.

15. Rabinow and Rose, "Biopower Today."

16. See Wolfe, "Ecologizing Biopolitics."

17. See Bingham and Hinchliffe, "Reconstituting Natures"; Latour, *Politics of Nature*.

18. For an introduction, see Pires, "Rewilding Ecological Communities"; Tylianakis et al., "Conservation of Species Interaction Networks."

19. Coyte, Schluter, and Foster, "Ecology of the Microbiome."

20. In relation to rewilding, see Monbiot, *Feral*; Rewilding Europe, *Vision for a Wilder Europe*; Donlan et al., "Pleistocene Rewilding"; Svenning et al., "Science for a Wilder Anthropocene." On helminths, see Elliott, Summers, and Weinstock, "Helminths as Governors"; Lukeš et al., "(Self-)Infections with Parasites."

21. Nilsen et al., "Wolf Reintroduction to Scotland."

22. This was pointed out to me in forthright terms in interview with John Scott. Parker discusses the potential of a modified version of the bovine tapeworm in Parker et al., "Prescription for Clinical Immunology."

23. Donlan et al., "Pleistocene Rewilding." There have also been speculative proposals to introduce African elephants into Australia to control invasive African grass species. Bowman, "Conservation."

24. Seddon et al., "Reversing Defaunation"; Griffiths and Harris, "Prevention of Secondary Extinctions"; Hansen et al., "Ecological History"; Hennessy, "Molecular Turn."

25. Fleming and Weinstock, "Clinical Trials of Helminth Therapy"; Schölmerich et al., "Randomised, Double-Blind, Placebo-Controlled Trial."

26. The cysticercoids of a rat tapeworm *(Hymenolepis diminuta)*, or HDC, are raised and sold by the helminth therapy company Biome Restoration. This company is advised by William Parker, whose lab has published a scientific protocol for producing HDC. Smyth et al., "Production and Use of *Hymenolepis diminuta* Cysticercoids."

27. See, e.g., Braverman, "Governing the Wild."

28. See Braverman, "Anticipating Endangerment"; Youatt, "Counting Species"; Biermann and Mansfield, "Biodiversity, Purity, and Death"; Holloway et al., "Biopower, Genetics"; Lulka, "Stabilizing the Herd"; Lulka, "Form and Formlessness."

29. See Lorimer, "Nonhuman Charisma."

30. This point is explored in relation to wolves in upland France in Buller, "Safe from the Wolf."

31. Hinchliffe et al., "Biosecurity"; Barker, "Biosecurity."

32. Soulé and Noss, "Rewilding and Biodiversity"; Foreman, *Rewilding North America.*

33. For discussions in European and North American contexts, see Schwartz, "Wild Horses"; Hintz, "Some Political Problems"; Jepson, Schepers, and Helmer, "Governing with Nature."

34. The original proposal for assisted colonization can be found in McLachlan, Hellmann, and Schwartz, "Framework for Debate." Its implications for rewilding are discussed in Sandom and MacDonald, "What Next?"; Seddon et al., "Reversing Defaunation"; Louys et al., "Rewilding the Tropics." The idea of assisted migration is contested by others, e.g., Ricciardi and Simberloff, "Assisted Colonization."

35. I develop the concept of animals' mobilities in Hodgetts and Lorimer, "Animals' Mobilities."

36. Lawton, *Making Space for Nature*; Crooks and Sanjayan, *Connectivity Conservation*; Goldman, "Constructing Connectivity"; Proctor et al., "Grizzly Bear Connectivity Mapping."

37. Seddon et al., "Reversing Defaunation."

38. Butterfield et al., "Prestoration," S155; emphasis in original.

39. The training of animals to be wild is a common part of animal rehabilitation programs. Its practices, paradoxes, and idiosyncrasies have been explored in Collard, "Putting Animals Back Together"; Bradshaw, *Elephants on the Edge*. A good example is provided by the Save China's Tigers project (http://www.savechinastigers.org/), which rewilds formerly captive Chinese tigers in South Africa. See also Quan, *Rewilded*.

40. For a review, see Thompson, "How Oral Vaccines Could Save Ethiopian Wolves."

41. Baerselman and Vera, *Nature Development*; Hootsman and Kampf, *Ecological Networks*.

42. As a result of political opposition from local farmers on the Flevoland, the corridor has yet to be constructed.

43. Drenthen, "Return of the Wild"; van Heel et al., "Analysing Stakeholders' Perceptions."

44. Schwartz, *Nature and National Identity*.

45. It is currently illegal to import hookworm and many other helminths into the United States; see "Import Alert 57-21," U.S. Food and Drug Administration, 2016, https://www.fda.gov/.

46. Paxson, "Post-Pasteurian Cultures."

47. The concept of microbial citizenship is developed by Spackman, "Formulating Citizenship." The mode of microbial citizenship associated with helminth therapy is similar to that described by Wolf-Meyer in his work on FMT. Wolf-Meyer, "Normal, Regular, and Standard."

48. Finlay and Arrieta, *Let Them Eat Dirt*; Gilbert and Knight, *Dirt Is Good*.

49. In her analysis of captive breeding of tortoises for reintroduction to the Galápagos Islands, Elizabeth Hennessy terms this dependence of rewilding on complex cutting-edge genetic technologies the "paradox" of producing prehistoric life. Hennessy, "Producing 'Prehistoric' Life," 71.

50. This argument is made by those involved in the Tauros Programme. See Goderie et al., *Aurochs*. These claims are contested by Frans Vera and others working with Heck cattle, who point to the animals' proven ability to survive harsh winters, as well as the likely heterogeneity of the aurochs phenotype.

51. Park et al., "Genome Sequencing."

52. Goderie et al., *Aurochs*.

53. This project seems to be in abeyance, but it was once promoted by the True Nature Foundation and received some support and publicity from Stuart Brand's Revive and Restore initiative (https://reviverestore.org/), which claims to be "building the 21st century genetic rescue toolkit for conservation."

54. See, e.g., Donlan, "De-extinction"; Svenning et al., "Science for a Wilder Anthropocene"; Zimov, "Pleistocene Park."

55. For a discussion of the many slippery meanings of domestication, see Cassidy and Mullin, *Where the Wild Things Are Now*; Swanson, Lien, and Ween, *Domestication Gone Wild*.

56. Biome Restoration, "The Science," https://biomerestoration.com/.

57. Cheng et al., "Overcoming Evolutionary Mismatch," 2.

58. Cheng et al., "Overcoming Evolutionary Mismatch," 2

59. Parker et al., "Prescription for Clinical Immunology," 1200.

60. Lorimer, "Scaring Crows"; Lorimer, Hodgetts, and Barua, "Animals' Atmospheres."

61. International Committee on the Management of Large Herbivores in the Oostvaardersplassen (ICMO), *Natural Processes* and *Reconciling Nature and Human Interests*.

62. Although they were also introduced to the reserve, the deer are considered wild and are not killed. Instead they are left to die of starvation.

63. Klaver et al., "Born to Be Wild"; Korthals et al., "Battles of Nature."

64. van Klink, Ruifrok, and Smit, "Rewilding with Large Herbivores."

65. Project Wolf is carried out by volunteers working for the organization Trees for Life at their rewilding project in the Scottish Highlands. I am grateful to Filipa Soares for bringing this project to my attention.

66. Cromsigt et al., "Hunting for Fear."

67. Navarro, Ferreira, and Loukas, "Hookworm Pharmacopoeia."

68. Parker et al., "Prescription for Clinical Immunology."

69. Hotez et al., "Human Hookworm Vaccine."

70. Smit et al., "Rewilding with Large Herbivores."

71. Seddon et al., "Reversing Defaunation."

72. Tree, *Wilding*.

73. Smit et al., "Rewilding with Large Herbivores"; van Klink, Ruifrok, and Smit, "Rewilding with Large Herbivores."

74. Battisti, Poeta, and Fanelli, *Introduction to Disturbance Ecology*.

75. There is no naturalistic landscape of fear at Knepp. Herbivores are occasionally scared by walkers' dogs, but hunting is done through conventional means.

76. This process and these plans are discussed in Tree, *Wilding*.

77. See Lorimer, "Rot."

78. Rewilding Europe, *Circle of Life*, promotes this as the reintroduction of the circle of life.

79. See Bocci, "Tangles of Care"; Wanderer, "Biologies of Betrayal."

80. For a discussion of mobility management through permeable fencing in the context of rewilding with horses in Portugal, see DeSilvey and Bartolini, "Where Horses Run Free?"

81. See discussion in Buller, "Introducing Aliens."

82. Perhaps the most notorious example is the cane toad, introduced to Australia to control an insect pest of sugarcane, which then went feral, spreading across the country with severe ecological consequences.

83. This triangular relationship is explored in Hodgetts, "Wildlife Conservation."

84. Reardon, "Phage Therapy."

85. Sassone-Corsi and Raffatellu, "No Vacancy"; Louys et al., "Rewilding the Tropics"; Griffiths et al., "Assessing the Potential."

86. Massumi, "National Enterprise Emergency," 154.

87. Tsing, "The Buck, the Bull," 6.

88. Wolfe, "Ecologizing Biopolitics."

89. Schweiger et al., "Importance of Ecological Memory," 12.

90. Here they reference and develop earlier reflections on this question by Carpenter et al., "From Metaphor to Measurement"; Carpenter and Folke, "Ecology for Transformation."

91. Keulartz takes this phrase from the work of Norbert Elias. See Keulartz, "Emergence of Enlightened Anthropocentrism," 58.

92. Tsing, *Mushroom*, 176.

93. Paxson, *Life of Cheese*, 161.

94. Paxson, *Life of Cheese*, 161.

95. Paxson, "Microbiopolitics," 118.

## 4. Wild Experiments

1. Latour, *Politics of Nature.*

2. Kohler, *Landscapes and Labscapes*; Gieryn, "City as Truth-Spot"; Livingstone, "Spaces of Knowledge"; Shapin, "Placing the View from Nowhere."

3. The character of experiments in nature and natural experiments is explored in some detail in Kohler, *Landscapes and Labscapes.*

4. The lack of science and monitoring is a criticism made by the International Committee on the Management of Large Herbivores in the Oostvaardersplassen (ICMO) in their two reports on the management of the OVP. See ICMO, *Natural Processes* and *Reconciling Nature and Human Interests.*

5. Tree makes this point in her account of rewilding at Knepp in *Wilding*; Sutherland, "Conservation Biology."

6. See van den Belt, "Networking Nature."

7. Flohr, Quinnell, and Britton, "Do Helminth Parasites Protect"; Stein et al., "Role of Helminth Infection."

8. Correale and Farez, "Impact of Parasite Infections."

9. Rook, *Hygiene Hypothesis*, 248.

10. Terborgh and Estes, *Trophic Cascades*, xv.

11. Paine, "Note on Trophic Complexity."

12. Smit et al., "Rewilding with Large Herbivores."

13. The processes of using lab techniques at field sites is discussed in Bockman and Eyal, "Eastern Europe as a Laboratory"; Greenhough, "Assembling an Island Laboratory."

14. In their review of existing scientific work on rewilding, Svenning et al. lament that "empirical studies are few whereas essays and opinion pieces predominate." Svenning et al., "Science for a Wilder Anthropocene," 900.

15. This issue is explored in Terborgh and Estes, *Trophic Cascades*. OVP provides a clear example where there are regular controversies about whether wild cattle, and especially horses, should be fed to survive the winter. See Lorimer and Driessen, "Bovine Biopolitics."

16. Gieryn, "City as Truth-Spot."

17. Elliott and Weinstock, "Nematodes and Human Therapeutic Trials"; Fleming and Weinstock, "Clinical Trials of Helminth Therapy."

18. Cheng et al., "Overcoming Evolutionary Mismatch," 20.

19. Elliott and Weinstock, "Nematodes and Human Therapeutic Trials."

20. Cheng et al., "Overcoming Evolutionary Mismatch."

21. Garner et al., "Introducing Therioepistemology," provide an introduction to this broader shift and to the study of how knowledge is gained from laboratory animals—what they term therioepistemology.

22. Babayan et al., "Wild Immunology."

23. Maizels and Nussey, "Into the Wild."

24. Leung et al., "Rapid Environmental Effects."

25. Graham, "Walk on the Wild Side"; Beans, "News Feature."

26. Leung et al., "Rapid Environmental Effects."

27. Maizels and Nussey, "Into the Wild."

28. Cheng et al., "Overcoming Evolutionary Mismatch"; Liu et al., "Practices and Outcomes."

29. Cheng et al., "Overcoming Evolutionary Mismatch," 2.

30. Harris, Wyatt, and Kelly, "Gift of Spit"; Hogarth and Saukko, "Market in the Making."

31. Cooper, "Pharmacology of Distributed Experiment," 18.

32. It is important to note that these shifts toward wild immunology and the revalorization of accidental ecologies are not universal or uncontested. Other scientists have flagged the challenge of reproducing these experiments to corroborate their findings. They bemoan their dependence on local specificities—or practices of place—and thus their inability to generate laws and rules with

application elsewhere. They also caution that the interests of some of the more zealous advocates of rewilding and biome restoration might shape the selection of study sites and populations, and blinker the interpretation of the gathered data. See, e.g., Briggs et al., "Hygiene Hypothesis."

33. Krohn and Weyer, "Society as a Laboratory."

34. Gross, "Public Proceduralization"; Gross and Hoffmann-Riem, "Ecological Restoration."

35. Braun, "From Critique to Experiment?," emphasis in original.

36. For a review, see Cooper, "Pharmacology of Distributed Experiment"; Muniesa and Callon, "Economic Experiments"; Last, "Experimental Geographies."

37. Chandler, *Ontopolitics*, 21.

38. Chandler, *Ontopolitics*, 20.

39. Gabrys, *Program Earth*.

40. Youatt, "Counting Species"; Braverman, "Governing the Wild"; Braverman, "Anticipating Endangerment"; Benson, *Wired Wilderness*; Fortun, "Biopolitics."

41. Hinchliffe and Lavau, "Differentiated Circuits"; Anderson, "Preemption, Precaution"; Anderson, "Affect and Biopower"; Keck, "Liberating Sick Birds"; Braun, "Biopolitics and the Molecularization of Life."

42. Chandler, *Ontopolitics*, 21.

43. Chandler, *Ontopolitics*, 36.

44. Chandler, *Ontopolitics*, 37.

45. Braun, "New Urban Dispositif?"

46. Chandler, *Ontopolitics*, 10. See also Latour, *Facing Gaia*, 82–83. Wakefield describes these approaches as being posttruth. Wakefield, "Inhabiting the Anthropocene Back Loop," 82.

47. Swyngedouw, "Apocalypse Forever?," 213; see also Grove and Chandler, "Introduction."

48. Chandler, *Ontopolitics*, 20–21.

49. For an overview, see Wilmers et al., "Golden Age"; Hodgson et al., "Precision Wildlife Monitoring."

50. Schulze et al., "How to Predict Molecular Interactions."

51. Pires, "Rewilding Ecological Communities."

52. Jordán, "Keystone Species."

53. Pires, "Rewilding Ecological Communities," 260.

54. Raimundo, Guimarães, and Evans, "Adaptive Networks."

55. Mackenzie, *An Engine, Not a Camera*; Edwards and Bowker, *Vast Machine*.

56. Latour, *Pandora's Hope*.

57. Bowker, "Biodiversity Datadiversity"; Robbins, "Fixed Categories."

58. Schweiger et al., "Importance of Ecological Memory."

59. See Keulartz, "Emergence of Enlightened Anthropocentrism"; Gunderson and Holling, *Panarchy*.

60. Chandler, *Ontopolitics*, 22.

61. Chandler, *Ontopolitics*, 51.

62. For example, Schweiger et al. explain: "As rewilding projects are generally planned to be open-ended, their key goals are: (i) the promotion of dynamic landscapes, natural processes, and, in some cases, concomitant ecosystem services, and (ii) the general maintenance of high levels of biodiversity rather than fixed reference states in species composition or habitat characteristics." Schweiger et al., "Importance of Ecological Memory."

63. I take the term "countermapping" from Nancy Peluso, for whom it describes a participatory technique for generating maps that challenge the status quo and to further progressive goals. Peluso, "Whose Woods Are These?" Countermapping forms part of the wider turn toward critical cartography in geography and cognate fields. For an introduction, see Harris and Hazen, "Power of Maps"; Crampton, "Cartography."

64. Wakefield develops the concept of the back-loop experiment from a revisionist reading of resilience theory and the heuristic of the adaptive cycle. She identifies "the possibility of new forms of life in its phase of release and reorganisation: the back loop. More than a brief, negative phase to govern or navigate, I argue that the back loop offers the possibility for a practical orientation to the Anthropocene based on experimentation with new uses, release of old frameworks, and allowance for the unknown." Wakefield, "Inhabiting the Anthropocene Back Loop," 77.

65. Cantrell, Martin, and Ellis, "Designing Autonomy," 156.

66. Adams, "Geographies of Conservation 2."

67. Cantrell, Martin, and Ellis, "Designing Autonomy," 162.

68. Cantrell, Martin, and Ellis, "Designing Autonomy," 163.

69. I take the term "beaver believer" from the working title of a documentary, *The Beaver Believers* (https://www.thebeaverbelievers.com/), which looks at a network of enthusiasts pioneering beaver reintroduction in the United States.

70. Chandler, *Ontopolitics*, 23.

71. See Tsing, *Mushroom*.

72. I do not provide names here because some of the activities I report are illegal. In each case, I am referring to the activities of four or five individuals.

73. Campbell-Palmer, *Eurasian Beaver Handbook*.

74. Cheng et al. explain that the field of self-treatment with helminths advanced with users "testing different helminth formulations much as chefs test various recipes." Cheng et al., "Overcoming Evolutionary Mismatch," 19.

75. Kohler terms these "practices of place," which he differentiates from the placeless practices of laboratory science. Kohler, *Landscapes and Labscapes*.

76. Tree, *Wilding*, 77.

77. I take this phrase from Despret, "Body We Care For"; Despret, "Responding Bodies." The processes of learning to be affected in the field sciences of conservation are addressed in Lorimer, "Counting Corncrakes." "Craft" is a loaded term. Its definition and relationships with science and with labor have been the subject of much analysis in science studies and philosophy. Compare with Sennett, *Craftsman*; Arendt, *Human Condition*; Ingold, *Making*.

78. Bubandt and Tsing, "Ethnoecology for the Anthropocene"; Tsing, *Mushroom*; Kirksey and Helmreich, "Emergence of Multispecies Ethnography"; Kirksey, *Emergent Ecologies*.

79. Hinchliffe, "Reconstituting Nature Conservation"; Hinchliffe et al., "Urban Wild Things."

80. See Keulartz and van den Belt, *DIY-Bio*.

81. Callon, Lascoumes, and Barthe, *Acting in an Uncertain World*.

82. Gammon, "Many Meanings of Rewilding."

83. For a discussion of anti-Pasteurian approaches and how they differ from post-Pasteurian microbiopolitics, see Paxson, "Microbiopolitics."

84. See Wark, *Hacker Manifesto*.

85. Flowers, "Chronic Disease," 326.

86. See Epstein, *Impure Science*.

87. Crowley, Hinchliffe, and McDonald, "Nonhuman Citizens on Trial."

88. Chandler, *Ontopolitics*; Wakefield, "Infrastructures of Liberal Life."

89. Lorimer and Driessen, "Wild Experiments."

90. Rheinberger, *Toward a History of Epistemic Things*. As Melinda Cooper argues, a successful real-world experiment involves "a transvaluation of the value of accident itself, from standardized, measurable risk to incalculable event. The real-life experiment imposes unknowable risks on society at large, but in the process it also has the potential to generate qualitatively new forms of experience, unobtainable under the controlled conditions of the trial." Cooper, "Pharmacology of Distributed Experiment," 36.

## 5. Geographies of Dysbiosis

1. For introductions, see Robbins, *Political Ecology*; Perreault, Bridge, and McCarthy, *Routledge Handbook of Political Ecology*; Peet and Watts, *Liberation Ecologies*.

2. Farmer, *Pathologies of Power*; Biehl and Petryna, *When People Come First*.

3. Hinchliffe et al., "Biosecurity"; Hinchliffe et al., *Pathological Lives*.

4. See also Ali and Keil, *Networked Disease*.

5. Hinchliffe et al., "Biosecurity," 538.

6. See also Singer, "Ecosyndemics."

7. Hinchliffe et al., *Pathological Lives*, 25.

8. Brown and Kelly, "Material Proximities"; Lowe, "Viral Clouds"; Loon, "Epidemic Space"; Blanchette, "Herding Species."

9. Wallace and Wallace, "Blowback." See also Shukin, *Animal Capital*; Wallace, *Big Farms Make Big Flu*.

10. There has been surprisingly little qualitative research in the humanities and critical social sciences on the autoimmune diseases that are currently associated with missing microbes; but see Mitman, *Breathing Space*; Brown, *Immunitary Life*.

11. For discussion, see Brown and Kelly, "Material Proximities"; Guthman and Mansfield, "Implications of Environmental Epigenetics"; Braun, "New Materialisms."

12. Hinchliffe et al., *Pathological Lives*.

13. Mutualism and symbiosis were once synonymous in ecology, but the latter is increasingly used to describe a wider range of relations, including parasitism.

14. Perry, "Parasites and Human Evolution."

15. Mitchell, "Origins of Human Parasites."

16. Smallwood et al., "Helminth Immunomodulation"; Pritchard and Brown, "Is *Necator americanus* Approaching."

17. Cox, "History of Human Parasitology"; Mitchell, "Origins of Human Parasites."

18. Velasquez-Manoff, *Epidemic of Absence*, 2.

19. Pullan et al., "Global Numbers."

20. Geissler, "Worms Are Our Life," parts 1 and 2; Parker, Allen, and Hastings, "Resisting Control"; Moran-Thomas, "Salvage Ethnography"; Bardosh et al., "Controlling Parasites."

21. Brooker, "Estimating the Global Distribution," 1141.

22. Peter Hotez, arguably the most recognized global authority on hookworm control, has proposed a worm index of human development. Hotez and Herricks, "Helminth Elimination."

23. Serpell, *Domestic Dog*.

24. Pierotti and Fogg, *First Domestication*.

25. Brooker, "Estimating the Global Distribution."

26. Brooker, Bethony, and Hotez, "Human Hookworm Infection."

27. Peter Hotez describes hookworm as part of what he terms the unholy trinity of soil-transmitted helminths, which comprise hookworm, roundworm,

and whipworm. Together, they are responsible for the most prevalent neglected tropical disease of helminthiasis. Hotez, *Forgotten People*.

28. Scott, *Against the Grain*.

29. Zuckerman et al., "Evolution of Disease."

30. I take this relational and geographic understanding of domestication from Helen Leach, who explores how the novel ecology of the agricultural village created new conditions for human–animal coevolution. Leach, "Human Domestication Reconsidered." See also Kelly and Lezaun, "Urban Mosquitoes."

31. Palmer, "Migrant Clinics," 679.

32. Palmer, "Migrant Clinics," 708.

33. Couacaud, "Hookworm Disease." The settlements of indigenous peoples forcibly confined to reserves were also necatoriasis hot spots (e.g., in Australia).

34. Jahiel and Babor, "Industrial Epidemics."

35. As various authors argue, the macroparasitic (i.e., asymmetrical and exploitative) character of plantation colonial capitalism exacerbates the microparasitic depredations of the hookworm. See Brown, "Microparasites and Macroparasites."

36. Tsing, "Threat to Holocene Resurgence," 52.

37. Ettling, *Germ of Laziness*.

38. Palmer, *Launching Global Health*; Farley, *To Cast Out Disease*.

39. This history and the influence of the Rockefeller Hookworm Eradication campaigns are celebrated in Farley, *To Cast Out Disease*.

40. Hotez, "Neglected Diseases"; McKenna et al., "Human Intestinal Parasite Burden."

41. Loukas et al., "Hookworm Infection."

42. Gunawardena et al., "Soil-Transmitted Helminth Infections."

43. Loukas et al., "Hookworm Infection."

44. The general idea of the poverty trap is outlined in Bonds et al., "Poverty Trap." It is developed in relation to hookworm in Hotez et al., "Rescuing the Bottom Billion."

45. World Health Organization, "Intestinal Worms," https://www.who.int/. The efficacy of school-based deworming programs is fiercely contested, especially in areas with poor sanitation and high reinfection risk. See World Bank, "Worm Wars: The Anthology," https://blogs.worldbank.org/; Parker and Allen, "De-politicizing Parasites."

46. Loukas et al., "Hookworm Infection."

47. Panic et al., "Repurposing Drugs."

48. Couacaud, "Hookworm Disease."

49. A review by key practitioners notes that "despite huge international efforts to control hookworm and other STH [soil-transmitted helminth] infections, little progress has been made." The authors suggest that "eliminating the public health problem that endemic hookworm infection causes will probably also require simultaneous and substantial economic development and sanitation improvements in affected areas." Loukas et al., "Hookworm Infection," 12.

50. Paxson, "Post-Pasteurian Cultures."

51. Bach, "Effect of Infections."

52. National Institutes of Health, National Institute of Environmental Health Sciences, "Autoimmune Diseases," https://www.niehs.nih.gov/.

53. American Autoimmune Related Disease Association (AARDA), "Autoimmune Disease Statistics," https://www.aarda.org/.

54. Ngo, Steyn, and McCombe, "Gender Differences."

55. Lerner, Jeremias, and Matthias, "World Incidence."

56. Velasquez-Manoff, *Epidemic of Absence*, 5–6.

57. Pollard, *Western Diseases*. Immunologists working on helminths describe these as diseases of modernity. Maizels, McSorley, and Smyth, "Helminths in the Hygiene Hypothesis."

58. Mitman, "Hay Fever Holiday"; Jackson, *Asthma*.

59. McKenna et al., "Human Intestinal Parasite Burden."

60. Berbudi et al., "Parasitic Helminths"; Cooke, "Review Series on Helminths."

61. Atkinson, Eisenbarth, and Michels, "Type 1 Diabetes"; DeFronzo et al., *International Textbook of Diabetes Mellitus*.

62. Niazi and Kalra, "Diabetes and Tuberculosis."

63. Beran and Yudkin, "Double Scandal."

64. Gale, "Dying of Diabetes." The life expectancy for a newly diagnosed child with type 1 diabetes is seven months. Atkinson, Eisenbarth, and Michels, "Type 1 Diabetes."

65. The inventor of insulin, Frederick Banting, and his colleagues made the patent for the drug available without charge and did not attempt to control commercial production. For discussion of the political economy of insulin, see Beran and Yudkin, "Double Scandal." The International Insulin Foundation 100 Campaign (http://www.access2insulin.org/) seeks to widen access to this drug and other diabetic technology.

66. But see Neel and Sargis, "Paradox of Progress."

67. Brennan et al., "Shifting Brucellosis Risk."

68. Craig, "Atopic Dermatitis."

69. von Mutius and Vercelli, "Farm Living"; Fujimura et al., "House Dust Exposure."

70. Fleming and Weinstock, "Clinical Trials of Helminth Therapy." See also Smallwood et al., "Helminth Immunomodulation."

71. Information on helminth users and their geography is taken from Cheng et al., "Overcoming Evolutionary Mismatch"; Liu et al., "Practices and Outcomes." The spatial patterns suggested in this survey work were confirmed by my own interviews with key informants among helminth users.

72. Fourie, Jackson, and Aveyard, "Living with Inflammatory Bowel Disease."

73. Ng et al., "Worldwide Incidence and Prevalence."

74. Clinical trials using helminths to treat IBD have tended to use TSO (Trichuris suis ova). See Elliott and Weinstock, "Nematodes and Human Therapeutic Trials." A Cochrane review finds limited evidence of their effectiveness. Garg, Croft, and Bager, "Helminth Therapy (Worms)."

75. I base this short account on my interviews with those using helminths to treat their IBD, supplemented by blogs and experiences collected on the Helminthic Therapy Wiki (https://helminthictherapywiki.org/). For further discussion, see Lorimer, "Gut Buddies."

76. Frohlich, "Social Construction."

77. See Richard Schiffman, "Are Pets the New Probiotic?," *New York Times*, June 6, 2017, https://www.nytimes.com/.

78. Lasanta et al., "Space-Time Process."

79. The contrasting geographies of defaunation and refaunation associated with this process are explored in Svenning, "Future Megafaunas."

80. For a discussion, see Robbins and Moore, "Ecological Anxiety Disorder."

81. Buller, "Safe from the Wolf."

82. Chakrabarty, *Provincializing Europe*; Anderson, *Colonial Pathologies*.

## 6. Future Pasts

1. Ecologists commonly talk of ecological baselines. The concept rose to prominence in conservation through discussion of shifting baselines—that is, the tendency for one generation of conservationists or resource managers to fix their understanding of a natural ecology to a point in their childhood. See Pauly, "Anecdotes." Raymond Williams describes a comparable "backward moving escalator" that shifts the idealized historic reference condition in literary understandings of the pastoral. Williams, *The Country and the City*, 9.

2. Rewilding Europe, *Rewilding Europe*.

3. Paxson, *Life of Cheese*. Paxson develops the idea of the postpastoral from the writings of Terry Gifford. Gifford, *Reconnecting with John Muir*.

4. Williams, *The Country and the City*.

5. Paxson's affirmative reading of postpastoral artisan cheesemaking resonates with Anna Tsing's valorization of the Matsutake Crusaders involved in the revitalization of *satoyama* forest landscapes and peasant forestry techniques in Japan. Tsing, *Mushroom*.

6. I take the term from Herf, *Reactionary Modernism*.

7. The two brothers published accounts of their breeding in international journals and popular books as part of their post–World War II efforts to curate and denazify their reputations. See Heck, "Breeding-Back of the Aurochs"; Heck, *Animals, My Adventures*.

8. Gritzbach, *Hermann Göring*; Gautschi, *Reichsjägermeister*.

9. Barnes and Minca, "Nazi Spatial Theory."

10. Heck, "Behördiche Landschaftsgestaltung."

11. In his archival research, Clemens Driessen found that the Nazi authorities seriously doubted Heinz's personal allegiance to Nazism. His application in 1936 for formal approval to be allowed to publish and give public lectures led to a lengthy inquiry into his moral and political reliability. Driessen and Lorimer, "Back-Breeding the Aurochs."

12. Kater, *Das "Ahnenerbe" der SS*.

13. Goderie et al., *Aurochs*.

14. For discussion of these and further examples of animal nationalisms, see Saraiva, *Fascist Pigs*; Sax, *Animals in the Third Reich*; Ritvo, *Animal Estate*; Ritvo, *Noble Cows*; Skabelund, "Breeding Racism"; Sharma, *Green and Saffron*; Jha, *Myth*. The Heck brothers' back-breeding projects were one of a number supported by the National Socialists. These included the project of Ahnenerbe explorer and zoologist Ernst Schaefer, who was breeding a hardy steppe horse for an imagined elite breed of frontier S.S. men. These so-called Wehrbauer were modeled on Teutonic knights. Schaefer hoped they would pioneer landscape development and farming while policing the Reich's extended borders on horseback. See Kater, *Das "Ahnenerbe" der SS*.

15. Stallins et al., "Geography and Postgenomics"; Benezra, "Datafying Microbes"; Leiper, "'Re-wilding' the Body."

16. Anderson, "Excremental Colonialism."

17. Driessen and Lorimer, "Back-Breeding the Aurochs."

18. Reichert, *Nibelungenlied*.

19. The notion of *Heimat* (which vaguely translates as "homeland") combines ideas of regional folk culture with personal rootedness through ancestry. It is unique to German culture and is often coded with nostalgic, pastoral associations.

20. Blackbourn, *Conquest of Nature*.

21. Vera, *Grazing Ecology*. Jørgensen, "Rethinking Rewilding," identifies a wider range of ecological baselines that inform rewilding projects in other parts of the world.

22. The Pleistocene overkill hypothesis was first put forward by Paul Martin in the 1960s. For a full introduction published much later in his life, see Martin, *Twilight of the Mammoths*. It remains controversial, not least because it has been used by some to advocate for a prehistorical start date for the Anthropocene that some see as excusing modern humans of responsibility. Ruddiman, "Anthropogenic Greenhouse Era"; Ruddiman et al., "Defining the Epoch." For a critique, see Hamilton, *Defiant Earth*.

23. See, e.g., Rewilding Europe, *Vision for a Wilder Europe*.

24. Paleo reenactment projects that describe processes of rewilding do exist, but they differ markedly from the bright green tenor of Rewilding Europe. See, e.g., Gammon, "Many Meanings of Rewilding"; Olson, *Unlearn, Rewild*; Taylor, *Spirit of Rewilding*; Scout, *Rewild or Die*.

25. Leach, *Rewild*.

26. Leach, "(Re)Becoming Human."

27. Ed Yong provides a useful overview of this research, identifying target groups in addition to the Hadza, including the Yanomami in Venezuela, the Matses in Peru, the Baku in the Central African Republic, and the Pygmies of Cameroon. See Yong, *I Contain Multitudes*, 131–34.

28. See Finlay and Arrieta, *Let Them Eat Dirt*; Gilbert and Knight, *Dirt Is Good*.

29. Paxson, *Life of Cheese*.

30. Paxson, *Life of Cheese*, 18. Tsing also identifies the centrality of work in the *satoyama* revitalization project of the Matsutake crusaders in Japan. Unlike Paxson's cheesemakers, she suggests that this restoration project aims for unintentional cultivation. Tsing, *Mushroom*.

31. Goderie et al. suggest that "it's our somewhat nostalgic craving for outcompeted, unprofitable old farming systems that makes the 'management' of European nature so expensive." Goderie et al., *Aurochs*, 148.

32. Balmford, Green, and Phalan, "What Conservationists Need to Know"; Phalan, "What Have We Learned"; Phalan et al., "Reconciling Food Production."

33. Heinz Heck suggests that "all specific domestic characteristics are faulty mutations that have been taken up by human breeders to change the animal for a particular unnatural purpose." Translation by Clemens Driessen. Heck, "Ueber die Rueckzuechtung des Urs," 11. Contemporary back-breeders talk of the "sad remains of the aurochs amongst domestic cattle." See Goderie et al., *Aurochs*, 56.

34. Cronon, "Trouble with Wilderness"; Hinchliffe, "Cities and Natures."

35. Radkau and Uekoetter, *Naturschutz und Nationalsozialismus*; Brügge-meier, Cioc, and Zeller, *How Green Were the Nazis?* This is an extreme model of romantic back-to-the-land movements that were common across Europe in the twentieth century. See Matless, *Landscape and Englishness.*

36. Rewilding Europe, *Vision for a Wilder Europe.*

37. See, e.g., Spector, *Diet Myth.*

38. Marx, *Machine in the Garden.*

39. See Cairns et al., *Wild Wonders of Europe.* In a striking example, those involved in producing a high-profile natural history documentary about the OVP—entitled *The New Wilderness*—went to great lengths to downplay its sub-urban location and visible urban and industrial infrastructure.

40. The concept of the heartland is a tongue-in-cheek reference to the geopolitical writings of Halford MacKinder. MacKinder, "Geographical Pivot."

41. See Schwartz, *Nature and National Identity.*

42. Driessen and Lorimer, "Back-Breeding the Aurochs."

43. Gross, *Inventing Nature.*

44. Creating spaces for biodiversity became an important requirement for joining the European Union. Waterton, "From Field to Fantasy"; Kay, "Europeanization."

45. Ceauşu et al., "Mapping Opportunities"; Merckx and Pereira, "Reshap-ing Agri-environmental Subsidies"; Pereira and Navarro, *Rewilding European Landscapes.*

46. This conference was organized by Wild Europe in 2010 in Brussels, Bel-gium. As of 2015, the organization works in seven marginal areas in nine countries: Spain, Portugal, Italy, Poland, Croatia, Slovakia, Ukraine, Bulgaria, and Romania.

47. See DeSilvey and Bartolini, "Where Horses Run Free?"; Pellis, Felder, and van der Duim, "Socio-political Conceptualization"; Vasile, "Vulnerable Bison"; Schwartz, *Nature and National Identity.*

48. For discussion, see Clemente et al., "Microbiome of Uncontacted Amer-indians"; Dominguez-Bello, "Microbial Anthropologist"; Ed Yong, "There Is No 'Healthy' Microbiome," *New York Times*, November 1, 2014, https://www.nytimes.com/; Yong, *I Contain Multitudes.*

49. Hayden, *When Nature Goes Public*; Neimark and Wilson, "Re-mining the Collections."

50. Heck, "Neuzüchtung des Auerochsen," 537.

51. Schama, *Landscape and Memory.*

52. Goderie et al., *Aurochs*, 4.

53. Schama, *Landscape and Memory.*

54. Goderie et al., *Aurochs.*

55. Goderie et al., *Aurochs.*

56. This argument is made by Jørgensen, "Rethinking Rewilding," and contested by Prior and Ward, "Rethinking Rewilding."

57. This argument is most commonly associated with Louv, *Last Child*.

58. Monbiot, *Feral*.

59. Gammon, "Many Meanings of Rewilding."

60. Tsing, "The Buck, the Bull," 16.

61. See Nash, *Wilderness*.

62. Deichmann, *Biologists under Hitler*; Sax, *Animals in the Third Reich*; Burkhardt, *Patterns of Behavior*; Kalikow, "Konrad Lorenz's Ethological Theory"; Sax, "What Is a 'Jewish Dog'?"

63. Heck, "Über die Neuzüchtung"; Heck, *Animals, My Adventures*.

64. Heck, "Breeding-Back of the Aurochs."

65. Harrington, *Reenchanted Science*.

66. Rewilding Europe describes this as selective science and explains that "the breeding programme is founded on a broad, multidisciplinary scientific base, including geneticists, ecologists, molecular biologists, archaeologists, archaeozoologists, historians, isotope experts, cattle experts and European cattle breeding organizations." See Rewilding Europe, "Rewilding in Action: The Aurochs—Europe's Defining Animal," https://rewildingeurope.com/.

67. Felius, *Cattle Breeds*.

68. U. K. T. Schulz, dir., *Wisente* (Berlin: Universum Film [UFA], 1941), Bundesfilmarchiv, Berlin. Translation by Clemens Driessen.

69. This work does not seem to have been written and published, but it formed part of a research project entitled "The Forest and Tree in the Aryan-Germanic History of Thought and Culture," headed by Lutz.

70. Lacoue-Labarthe, Nancy, and Holmes, "Nazi Myth."

71. Goderie et al., *Aurochs*, 74.

72. Lorimer, "Nonhuman Charisma."

73. Sunstein, "Idea of a Useable Past," 603.

74. Herf, *Reactionary Modernism*.

75. Barnes and Minca, "Nazi Spatial Theory," 671.

76. Latour uses the term "reactionary" in an unpublished piece commentating on Nordhaus and Schellenberger's 2007 book *Break Through*. See Latour, "It's Development, Stupid!," 7. A condensed version, without the term "reactionary," is Latour, "Love Your Monsters." In another essay, he cautions against the use of the term "reactionary" and the linear model of time, progress, and critique that it implies. Latour, "Attempt." Latour subsequently distances himself from ecomodernism, as it is put forward by the Breakthrough Institute. See Latour, "Fifty Shades of Green."

77. See Svenning et al., "Science for a Wilder Anthropocene."

## 7. Probiotic Value

1. Osborne, "Leave It to Beavers."

2. Poliquin, *Beaver*. Poliquin explains how this imaginary was used to naturalize kinds of work in settler colonialism: outdoor and physical for men, indoor and reproductive for women.

3. Campbell-Palmer, Gow, and Needham, *Eurasian Beaver*.

4. Mike's blog is now dormant, but it can still be accessed at Colon Comrades, https://coloncomrades.wordpress.com/. A translation of the Spanish reads, "Thanks my beautiful helminths. Always in my guts and my heart."

5. I take the concept of "lively commodities" from Collard and Dempsey, who define it as those "whose capitalist value is derived from their status as living beings." Collard and Dempsey, "Life for Sale?," 2684. See also Barua, "Lively Commodities."

6. Worster, *Nature's Economy*. I take the concept of the naturalization of work from Besky and Blanchette, eds., *How Nature Works*. See also Besky and Blanchette, "Naturalization of Work."

7. Battistoni, "Bringing in the Work of Nature," 21.

8. Barua, "Nonhuman Labour"; Barua, "Lively Commodities"; Porcher, *Ethics of Animal Labor*; Haraway, *When Species Meet*.

9. Marx, *Economic and Philosophic Manuscripts*.

10. Marx, *Economic and Philosophic Manuscripts*, 24.

11. Marx, *Capital*, 284.

12. Ingold, "The Architect and the Bee"; Ingold, *Hunters, Pastoralists and Ranchers*; Haraway, *When Species Meet*.

13. Haraway, *When Species Meet*. See also Barua, "Encounter."

14. Barua, "Nonhuman Labour," 281.

15. In addition to the work of Maan Barua, see Collard, "Putting Animals Back Together"; Collard and Dempsey, "Life for Sale?"; Dempsey, *Enterprising Nature*; Porcher, *Ethics of Animal Labor*; Johnson, "At the Limits"; Johnson and Goldstein, "Biomimetic Futures"; Coulter, *Animals, Work*; Besky and Blanchette, "Naturalization of Work."

16. Barua, "Animal Work." See also Blanchette, "Herding Species"; Wadiwel, "Chicken Harvesting Machine"; Beldo, "Metabolic Labor." The concept of sentient commodities is developed in Wilkie, "Sentient Commodities."

17. Adams, *Sexual Politics of Meat*; Haraway, *When Species Meet*; Federici, *Caliban and the Witch*.

18. Donovan and Adams, *Feminist Care Tradition*.

19. See Lorimer, *Wildlife in the Anthropocene*; Barua, "Nonhuman Labour."

20. Collard and Dempsey explore the concept of ecological labor and distinguish two types of lively commodities to which it gives rise according to "two

sets of distinctions: animate, encounterable life versus reproductive life; and individual versus 'massified' or aggregate life." Collard and Dempsey, "Life for Sale?," 2688.

21. Barua, "Animal Work." We can differentiate the idea of ecological work from what DiNovelli-Lang and Hebert describe as the ecological labor performed by humans: "By *ecological labor*, we refer not merely to the labor demanded by projects of environmental protection and conservation as opposed to the better known labor of resource extraction, but also to how many engagements with the environment are today being reconfigured as ecologically reproductive 'care work,' the affective and performative dimensions of which are often valued more than objective results." DiNovelli-Lang and Hébert, "Ecological Labor."

22. Crumley, *Nature's Architect*. In their promotional materials, the Devon Wildlife Trust presents beavers as nature's water engineers; see Devon Wildlife Trust, *Beavers—Nature's Water Engineers*.

23. Puttock et al., "Eurasian Beaver Activity"; Law et al., "Using Ecosystem Engineers."

24. Crumley, *Nature's Architect*, 195.

25. I take the concept of ecomanagerialism from Timothy Luke, for whom it describes a mode of government manifest in the training of professional ecologists in which "nature loses its transcendental qualities and its locales, resources, and systems become objects of capitalist manipulation." Luke, "Ecomanagerialism," 120.

26. Paxson, "Naturalization of Nature."

27. Johnson and Goldstein, "Biomimetic Futures," 388.

28. Department for Environment, Food and Rural Affairs (DEFRA), *Working with the Grain of Nature*; Rewilding Britain, "How Rewilding Reduces Flood Risk," 2016, https://www.rewildingbritain.org.uk/assets/uploads/files/publications/Rewilding_FloodReport_AUG2016_FINAL.pdf.

29. For an overview, see Campbell-Palmer, *Eurasian Beaver Handbook*.

30. Crumley, *Nature's Architect*; Osborne, "Leave It to Beavers"; Hood, *Beaver Manifesto*.

31. For an introduction to the concept of animal clinical labor, see Clark, "Labourers or Lab Tools."

32. Melinda Cooper and Catherine Waldby introduce the concept of human clinical labor in Cooper and Waldby, *Clinical Labor*.

33. I take this four-part categorization of work from economic history and sociology, where it is associated with the writings of figures like Colin Clark and Jean Fourastié.

34. The history of the hookworm as a parasitic productivity problem is detailed in chapter 5. For an analysis of the beaver as a resource that shaped

North American settler colonialism, see Wolf, Eriksen, and Diaz, *Europe and the People without History*; Ogden, "Beaver Diaspora."

35. I take this understanding of commoning from Gibson-Graham, Cameron, and Healy, *Take Back the Economy*. Ginn and Ascensão identify three types of commoning within this work: "The first is commoning enclosed resources. This moves narrow access to wide, and expands the use and benefit—as well as responsibilities of care—from owner to a wider community. The second is maintaining existing commons. The third is bringing resources that are not yet managed (or may not yet exist) from open, unregulated access into some form of commons." Ginn and Ascensão, "Autonomy, Erasure, and Persistence," 933. It is the third type of commoning that is at work in this example.

36. Bresnihan, "More-than-Human Commons." See also Johnson, "At the Limits"; Nelson and Braun, "Autonomia in the Anthropocene."

37. Bresnihan, "More-than-Human Commons," 93.

38. Bresnihan, "More-than-Human Commons," 94.

39. Bresnihan, "More-than-Human Commons," 95. See also Woelfle-Erskine, "Beavers as Commoners?"

40. Bresnihan, "More-than-Human Commons," 94.

41. Symmbio, "About Us," http://www.symmbio.com/.

42. Three doses of *Necator americanus* hookworm larvae cost $450, valid for two years from the date of purchase, or $200 for a single dose. For *Trichuris trichiura* whipworm ova, prices start at $1 per ovum for doses of 25, 50, 100, or 250, with a minimum order of $300. Symmbio, http://www.symmbio.com/.

43. Tsing, *Mushroom*.

44. I develop the concept of biosocial capital from Pierre Bourdieu's well-known idea of social capital. In this context, it describes the skills, connections, and experience required to navigate virtual communities, to solicit relevant and trustworthy health advice, and to successfully acquire and administer therapeutics.

45. Tsing, *Mushroom*, 180.

46. Woelfle-Erskine, "Beavers as Commoners?"

47. Battistoni, "Bringing in the Work of Nature," 22.

48. Some, like Porcher, make a case for animal work that is founded on a sense of responsibility and reciprocity and that ensures ongoing coevolved relationships. See Porcher, *Ethics of Animal Labor*. Others—for example, those invested in animal liberation—see virtually all working relations with animals as exploitative. See, e.g., Nibert, *Animal Oppression*; Wadiwel, *War against Animals*; Noske, *Humans and Other Animals*.

49. This conditionality is explored in different contexts by Crowley, Hinchliffe, and McDonald, "Nonhuman Citizens on Trial"; Woelfle-Erskine and Cole, "Transfiguring the Anthropocene."

50. Campbell-Palmer, Gow, and Needham, *Eurasian Beaver.*

51. Dempsey, *Enterprising Nature.* For a discussion of the close connections between the discourse of ecosystem services and resilience theory, see Nelson, "Resilience."

52. Buscher, Dressler, and Fletcher, *Nature Inc.*; Heynen, *Neoliberal Environments*; Cooper, *Life as Surplus.*

53. Ditgen et al., "Harnessing the Helminth Secretome."

54. One market research report estimates that the "global autoimmune diseases treatment market value" was $36.41 billion in 2016 and was estimated to grow at 3.8 percent per year to reach $45.54 billion by 2022. Research and Markets, "Global Autoimmune Disease Treatment Market," https://www.researchandmarkets.com/.

55. See Harnett and Harnett, "Can Parasitic Worms Cure the Modern World's Ills?"

56. Seres Therapeutics, "Microbiome Overview," https://www.serestherapeutics.com/.

57. For a discussion of the initial failure of this trial (ClinicalTrials.gov identifier NCT02437500) and its effect on the microbiome investment climate, see Ratner, "Seres's Pioneering Microbiome Drug Fails"; Ed Yong, "Sham Poo Washes Out," *Atlantic*, August 1, 2016, https://www.theatlantic.com/. The social, legal, and ethical implications of the classification of feces as drugs are discussed in Ma et al., "Ethical Issues"; Sachs and Edelstein, "Ensuring the Safe and Effective FDA Regulation."

58. I take the concepts of lively and biocapital from Rajan, ed., *Lively Capital*; Sunder Rajan, *Biocapital.*

59. Johnson and Goldstein, "Biomimetic Futures"; Cooper, *Life as Surplus.*

60. See, e.g., Dempsey, *Enterprising Nature*; Robertson, "Measurement and Alienation"; Sullivan, "Banking Nature?"

61. The Natural Capital Coalition defines natural capital as "the stock of renewable and nonrenewable resources (e.g. plants, animals, air, water, soils, minerals) that combine to yield a flow of benefits to people. All this means is that any part of the natural world that benefits people, or that underpins the provision of benefits to people, is a form of natural capital. Natural capital is a stock, and from it flows ecosystem services or benefits." National Capital Coalition, "What Is Natural Capital?," https://naturalcapitalcoalition.org/.

62. Department for Environment, Food, and Rural Affairs, Natural Capital Committee (https://www.gov.uk/), chaired by Dieter Helm.

63. Campbell-Palmer, *Eurasian Beaver Handbook*, 117.

64. The phrase "public money for public goods" has been central to the U.K. government's discourse promoting a new regime for agricultural subsidies

that would replace the Common Agricultural Policy after the United Kingdom leaves the European Union.

65. Papadopoulos, *Insurgent Posthumanism*.

66. Smith, "Nature as an Accumulation Strategy," 28–29.

67. Smith, "Nature as an Accumulation Strategy," 21.

68. Wammes et al., "Helminth Therapy or Elimination."

69. Parker et al., "Prescription for Clinical Immunology."

70. The idea that the generative potentials of life itself become an accumulation strategy is first discussed by Haraway in *Modest Witness*.

71. Robertson, "Nature that Capital Can See."

72. The biopolitical consequences of this enterprising of nature have been explored in Johnson and Goldstein, "Biomimetic Futures"; Collard and Dempsey, "Life for Sale?"; Collard and Dempsey, "Capitalist Natures."

73. Helm, *Green and Prosperous Land*, 36.

74. A similar analysis is offered by Cleo Woelfle-Erskine in work on the use of beavers for ecological restoration in Oregon. See Woelfle-Erskine, "Beavers as Commoners?"

75. For a discussion of spectacular accumulation, see Barua, "Nonhuman Labour."

76. This point is made by Genes et al., "Why We Should Let Rewilding Be Wild." Their study critiques a previous manifesto for making "rewilding fit for policy," which defines rewilding as "the reorganisation of biota and ecosystem processes to set an identified social–ecological system on a preferred trajectory, leading to the self-sustaining provision of ecosystem services with minimal ongoing management." Pettorelli et al., "Making Rewilding Fit for Policy," 1117.

77. I take the concept of a socioecological fix from Castree, "Neoliberalising Nature."

78. Hribal, "'Animals Are Part of the Working Class'"; Hribal and Clair, *Fear of the Animal Planet*. Chris Philo, one of the architects of the revival of animal geography in the 1990s, cautions against the use of the word "resistance" in relation to animals, suggesting that it is linked too strongly to intentionality. He suggests "transgression" be used in its place. See Philo, "Animals, Geography." For a wider discussion, see Palmer, "Taming the Wild Profusion."

79. The history of biocontrol, and the rare but disastrous examples of introduced species that subsequently became invasive are reviewed in Barratt et al., "Status of Biological Control."

80. Cane toads were introduced into Australia in 1935 for the biological control of agricultural pests, especially beetles that eat sugarcane. The toads subsequently spread across much of Northeast Australia, eating local fauna. The toads are poisonous and have killed numerous indigenous animals. Low, *Feral*

*Future*; Shine and Greene, *Cane Toad Wars*. An example of invasive beavers can be found in Ogden, *Beaver Diaspora*.

81. I take the concepts of emergent ecologies from Kirksey, *Emergent Ecologies*.

82. Rutherford, "Anthropocene's Animal?," 207.

83. See Collard, "Cougar–Human Entanglements"; Nagy, Johnson, and Malamud, *Trash Animals*; Barua and Sinha, "Animating the Urban."

84. Hearn, Watkins, and Balzaretti, "Cultural and Land Use Implications"; Boonman-Berson, Driessen, and Turnhout, "Managing Wild Minds"; Tack, *Wild Boar* (Sus scrofa) *Populations in Europe*.

85. Gavier-Widén et al., "African Swine Fever."

86. Woelfle-Erskine and Cole, "Transfiguring the Anthropocene."

87. Stillfried et al., "Wild Inside."

88. I take the concept of the globalized diseasescape from Harper and Armelagos, "Changing Disease-scape." See also Ali and Keil, *Networked Disease*.

89. For a wider discussion of the intensities of industrial pig production, see Blanchette, "Herding Species"; Hinchliffe and Ward, "Geographies of Folded Life"; White, "From Globalized Pig Breeds to Capitalist Pigs."

90. I take the concept of animal capital from Shukin, *Animal Capital*.

91. Examples of alternative visioning of postantibiotic futures include Ashby et al., *Infectious Futures*; Emanuel, *Truth about Hawaii*.

92. Grant, *Parasite*, *Chimera*, and *Symbiont*.

93. Dempsey, *Enterprising Nature*, 10.

94. Johnson, "At the Limits," 289. Comparable arguments are made by Sara Holiday Nelson in her reworking of the theory of resilience for the purposes of socioecological transformation. Nelson, "Resilience." See also Sullivan, "Nature on the Move 3."

95. I take the concept of the uncommons from Marisol de la Cadena. See Blaser and de la Cadena, "Uncommons"; de la Cadena, "Uncommoning Nature."

96. Arendt, *Human Condition*.

97. Arendt, *Human Condition*.

98. Arendt is not commonly considered a good place to start for more-than-humanism, given her antipathy toward animals and nature, but the beginnings of such a revisionist reading are offered in Bowring, "Arendt after Marx."

99. McMullen, "Rewilding Scotlands Highlands." See also Porcher, *Ethics of Animal Labor*.

100. This phrase appeared on the entrance of Auschwitz and other Nazi concentration camps.

101. Weeks, *Problem with Work*; Papadopoulos, "Insurgent Posthumanism."

102. Crumley, *Nature's Architect*.

103. For a discussion of the animal as artist, see Baker, *Artist Animal*.

104. The character of animal play and its potential to ground a multispecies ethics for rewilding is promoted by Bekoff in *Minding Animals* and *Rewilding Our Hearts*.

## Conclusions

1. Keulartz, "Emergence of Enlightened Anthropocentrism."

2. I explore this tension in more detail in Lorimer, "Decoupling without Disconnection."

3. Svenning et al., "Science for a Wilder Anthropocene."

4. Parker and Ollerton, "Evolutionary Biology and Anthropology."

5. Lorimer and Driessen, "Bovine Biopolitics"; Klaver et al., "Born to Be Wild."

6. Latour, *Down to Earth*.

7. Herf, *Reactionary Modernism*.

8. It is worth noting that many of the landscapes currently targeted for rewilding interventions across Europe and the Americas maintain their wildlife as they were cleared of people through violent acts of colonial dispossession, in some cases legitimated through ecological claims to protect nature. These dark histories are used by some nationalist groups as grounds to rewilding projects proposed by international or urban environmental organizations. For examples in the Baltic region, see Schwartz, *Nature and National Identity*.

9. Monbiot, *Feral*.

10. For a discussion, see Granjou and Phillips, "Living and Labouring Soils"; Krzywoszynska, "Caring for Soil Life"; Spackman, "Formulating Citizenship."

11. The potential of this type of analysis can be seen in Münster, "Performing Alternative Agriculture"; Hinchliffe, Butcher, and Rahman, "AMR Problem"; Jasarevic, "Thing in a Jar."

12. The intersections between race and probiotic thinking are evident from existing work on the microbiome and epigenetics. See Stallins et al., "Geography and Postgenomics"; Mansfield and Guthman, "Epigenetic Life"; Benezra, "Datafying Microbes."

13. Hörl and Burton, eds., *General Ecology*. The title of this book is taken from an earlier claim for the rise of ecological social thought by Félix Guattari, *Three Ecologies*. This is different from the new ecological order feared by Ferry, *New Ecological Order*.

14. Worster, *Nature's Economy*.

15. This high-profile workshop was conceived by Déborah Danowski, Eduardo Viveiros de Castro, and Bruno Latour, and was held in Rio de Janeiro

in 2014. See the Thousand Names of Gaia (https://thethousandnamesofgaia
.wordpress.com/).

16. Stengers, "Gaia," 3.

17. Paxson, "Microbiopolitics."

18. Paxson and Helmreich, "Perils and Promises."

19. The history of vitalist and holistic thinking and its intersections with
fascism are explored in Harrington, *Reenchanted Science*; Klinke, "Vitalist Temptations"; Abrahamsson, review of *Vibrant Matter* by Jane Bennett.

20. In *Primate Visions*, Haraway writes of the importance for science studies
scholars of resisting the siren call of the scientists they study.

21. I take my understanding of strong objectivity from Sandra Harding and
other feminist philosophers of science.

22. For a summary of these critiques, see Wakefield, "Infrastructures of
Liberal Life."

23. Nelson, "Resilience," 3.

24. Braun, "New Materialisms"; Johnson and Goldstein, "Biomimetic
Futures."

25. Esposito, *Immunitas*.

26. Wolfe, *Before the Law*; Clarke, "Planetary Immunity." I explore this argument in more detail in relation to helminth therapy in Lorimer, "Gut Buddies."

27. Staying with the trouble is Haraway's exhortation to those working in
multispecies trouble to avoid making recourse to simple, pure, or final solutions
to the inexorable challenges of human–nonhuman relations. See Haraway, *Staying with the Trouble*.

28. David Chandler and Jonathan Pugh outline this three-part typology of
forms of hope in a call for papers for a session at the RGS-IBG Conference in
London in August 2019.

29. Latour, *Down to Earth*.

30. For a wider discussion of the role of grief in the Anthropocene, see
Head, *Hope and Grief*.

31. Jepson, "Recoverable Earth."

32. Scranton, *Learning to Die in the Anthropocene*.

# BIBLIOGRAPHY

Abrahamsson, Christian. Review of *Vibrant Matter: A Political Ecology of Things*, by Jane Bennett. *Dialogues in Human Geography* 1, no. 3 (2011): 399–402.

Ackert, Lloyd. *Sergei Vinogradskii and the Cycle of Life: From the Thermodynamics of Life to Ecological Microbiology, 1850–1950*. Amsterdam: Springer, 2012.

Adams, Carol J. *The Sexual Politics of Meat: A Feminist-Vegetarian Critical Theory*. New York: Continuum, 1990.

Adams, William. *Against Extinction: The Story of Conservation*. London: Taylor & Francis, 2013.

Adams, William. "Geographies of Conservation 2: Technology, Surveillance and Conservation by Algorithm." *Progress in Human Geography* 43, no. 2 (2019): 337–50.

Albuquerque, Ulysses Paulino, Paulo Henrique Santos Gonçalves, Washington Soares Ferreira Júnior, et al. "Humans as Niche Constructors: Revisiting the Concept of Chronic Anthropogenic Disturbances in Ecology." *Perspectives in Ecology and Conservation* 16, no. 1 (2018): 1–11.

Ali, Harris, and Roger Keil. *Networked Disease: Emerging Infections in the Global City*. New York: Wiley, 2011.

Allen, Judith, and Rick Maizels. "Diversity and Dialogue in Immunity to Helminths." *Nature Reviews Immunology* 11, no. 6 (2011): 375–88.

Anderson, Ben. "Affect and Biopower: Towards a Politics of Life." *Transactions of the Institute of British Geographers* 37, no. 1 (2012): 28–43.

Anderson, Ben. "Preemption, Precaution, Preparedness: Anticipatory Action and Future Geographies." *Progress in Human Geography* 34, no. 6 (2010): 777–98.

Anderson, Warwick. *Colonial Pathologies: American Tropical Medicine, Race, and Hygiene in the Philippines*. Durham, N.C.: Duke University Press, 2006.

Anderson, Warwick. "Excremental Colonialism: Public Health and the Poetics of Pollution." *Critical Inquiry* 21, no. 3 (1995): 640–69.

Anderson, Warwick. "Natural Histories of Infectious Disease: Ecological Vision in Twentieth-Century Biomedical Science." *Osiris* 19 (2004): 39–61.

Anderson, Warwick, and Ian Mackay. *Intolerant Bodies: A Short History of Autoimmunity.* Baltimore, Md.: Johns Hopkins University Press, 2014.

Arendt, H. *The Human Condition.* Chicago: University of Chicago Press, 1958.

Arènes, Alexandra, Bruno Latour, and Jérôme Gaillardet. "Giving Depth to the Surface: An Exercise in the Gaia-graphy of Critical Zones." *Anthropocene Review* 5, no. 2 (2018): 120–35.

Asafu-Adjaye, J., L. Blomqvist, S. Brand, et al. *An Ecomodernist Manifesto.* Oakland, Calif.: Oakland, Calif.: Breakthrough Institute, 2015.

Ashby, Madeline, A. S. Fields, Jenni Hill, et al. *Infectious Futures: Stories of the Post-antibiotic Apocalypse.* London: Nesta, 2015.

Atkinson, Mark A., George S. Eisenbarth, and Aaron W. Michels. "Type 1 Diabetes." *Lancet* 383, no. 9911 (2014): 69–82.

Babayan, Simon A., Judith E. Allen, Jan E. Bradley, et al. "Wild Immunology: Converging on the Real World." *Annals of the New York Academy of Sciences* 1236, no. 1 (2011): 17–29.

Bach, Jean-François. "The Effect of Infections on Susceptibility to Autoimmune and Allergic Diseases." *New England Journal of Medicine* 347, no. 12 (2002): 911–20.

Baerselman, Fred, and Frans Vera. *Nature Development: An Exploratory Study for the Construction of Ecological Networks.* The Hague: Ministry of Agriculture, Nature Management and Fisheries, 1995.

Baker, Steve. *Artist Animal.* Minneapolis: University of Minnesota Press, 2013.

Balmford, Andrew, Rhys Green, and Ben Phalan. "What Conservationists Need to Know about Farming." *Proceedings of the Royal Society B: Biological Sciences* 279, no. 1739 (2012): 2714–24.

Bardosh, Kevin, Phouth Inthavong, Sivilai Xayaheuang, and Anna L. Okello. "Controlling Parasites, Understanding Practices: The Biosocial Complexity of a One Health Intervention for Neglected Zoonotic Helminths in Northern Lao PDR." *Social Science and Medicine* 120 (2014): 215–23.

Barker, Kezia. "Biosecurity: Securing Circulations from the Microbe to the Macrocosm." *Geographical Journal* 181, no. 4 (2015): 357–65.

Barnes, Trevor, and Claudio Minca. "Nazi Spatial Theory: The Dark Geographies of Carl Schmitt and Walter Christaller." *Annals of the Association of American Geographers* 103, no. 3 (2013): 669–87.

Barratt, B. I. P., V. C. Moran, F. Bigler, and J. C. van Lenteren. "The Status of Biological Control and Recommendations for Improving Uptake for the Future." *BioControl* 63, no. 1 (2018): 155–67.

Barua, Maan. "Animal Work: Metabolic, Ecological, Affective." Theorizing the Contemporary, Fieldsights, July 26, 2018. https://culanth.org/.

Barua, Maan. "Encounter." *Environmental Humanities* 7, no. 1 (2016): 265–70.

Barua, Maan. "Lively Commodities and Encounter Value." *Environment and Planning D: Society and Space* 34, no. 4 (2016): 725–44.

Barua, Maan. "Nonhuman Labour, Encounter Value, Spectacular Accumulation: The Geographies of a Lively Commodity." *Transactions of the Institute of British Geographers* 42, no. 2 (2017): 274–88.

Barua, Maan, and Anindya Sinha. "Animating the Urban: An Ethological and Geographical Conversation." *Social and Cultural Geography* (2017): 1–21.

Battisti, Corrado, Gianluca Poeta, and Giuliano Fanelli. *An Introduction to Disturbance Ecology: A Road Map for Wildlife Management and Conservation.* Amsterdam: Springer, 2016.

Battistoni, Alyssa. "Bringing in the Work of Nature: From Natural Capital to Hybrid Labor." *Political Theory* 45, no. 1 (2016): 5–31.

Beans, Carolyn. "News Feature: What Happens When Lab Animals Go Wild." *Proceedings of the National Academy of Sciences of the United States of America* 115, no. 13 (2018): 3196–99.

Beinart, William. *The Rise of Conservation in South Africa: Settlers, Livestock, and the Environment 1770–1950.* Oxford: Oxford University Press, 2008.

Bekoff, Marc. *Minding Animals: Awareness, Emotions, and Heart.* Oxford: Oxford University Press, 2002.

Bekoff, Marc. *Rewilding Our Hearts: Building Pathways of Compassion and Coexistence.* Novato, Calif.: New World Library, 2014.

Beldo, Les. "Metabolic Labor: Broiler Chickens and the Exploitation of Vitality." *Environmental Humanities* 9, no. 1 (2017): 108–28.

Benezra, Amber. "Datafying Microbes: Malnutrition at the Intersection of Genomics and Global Health." *BioSocieties* 11, no. 3 (2016): 334–51.

Bennett, Jane. "The Force of Things—Steps toward an Ecology of Matter." *Political Theory* 32, no. 3 (2004): 347–72.

Benson, Etienne. *Wired Wilderness: Technologies of Tracking and the Making of Modern Wildlife.* Baltimore, Md.: Johns Hopkins University Press, 2010.

Beran, D., and J. S. Yudkin. "The Double Scandal of Insulin." Editorial. *Journal of the Royal College of Physicians of Edinburgh* 43, no. 3 (2013): 194–96.

Berbudi, Afiat, Jesuthas Ajendra, Ajeng P. F. Wardani, Achim Hoerauf, and Marc P. Hübner. "Parasitic Helminths and Their Beneficial Impact on Type 1 and Type 2 Diabetes." *Diabetes/Metabolism Research and Reviews* 32, no. 3 (2016): 238–50.

Besky, Sarah, and Alex Blanchette. "The Naturalization of Work." Theorizing the Contemporary, Fieldsights, July 26, 2018. https://culanth.org/.

Besky, Sarah, and Alex Blanchette, eds. *How Nature Works: Rethinking Labor on a Troubled Planet*. Santa Fe, N.M.: SAR Press, 2019.

Bested, Alison C., Alan C. Logan, and Eva M. Selhub. "Intestinal Microbiota, Probiotics and Mental Health: From Metchnikoff to Modern Advances: Part I—Autointoxication Revisited." *Gut Pathogens* 5, no. 1 (2013): 5.

Biehl, João, and Adriana Petryna. *When People Come First: Critical Studies in Global Health*. Princeton, N.J.: Princeton University Press, 2013.

Biermann, Christine, and Becky Mansfield. "Biodiversity, Purity, and Death: Conservation Biology as Biopolitics." *Environment and Planning D: Society and Space* 32, no. 2 (2014): 257–73.

Biermann, F., K. Abbott, S. Andresen, et al. "Navigating the Anthropocene: Improving Earth System Governance." *Science* 335, no. 6074 (2012): 1306–7.

Bingham, Nick, and Steve Hinchliffe. "Reconstituting Natures: Articulating Other Modes of Living Together." *Geoforum* 39, no. 1 (2008): 83–87.

Birks, H. J. B. "Mind the Gap: How Open Were European Primeval Forests?" *Trends in Ecology and Evolution* 20, no. 4 (2005): 154–56.

Blackbourn, David. *The Conquest of Nature: Water, Landscape, and the Making of Modern Germany*. New York: Random House, 2011.

Blanchette, Alex. "Herding Species: Biosecurity, Posthuman Labor, and the American Industrial Pig." *Cultural Anthropology* 30, no. 4 (2015): 640–69.

Blaser, Mario, and Marisol de la Cadena. "The Uncommons: An Introduction." *Anthropologica* 59, no. 2 (2017): 185–93.

Blaser, Martin. *Missing Microbes: How Killing Bacteria Creates Modern Plagues*. London: Oneworld Publications, 2014.

Blaser, Martin J., Zoe G. Cardon, Mildred K. Cho, et al. "Toward a Predictive Understanding of Earth's Microbiomes to Address 21st Century Challenges." Editorial. *mBio* 7, no. 3 (2016).

Blaser, Martin, and Stanley Falkow. "What Are the Consequences of the Disappearing Human Microbiota?" *Nature Reviews Microbiology* 7, no. 12 (2009): 887–94.

Blomqvist, Linus, Ted Nordhaus, and Michael Shellenberger. *Nature Unbound: Decoupling for Conservation*. Oakland, Calif.: Breakthrough Institute, 2015.

Böbel, Till S., Sascha B. Hackl, et al. "Less Immune Activation Following Social Stress in Rural vs. Urban Participants Raised with Regular or No Animal Contact, Respectively." *Proceedings of the National Academy of Sciences of the United States of America* 115, no. 20 (2018): 5259–64.

Bocci, Paolo. "Tangles of Care: Killing Goats to Save Tortoises on the Galápagos Islands." *Cultural Anthropology* 32, no. 3 (2017): 424–49.

Bockman, Johanna, and Gil Eyal. "Eastern Europe as a Laboratory for Economic Knowledge: The Transnational Roots of Neoliberalism." *American Journal of Sociology* 108, no. 2 (2002): 310–52.

Bonds, Matthew H., Donald C. Keenan, Pejman Rohani, and Jeffrey D. Sachs. "Poverty Trap Formed by the Ecology of Infectious Diseases." *Proceedings of the Royal Society B: Biological Sciences* 277, no. 1685 (2010): 1185–92.

Bono-Lunn, Dillan, Chantal Villeneuve, Nour J. Abdulhay, Matthew Harker, and William Parker. "Policy and Regulations in Light of the Human Body as a 'Superorganism' Containing Multiple, Intertwined Symbiotic Relationships." *Clinical Research and Regulatory Affairs* 33, no. 2–4 (2016): 39–48.

Boonman-Berson, Susan, Clemens Driessen, and Esther Turnhout. "Managing Wild Minds: From Control by Numbers to a Multinatural Approach in Wild Boar Management in the Veluwe, the Netherlands." *Transactions of the Institute of British Geographers* 44, no. 1 (2019): 2–15.

Bordenstein, Seth, and Kevin Theis. "Host Biology in Light of the Microbiome: Ten Principles of Holobionts and Hologenomes." *PLoS Biology* 13, no. 8 (2015): A005.

Botkin, Daniel. *Discordant Harmonies: A New Ecology for the Twenty-First Century.* Oxford: Oxford University Press, 1990.

Bowker, Geoffrey. "Biodiversity Datadiversity." *Social Studies of Science* 30, no. 5 (2000): 643–83.

Bowman, David. "Conservation: Bring Elephants to Australia?" *Nature* 482, no. 7383 (2012): 30.

Bowman, David M. J. S., Jennifer Balch, Paulo Artaxo, et al. "The Human Dimension of Fire Regimes on Earth." *Journal of Biogeography* 38, no. 12 (2011): 2223–36.

Bowring, Finn. "Arendt after Marx: Rethinking the Dualism of Nature and World." *Rethinking Marxism* 26, no. 2 (2014): 278–90.

Bradshaw, Gay. *Elephants on the Edge: What Animals Teach Us about Humanity.* New Haven, Conn.: Yale University Press, 2009.

Braun, Bruce. "Biopolitics and the Molecularization of Life." *Cultural Geographies* 14, no. 1 (2007): 6–28.

Braun, Bruce. "From Critique to Experiment? Rethinking Political Ecology for the Anthropocene." In *The Routledge Handbook of Political Ecology,* edited by Tom Perreault, Gavin Bridge, and James McCarthy, 102–14. London: Taylor & Francis, 2015.

Braun, Bruce. "New Materialisms and Neoliberal Natures." *Antipode* 47, no. 1 (2015): 1–14.

Braun, Bruce. "A New Urban Dispositif? Governing Life in an Age of Climate Change." *Environment and Planning D: Society and Space* 32, no. 1 (2014): 49–64.

Braverman, Irus. "Anticipating Endangerment: The Biopolitics of Threatened Species Lists." *BioSocieties* 12, no. 1 (2017): 132–57.

Braverman, Irus. "Governing the Wild: Databases, Algorithms, and Population Models as Biopolitics." *Surveillance and Society* 12, no. 1 (2014): 15–37.

Brennan, Angela, Paul C. Cross, Katie Portacci, Brandon M. Scurlock, and William H. Edwards. "Shifting Brucellosis Risk in Livestock Coincides with Spreading Seroprevalence in Elk." *PLoS One* 12, no. 6 (2017): e0178780.

Bresnihan, Patrick. "The More-than-Human Commons: From Commons to Commoning." In *Space, Power, and the Commons: The Struggle for Alternative Futures*, edited by S. Kirwan, L. Dawney, and J. Brigstocke, 93–112. New York: Routledge, 2015.

Bresnihan, Patrick. *Transforming the Fisheries: Neoliberalism, Nature, and the Commons*. Lincoln: University of Nebraska Press, 2016.

Briggs, Neima, Jill Weatherhead, K. Jagannadha Sastry, and Peter J. Hotez. "The Hygiene Hypothesis and Its Inconvenient Truths about Helminth Infections." *PLoS Neglected Tropical Diseases* 10, no. 9 (2016): e0004944.

Brooker, Simon. "Estimating the Global Distribution and Disease Burden of Intestinal Nematode Infections: Adding up the Numbers—A Review." *International Journal for Parasitology* 40, no. 10 (2010): 1137–44.

Brooker, Simon, Jeffrey Bethony, and Peter Hotez. "Human Hookworm Infection in the 21st Century." *Advances in Parasitology* 58 (2004): 197–288.

Brosschot, Tara P., and Lisa A. Reynolds. "The Impact of a Helminth-Modified Microbiome on Host Immunity." *Mucosal Immunology* 11, no. 4 (2018): 1039–46.

Brown, Hannah, and Ann H. Kelly. "Material Proximities and Hotspots: Toward an Anthropology of Viral Hemorrhagic Fevers." *Medical Anthropology Quarterly* 28, no. 2 (2014): 280–303.

Brown, Joel S., John W. Laundré, and Mahesh Gurung. "The Ecology of Fear: Optimal Foraging, Game Theory, and Trophic Interactions." *Journal of Mammalogy* 80, no. 2 (1999): 385–99.

Brown, Nik. *Immunitary Life*. London: Palgrave Macmillan, 2018.

Brown, Peter J. "Microparasites and Macroparasites." *Cultural Anthropology* 2, no. 1 (1987): 155–71.

Brüggemeier, F. J., M. Cioc, and T. Zeller. *How Green Were the Nazis? Nature, Environment, and Nation in the Third Reich*. Athens: Ohio University Press, 2005.

Bubandt, Nils, and Anna Tsing. "An Ethnoecology for the Anthropocene: How a Former Brown-Coal Mine in Denmark Shows Us the Feral Dynamics of

Post-industrial Ruin: Online Supplement." *Journal of Ethnobiology* 38, no. 1 (2018): 1–7.

Buller, Henry. "Introducing Aliens, Re-introducing Natives: A Conflict of Interest for Biosecurity." In *Biosecurity: The Socio-politics of Invasive Species and Infectious Diseases*, edited by Andrew Dobson, Kezia Barker, and Sarah L. Taylor, 183–98. London: Routledge, 2013.

Buller, Henry. "Safe from the Wolf: Biosecurity, Biodiversity, and Competing Philosophies of Nature." *Environment and Planning A* 40, no. 7 (2008): 1583–97.

Burkhardt, Richard W. *Patterns of Behavior: Konrad Lorenz, Niko Tinbergen, and the Founding of Ethology*. Chicago: University of Chicago Press, 2005.

Buscher, B., W. Dressler, and R. Fletcher. *Nature Inc.: Environmental Conservation in the Neoliberal Age*. Tucson: University of Arizona Press, 2014.

Butterfield, Bradley, Stella Copeland, Seth Munson, Carla Roybal, and Troy Wood. "Prestoration: Using Species in Restoration that Will Persist Now and into the Future." *Restoration Ecology* 25, S2 (2017): S155-S63.

Byers, James E., Kim Cuddington, Clive G. Jones, et al. "Using Ecosystem Engineers to Restore Ecological Systems." *Trends in Ecology and Evolution* 21, no. 9 (2006): 493–500.

Cairns, P., S. Widstrand, F. Möllers, and B. Wijnberg. *Wild Wonders of Europe*. New York: Harry N. Abrams, 2010.

Callicott, J. Baird. *In Defense of the Land Ethic: Essays in Environmental Philosophy*. Albany: State University of New York Press, 1989.

Callon, Michel, Pierre Lascoumes, and Yannick Barthe. *Acting in an Uncertain World: An Essay on Technical Democracy*. Cambridge, Mass.: MIT Press, 2009.

Campbell-Palmer, R. *The Eurasian Beaver Handbook: Ecology and Management of Castor Fiber*. Exeter, U.K.: Pelagic, 2016.

Campbell-Palmer, R., D. Gow, and R. Needham. *The Eurasian Beaver*. Exeter, U.K.: Pelagic, 2015.

Cantrell, Bradley, Laura J. Martin, and Erle C. Ellis. "Designing Autonomy: Opportunities for New Wildness in the Anthropocene." *Trends in Ecology and Evolution* 32, no. 3 (2017): 156–66.

Caro, T., and S. Girling. *Conservation by Proxy: Indicator, Umbrella, Keystone, Flagship, and Other Surrogate Species*. Washington, D.C.: Island Press, 2010.

Caro, Tim, and Paul Sherman. "Rewilding Can Cause Rather than Solve Ecological Problems." Letter. *Nature* 462, no. 24 (2009): 985.

Carpenter, Stephen R., and Carl Folke. "Ecology for Transformation." *Trends in Ecology and Evolution* 21, no. 6 (2006): 309–15.

Carpenter, Steve, Brian Walker, J. Marty Anderies, and Nick Abel. "From Metaphor to Measurement: Resilience of What to What?" *Ecosystems* 4, no. 8 (2001): 765–81.

Cassidy, Rebecca, and Molly H. Mullin. *Where the Wild Things Are Now: Domestication Reconsidered*. Oxford: Berg, 2007.

Castree, Noel. "Neoliberalising Nature: The Logics of Deregulation and Reregulation." *Environment and Planning A* 40, no. 1 (2008): 131–52.

Ceauşu, Silvia, Max Hofmann, Laetitia M. Navarro, Steve Carver, Peter H. Verburg, and Henrique M. Pereira. "Mapping Opportunities and Challenges for Rewilding in Europe." *Conservation Biology* 29, no. 4 (2015): 1017–27.

Chakrabarty, Dipesh. *Provincializing Europe: Postcolonial Thought and Historical Difference*. New ed. Princeton, N.J.: Princeton University Press, 2009.

Chandler, David. *Ontopolitics in the Anthropocene: An Introduction to Mapping, Sensing, and Hacking*. London: Taylor & Francis, 2018.

Chapron, Guillaume, Petra Kaczensky, John D. C. Linnell, et al. "Recovery of Large Carnivores in Europe's Modern Human-Dominated Landscapes." *Science* 346, no. 6216 (2014): 1517–19.

Cheng, Anna, Darshana Jaint, Steven Thomas, Janet Wilson, and William Parker. "Overcoming Evolutionary Mismatch by Self-Treatment with Helminths: Current Practices and Experience." *Journal of Evolutionary Medicine* 3 (2015): 1–22.

Chien, Shiuh-Shen, Dong-Li Hong, and Po-Hsiung Lin. "Ideological and Volume Politics behind Cloud Water Resource Governance—Weather Modification in China." *Geoforum* 85 (2017): 225–33.

Chilvers, J., and M. Kearnes. *Remaking Participation: Science, Environment and Emergent Publics*. London: Taylor & Francis, 2015.

Clark, Jonathon. "Labourers or Lab Tools? Rethinking the Role of Lab Animals in Clinical Trials." In *The Rise of Critical Animal Studies: From the Margins to the Centre*, edited by N. Taylor and R. Twine, 139–64. London: Taylor & Francis, 2014.

Clark, Nigel. *Inhuman Nature: Sociable Living on a Dynamic Planet*. Thousand Oaks, Calif.: SAGE Publications, 2011.

Clark, Nigel. "Mobile Life: Biosecurity Practices and Insect Globalization." *Science as Culture* 22, no. 1 (2013): 16–37.

Clarke, Bruce. "Planetary Immunity: Biopolitics, Gaia Theory, the Holobiont, and the Systems Counterculture." In *General Ecology: The New Ecological Paradigm*, edited by Erich Hörl and James Burton, 193–216. London: Bloomsbury, 2018.

Clarke, Bruce. "Rethinking Gaia: Stengers, Latour, Margulis." *Theory, Culture, and Society* 34, no. 4 (2017): 3–26.

Clemente, Jose C., Erica C. Pehrsson, Martin J. Blaser, et al. "The Microbiome of Uncontacted Amerindians." *Science Advances* 1, no. 3 (2015): e1500183.

Cocker, Mark. *Our Place: Can We Save Britain's Wildlife before It Is Too Late?* New York: Random House, 2018.

Colebrook, Clare. *Death of the Posthuman: Essays on Extinction.* London: Open Humanities Press, 2015.

Collard, Rosemary-Claire. "Cougar–Human Entanglements and the Biopolitical Un/Making of Safe Space." *Environment and Planning D: Society and Space* 30, no. 1 (2012): 23–42.

Collard, Rosemary-Claire. "Putting Animals Back Together, Taking Commodities Apart." *Annals of the Association of American Geographers* 104, no. 1 (2014): 151–65.

Collard, Rosemary-Claire, and Jessica Dempsey. "Capitalist Natures in Five Orientations." *Capitalism Nature Socialism* 28, no. 1 (2017): 78–97.

Collard, Rosemary-Claire, and Jessica Dempsey. "Life for Sale? The Politics of Lively Commodities." *Environment and Planning A* 45, no. 11 (2013): 2682–99.

Cooke, Anne. "Review Series on Helminths, Immune Modulation and the Hygiene Hypothesis: How Might Infection Modulate the Onset of Type 1 Diabetes?" *Immunology* 126, no. 1 (2009): 12–17.

Cooper, Melinda. *Life as Surplus: Biotechnology and Capitalism in the Neoliberal Era.* Seattle: University of Washington Press, 2008.

Cooper, Melinda. "The Pharmacology of Distributed Experiment—User-Generated Drug Innovation." *Body and Society* 18, no. 3–4 (2012): 18–43.

Cooper, Melinda, and Catherine Waldby. *Clinical Labor: Tissue Donors and Research Subjects in the Global Bioeconomy.* Durham, N.C.: Duke University Press, 2014.

Cornelissen, Perry. "Large Herbivores as a Driving Force of Woodland–Grassland Cycles." PhD diss., Wageningen University, 2017. http://edepot.wur.nl/396698.

Correale, Jorge, and Mauricio F. Farez. "The Impact of Parasite Infections on the Course of Multiple Sclerosis." *Journal of Neuroimmunology* 233, no. 1–2 (2011): 6–11.

Couacaud, L. "Hookworm Disease and Its Relationship to Capitalism and Urban Development." *Journal of Political Ecology* 21, no. 1 (2014): 349–71.

Coulter, K. *Animals, Work, and the Promise of Interspecies Solidarity.* New York: Palgrave Macmillan U.S., 2016.

Cox, Frank. "History of Human Parasitology." In *Topley and Wilson's Microbiology and Microbial Infections*, 10th ed., vol. 6, *Parasitology*, edited by F. E. G. Cox, Derek Wakelin, Stephen Gillespie, and Dickson D. Despommier, 6:3–23. New York: Wiley, 2007.

Coyte, Katharine Z., Jonas Schluter, and Kevin R. Foster. "The Ecology of the Microbiome: Networks, Competition, and Stability." *Science* 350, no. 6261 (2015): 663–66.

Craig, J. Mark. "Atopic Dermatitis and the Intestinal Microbiota in Humans and Dogs." *Veterinary Medicine and Science* 2, no. 2 (2016): 95–105.

Crampton, Jeremy W. "Cartography: Performative, Participatory, Political." *Progress in Human Geography* 33, no. 6 (2009): 840–48.

Crawford, R. L., and D. L. Crawford. *Bioremediation: Principles and Applications.* Cambridge: Cambridge University Press, 2005.

Cromsigt, Joris P. G. M., Dries P. J. Kuijper, Marius Adam, et al. "Hunting for Fear: Innovating Management of Human–Wildlife Conflicts." *Journal of Applied Ecology* 50, no. 3 (2013): 544–49.

Cronon, William. "The Trouble with Wilderness; or, Getting Back to the Wrong Nature." In *Uncommon Ground: Rethinking the Human Place in Nature*, edited by W. Cronon, 69–90. New York: Norton, 1996.

Crooks, K. R., and M. Sanjayan. *Connectivity Conservation.* Cambridge: Cambridge University Press, 2006.

Crowley, S. L., S. Hinchliffe, and R. A. McDonald. "Nonhuman Citizens on Trial: The Ecological Politics of a Beaver Reintroduction." *Environment and Planning A* 49, no. 8 (2017): 1846–66.

Crumley, Jim. *Nature's Architect: The Beaver's Return to Our Wild Landscapes.* Manchester: Saraband, 2015.

Darwin, Charles. *The Formation of Vegetable Mould: Through the Action of Worms, with Observations on Their Habits.* London: John Murray, 1881.

Daszkiewiez, P., and J. Aikhenbaum. *Aurochs: Le Retour d'une Supercherie Nazie.* Paris: L'Association, 2000.

Davies, P. D. S., J. Grant, and M. Catchpole. *The Drugs Don't Work: A Global Threat.* London: Penguin, 2013.

DeFronzo, R. A., E. Ferrannini, P. Zimmet, and G. Alberti. *International Textbook of Diabetes Mellitus.* New York: Wiley, 2015.

Deichmann, Ute. *Biologists under Hitler.* Cambridge, Mass.: Harvard University Press, 1996.

de la Cadena, Marisol. "Uncommoning Nature." *e-flux* 65 (2015).

Dempsey, Jessica. *Enterprising Nature: Economics, Markets, and Finance in Global Biodiversity Politics.* New York: Wiley, 2016.

Department for Environment, Food and Rural Affairs (DEFRA). *Working with the Grain of Nature: A Biodiversity Strategy for England.* London: DEFRA, 2002.

DeSilvey, Caitlin. *Curated Decay: Heritage beyond Saving.* Minneapolis: University of Minnesota Press, 2017.

DeSilvey, Caitlin, and Nadia Bartolini. "Where Horses Run Free? Autonomy, Temporality and Rewilding in the Côa Valley, Portugal." *Transactions of the Institute of British Geographers* 44, no. 1 (2019): 94–109.

Despret, Vinciane. "The Body We Care For: Figures of Anthropo-zoo-genesis." *Body and Society* 10, no. 2–3 (2004): 111–34.

Despret, Vinciane. "Responding Bodies and Partial Affinities in Human–Animal Worlds." *Theory, Culture, and Society* 30, no. 7/8 (2013): 51–76.

Devon Wildlife Trust. *Beavers—Nature's Water Engineers: A Summary of Initial Findings from the Devon Beaver Projects.* Exeter, U.K.: Devon Wildlife Trust, 2018.

de Vrieze, J. "The Promise of Poop." *Science* 341, no. 6149 (2013): 954–57.

Di Gioia, D., and B. Biavati. *Probiotics and Prebiotics in Animal Health and Food Safety.* Amsterdam: Springer, 2018.

DiNovelli-Lang, Danielle, and Karen Hébert. "Ecological Labor." Theorizing the Contemporary, Fieldsights, July 26, 2018. https://culanth.org/.

Diprose, Rosalyn. *Corporeal Generosity: On Giving with Nietzsche, Merleau-Ponty, and Levinas.* Albany: State University of New York Press, 2002.

Ditgen, Dana, Emmanuela Anandarajah, Kamila Meissner, Norbert Brattig, Carsten Wrenger, and Eva Liebau. "Harnessing the Helminth Secretome for Therapeutic Immunomodulators." *BioMed Research International* 2014 (2014): 964350.

Dobson, Andrew, Kezia Barker, and Sarah Tayor. *Biosecurity: The Socio-politics of Invasive Species and Infectious Diseases.* London: Routledge, 2013.

Dominguez-Bello, Maria Gloria. "A Microbial Anthropologist in the Jungle." *Cell* 167, no. 3 (2016): 588–94.

Donlan, C. Josh. "De-extinction in a Crisis Discipline." *Frontiers of Biogeography* 6, no. 1 (2014): 25–28.

Donlan, C. J., J. Berger, C. E. Bock, et al. "Pleistocene Rewilding: An Optimistic Agenda for Twenty-First Century Conservation." *American Naturalist* 168, no. 5 (2006): 660–81.

Donovan, Josephine, and Carol J. Adams. *The Feminist Care Tradition in Animal Ethics: A Reader.* New York: Columbia University Press, 2007.

Drenthen, Martin. "Developing Nature along Dutch Rivers: Place or Nonplace." In *New Visions of Nature*, edited by Martin A. M. Drenthen, F. W. Jozef Keulartz, and James Proctor, 205–28. Amsterdam: Springer, 2009.

Drenthen, Martin. "The Return of the Wild in the Anthropocene: Wolf Resurgence in the Netherlands." *Ethics, Policy, and Environment* 18, no. 3 (2015): 318–37.

Drenthen, Martin. "Rewilding in Layered Landscapes as a Challenge to Place Identity." *Environmental Values* 27, no. 4 (2018): 405–25.

Driessen, Clemens, and Jamie Lorimer. "Back-Breeding the Aurochs: The Heck Brothers, National Socialism and Imagined Geographies for Nonhuman Lebensraum." In *Hitler's Geographies: The Spatialities of the Third Reich*, edited by P. Giaccaria and C. Minca, 138–60. Chicago: University of Chicago Press, 2016.

Edwards, P. N., and G. C. Bowker. *A Vast Machine: Computer Models, Climate Data, and the Politics of Global Warming*. Cambridge, Mass.: MIT Press, 2010.

Eisenberg, C. *The Wolf's Tooth: Keystone Predators, Trophic Cascades, and Biodiversity*. Washington, D.C.: Island Press, 2010.

Elliott, David E., Robert W. Summers, and Joel V. Weinstock. "Helminths as Governors of Immune-Mediated Inflammation." *International Journal for Parasitology* 37, no. 5 (2007): 457–64.

Elliott, D. E., and J. V. Weinstock. "Nematodes and Human Therapeutic Trials for Inflammatory Disease." *Parasite Immunology* 39, no. 5 (2017): 10.1111/pim.12407.

Emanuel, Oliver. *The Truth about Hawaii*. Prod./dir. Kirsty Williams. BBC Radio 4. https://www.bbc.co.uk/.

Enticott, Gareth. "Lay Immunology, Local Foods and Rural Identity: Defending Unpasteurised Milk in England." *Sociologia Ruralis* 43, no. 3 (2003): 257–70.

Epstein, S. *Impure Science: AIDS, Activism, and the Politics of Knowledge*. Berkeley: University of California Press, 1996.

Esposito, Roberto. *Immunitas: The Protection and Negation of Life*. Translated by Zakiya Hanafi. 2002. Malden, Mass.: Polity, 2011.

Estes, J. A., J. Terborgh, J. S. Brashares, et al. "Trophic Downgrading of Planet Earth." *Science* 333, no. 6040 (2011): 301–6.

Esteves, L. *Managed Realignment: A Viable Long-Term Coastal Management Strategy?* Amsterdam: Springer, 2014.

Ettling, J. *The Germ of Laziness: Rockefeller Philanthropy and Public Health in the New South*. Cambridge, Mass.: Harvard University Press, 2013.

Evans, David. *A History of Nature Conservation in Britain*. London: Routledge, 1997.

Farley, John. *To Cast Out Disease: A History of the International Health Division of Rockefeller Foundation (1913–1951)*. Oxford: Oxford University Press, 2003.

Farmer, Paul. *Pathologies of Power: Health, Human Rights, and the New War on the Poor*. Berkeley: University of California Press, 2004.

Federici, S. *Caliban and the Witch*. Brooklyn, N.Y.: Autonomedia, 2004.

Felius, M. *Cattle Breeds: An Encyclopedia*. London: Trafalgar Square Publishing, 2007.

Ferry, L. *The New Ecological Order*. Chicago: University of Chicago Press, 1995.

Filyk, Heather A., and Lisa C. Osborne. "The Multibiome: The Intestinal Ecosystem's Influence on Immune Homeostasis, Health, and Disease." *EBioMedicine* 13 (2016): 46–54.

Finlay, B., and M. C. Arrieta. *Let Them Eat Dirt: How Microbes Can Make Your Child Healthier*. New York: Random House, 2016.

Fisher, Mark. "Self-Willed Land; Can Nature Ever Be Free?" *ECOS* 25, no. 1 (2004): 6–11.

Fleming, J. O., and J. V. Weinstock. "Clinical Trials of Helminth Therapy in Autoimmune Diseases: Rationale and Findings." *Parasite Immunology* 37, no. 6 (2015): 277–92.

Flies, Emily J., Chris Skelly, Rebecca Lovell, Martin F. Breed, David Phillips, and Philip Weinstein. "Cities, Biodiversity and Health: We Need Healthy Urban Microbiome Initiatives." *Cities and Health* 2, no. 2 (2018): 143–50.

Flohr, Carsten, Rupert Quinnell, and John Britton. "Do Helminth Parasites Protect against Atopy and Allergic Disease?" *Clinical and Experimental Allergy* 39, no. 1 (2009): 20–32.

Flowers, Stephen. "Chronic Disease, New Thinking, and Outlaw Innovation: Patients on the Edge in the Knowledge Commons." In *Governing Medical Knowledge Commons*, edited by Brett M. Frischmann, Katherine J. Strandburg, and Michael J. Madison, 326–47. Cambridge: Cambridge University Press, 2017.

Foreman, D. *Rewilding North America: A Vision for Conservation in the 21st Century*. Washington, D.C.: Island Press, 2004.

Fortun, Kim. "Biopolitics and the Informating of Environmentalism." In *Lively Capital: Biotechnologies, Ethics, and Governance in Global Markets*, edited by Kaushik Sunder Rajan, 306–27. Durham, N.C.: Duke University Press, 2012.

Foucault, Michel. *The Birth of Biopolitics: Lectures at the Collège de France, 1978–1979*. New York: Picador, 2010.

Foucault, Michel. *Society Must Be Defended: Lectures at the Collège de France, 1975–76*. New York: Picador, 2003.

Foucault, Michel, Michel Senellart, François Ewald, and Alessandro Fontana. *Security, Territory, Population: Lectures at the Collège de France, 1977–78*. New York: Palgrave Macmillan, République Française, 2007.

Fourie, Simona, Debra Jackson, and Helen Aveyard. "Living with Inflammatory Bowel Disease: A Review of Qualitative Research Studies." *International Journal of Nursing Studies* 87 (2018): 149–56.

Fraser, Caroline. *Rewilding the World: Dispatches from the Conservation Revolution*. New York: Metropolitan Books, 2009.

Frohlich, Dennis Owen. "The Social Construction of Inflammatory Bowel Disease Using Social Media Technologies." *Health Communication* 31, no. 11 (2016): 1412–20.

Fuhlendorf, S. D., D. M. Engle, J. Kerby, and R. Hamilton. "Pyric Herbivory: Rewilding Landscapes through the Recoupling of Fire and Grazing." *Conservation Biology* 23, no. 3 (2009): 588–98.

Fujimura, Kei E., Tine Demoor, Marcus Rauch, et al. "House Dust Exposure Mediates Gut Microbiome *Lactobacillus* Enrichment and Airway Immune Defense against Allergens and Virus Infection." *Proceedings of the National Academy of Sciences of the United States of America* 111, no. 2 (2014): 805–10.

Gabrys, Jennifer. *Program Earth: Environmental Sensing Technology and the Making of a Computational Planet*. Minneapolis: University of Minnesota Press, 2016.

Gabrys, Jennifer. "Programming Environments: Environmentality and Citizen Sensing in the Smart City." *Environment and Planning D: Society and Space* 32, no. 1 (2014): 30–48.

Gale, Edwin A. M. "Dying of Diabetes." *Lancet* 368, no. 9548 (2006): 1626–28.

Galetti, Mauro, Marcos Moleón, Pedro Jordano, et al. "Ecological and Evolutionary Legacy of Megafauna Extinctions." *Biological Reviews* 93, no. 2 (2018): 845–62.

Gammon, Andrea R. "The Many Meanings of Rewilding: An Introduction and the Case for a Broad Conceptualisation." *Environmental Values* 27, no. 4 (2018): 331–50.

Gan, Elaine, Anna Tsing, and Daniel Sullivan. "Using Natural History in the Study of Industrial Ruins." *BioOne* 38, no. 1 (2018): 39–54.

Garg, Sushil K., Ashley M. Croft, and Peter Bager. "Helminth Therapy (Worms) for Induction of Remission in Inflammatory Bowel Disease." *Cochrane Database of Systematic Reviews* (1) (2014): CD009400.

Garner, Joseph P., Brianna N. Gaskill, Elin M. Weber, Jamie Ahloy-Dallaire, and Kathleen R. Pritchett-Corning. "Introducing Therioepistemology: The Study of How Knowledge Is Gained from Animal Research." *Lab Animal* 46 (2017): 103.

Gautschi, A. *Der Reichsjägermeister: Fakten und Legenden um Hermann Göring*. Melsungen, Germany: Verlag J. Neumann-Neudamm, 2010.

Gavier-Widén, D., K. Ståhl, A. S. Neimanis, et al. "African Swine Fever in Wild Boar in Europe: A Notable Challenge." *Veterinary Record* 176, no. 8 (2015): 199–200.

Geissler, P. Wenzel. "'Worms Are Our Life,' Part 1: Understandings of Worms and the Body among the Luo of Western Kenya." *Anthropology and Medicine* 5, no. 1 (1998): 63–79.

Geissler, P. Wenzel. "'Worms Are Our Life,' Part 2: Luo Children's Thoughts about Worms and Illness." *Anthropology and Medicine* 5, no. 2 (1998): 133–44.

Genes, Luísa, Jens-Christian Svenning, Alexandra S. Pires, and Fernando A. S. Fernandez. "Why We Should Let Rewilding Be Wild and Biodiverse." *Biodiversity and Conservation* 28, no. 5 (2019): 1285–89.

Giacomin, Paul, John Croese, Lutz Krause, Alex Loukas, and Cinzia Cantacessi. "Suppression of Inflammation by Helminths: A Role for the Gut Microbiota?" *Philosophical Transactions of the Royal Society of London, Series B, Biological Sciences* 370 (2015): 20140296.

Gibson, Katherine, Deborah Bird Rose, and Ruth Fincher. *Manifesto for Living in the Anthropocene.* New York: Punctum Books, 2015.

Gibson-Graham, J. K., J. Cameron, and S. Healy. *Take Back the Economy: An Ethical Guide for Transforming Our Communities.* Minneapolis: University of Minnesota Press, 2013.

Gieryn, Thomas F. "City as Truth-Spot." *Social Studies of Science* 36, no. 1 (2006): 5–38.

Gifford, T. *Reconnecting with John Muir: Essays in Post-pastoral Practice.* Athens: University of Georgia Press, 2006.

Gilbert, J., and R. Knight. *Dirt Is Good: The Advantage of Germs for Your Child's Developing Immune System.* New York: St. Martin's Press, 2017.

Gilbert, Scott. *Ecological Developmental Biology.* Sunderland, Mass.: Sinauer, 2015.

Gilbert, Scott F., Thomas C. G. Bosch, and Cristina Ledon-Rettig. "Eco-Evo-Devo: Developmental Symbiosis and Developmental Plasticity as Evolutionary Agents." Review. *Nat Rev Genet* 16, no. 10 (2015): 611–22.

Gilbert, Scott, Jan Sapp, and Alfred Tauber. "A Symbiotic View of Life: We Have Never Been Individuals." *Quarterly Review of Biology* 87, no. 4 (2012): 325–41.

Ginn, Franklin, and Eduardo Ascensão. "Autonomy, Erasure, and Persistence in the Urban Gardening Commons." *Antipode* 50, no. 4 (2018): 929–52.

Giphart, R., and M. van Vugt. *Mismatch: How Our Stone Age Brain Deceives Us Every Day (And What We Can Do about It).* New York: Little, Brown, 2018.

Gluckman, P., and M. Hanson. *Mismatch: The Lifestyle Diseases Timebomb.* Oxford: Oxford University Press, 2008.

Goderie, R., W. Helmer, H. Kerkdijk-Otten, and S. Widstrand. *The Aurochs: Born to Be Wild.* Zutphen, Netherlands: Roodbont, 2013.

Goldman, M. "Constructing Connectivity: Conservation Corridors and Conservation Politics in East African Rangelands." *Annals of the Association of American Geographers* 99, no. 2 (2009): 335–59.

Graham, Dustin M. "A Walk on the Wild Side." *Lab Animal* 46 (2017): 423.

Granjou, Céline, and Catherine Phillips. "Living and Labouring Soils: Metagenomic Ecology and a New Agricultural Revolution?" *BioSocieties* 14 (2019): 393–415.

Grant, Mira. *Chimera*. London: Orbit, 2015.

Grant, Mira. *Parasite*. London: Orbit, 2013.

Grant, Mira. *Symbiont*. London: Orbit, 2014.

Grassberger, M., R. A. Sherman, O. S. Gileva, C. M. H. Kim, and K. Y. Mumcuoglu. *Biotherapy—History, Principles and Practice: A Practical Guide to the Diagnosis and Treatment of Disease Using Living Organisms*. Amsterdam: Springer, 2013.

Greenhough, Beth. "Assembling an Island Laboratory." *Area* 43, no. 2 (2011): 134–38.

Griffiths, Christine J., and Stephen Harris. "Prevention of Secondary Extinctions through Taxon Substitution." *Conservation Biology* 24, no. 3 (2010): 645–46.

Griffiths, Christine J., Carl G. Jones, Dennis M. Hansen, et al. "The Use of Extant Non-indigenous Tortoises as a Restoration Tool to Replace Extinct Ecosystem Engineers." *Restoration Ecology* 18, no. 1 (2010): 1–7.

Griffiths, C. J., N. Zuël, C. G. Jones, Z. Ahamud, and S. Harris. "Assessing the Potential to Restore Historic Grazing Ecosystems with Tortoise Ecological Replacements." *Conservation Biology* 27, no. 4 (2013): 690–700.

Grinevald, Jacques, and Giulia Rispoli. "Vladimir Vernadsky and the Co-evolution of the Biosphere, the Noosphere, and the Technosphere." *Technosphere Magazine*, June 20, 2018.

Gritzbach, E. *Hermann Göring, Werk und Mensch*. Munich: Eher Nachf, 1942.

Gross, Matthias. *Inventing Nature: Ecological Restoration by Public Experiments*. Lanham, Md.: Lexington Books, 2003.

Gross, Matthias. "The Public Proceduralization of Contingency: Bruno Latour and the Formation of Collective Experiments." *Social Epistemology* 24, no. 1 (2010): 63–74.

Gross, Matthias, and Holger Hoffmann-Riem. "Ecological Restoration as a Real-World Experiment: Designing Robust Implementation Strategies in an Urban Environment." *Public Understanding of Science* 14, no. 3 (2005): 269–84.

Grove, Kevin. "Agency, Affect, and the Immunological Politics of Disaster Resilience." *Environment and Planning D: Society and Space* 32, no. 2 (2014): 240–56.

Grove, Kevin. *Resilience*. London: Taylor & Francis, 2018.

Grove, Kevin. "Security beyond Resilience." *Environment and Planning D: Society and Space* 35, no. 1 (2017): 184–94.

Grove, Kevin, and David Chandler. "Introduction: Resilience and the Anthropocene: The Stakes of 'Renaturalising' Politics." *Resilience* 5, no. 2 (2017): 79–91.

Grove, Richard. *Green Imperialism: Colonial Expansion, Tropical Island Edens, and the Origins of Environmentalism, 1600–1860*. Cambridge: Cambridge University Press, 1995.

Guarner, Francisco, Raphaëlle Bourdet-Sicard, Per Brandtzaeg, et al. "Mechanisms of Disease: The Hygiene Hypothesis Revisited." *Nature Clinical Practice Gastroenterology and Hepatology* 3 (2006): 275.

Guattari, Félix. *The Three Ecologies*. London: Athlone Press, 2000.

Gunawardena, Kithsiri, Balachandran Kumarendran, Roshini Ebenezer, Muditha Sanjeewa Gunasingha, Arunasalam Pathmeswaran, and Nilanthi de Silva. "Soil-Transmitted Helminth Infections among Plantation Sector Schoolchildren in Sri Lanka: Prevalence after Ten Years of Preventive Chemotherapy." *PLoS Neglected Tropical Diseases* 5, no. 9 (2011): e1341.

Gunderson, Lance H., and C. S. Holling. *Panarchy: Understanding Transformations in Human and Natural Systems*. Washington, D.C.: Island Press, 2002.

Guthman, Julie, and Becky Mansfield. "The Implications of Environmental Epigenetics: A New Direction for Geographic Inquiry on Health, Space, and Nature–Society Relations." *Progress in Human Geography* 37, no. 4 (2013): 486–504.

Haff, Peter. "Humans and Technology in the Anthropocene: Six Rules." *Anthropocene Review* 1, no. 2 (2014): 126–36.

Hajek, A. E. *Natural Enemies: An Introduction to Biological Control*. Cambridge: Cambridge University Press, 2004.

Hamilton, Clive. *Defiant Earth: The Fate of Humans in the Anthropocene*. New York: Wiley, 2017.

Hansen, Dennis M., C. Josh Donlan, Christine J. Griffiths, and Karl J. Campbell. "Ecological History and Latent Conservation Potential: Large and Giant Tortoises as a Model for Taxon Substitutions." *Ecography* 33, no. 2 (2010): 272–84.

Hanski, Ilkka, Leena von Hertzen, Nanna Fyhrquist, et al. "Environmental Biodiversity, Human Microbiota, and Allergy Are Interrelated." *Proceedings of the National Academy of Sciences of the United States of America* 109, no. 21 (2012): 8334–39.

Haraway, Donna. *Modest Witness: Feminism and Technoscience*. New York: Routledge, 1997.

Haraway, Donna. *Primate Visions: Gender, Race, and Nature in the World of Modern Science*. New York: Routledge, 1989.

Haraway, Donna. *Simians, Cyborgs, and Women: The Reinvention of Nature*. New York: Routledge, 1991.

Haraway, Donna. *Staying with the Trouble: Making Kin in the Chthulucene*. Durham, N.C.: Duke University Press, 2016.

Haraway, Donna. *When Species Meet*. Minneapolis: University of Minnesota Press, 2008.

Haraway, Donna, Noboru Ishikawa, Scott F. Gilbert, Kenneth Olwig, Anna L. Tsing, and Nils Bubandt. "Anthropologists Are Talking—About the Anthropocene." Roundtable. *Ethnos* 81, no. 3 (2016): 535–64.

Harnett, Margaret M., and William Harnett. "Can Parasitic Worms Cure the Modern World's Ills?" *Trends in Parasitology* 33, no. 9 (2017): 694–705.

Harper, Kristin, and George Armelagos. "The Changing Disease-scape in the Third Epidemiological Transition." *International Journal of Environmental Research and Public Health* 7, no. 2 (2010): 675.

Harrington, A. *Reenchanted Science: Holism in German Culture from Wilhelm II to Hitler*. Princeton, N.J.: Princeton University Press, 1999.

Harris, Anna, Sally Wyatt, and Susan E. Kelly. "The Gift of Spit (And the Obligation to Return It)." *Information, Communication, and Society* 16, no. 2 (2013): 236–57.

Harris, L. M., and H. D. Hazen. "Power of Maps: (Counter)Mapping for Conservation." Review. *ACME* 4, no. 1 (2006): 99–130.

Hastings, Alan, James E. Byers, Jeffrey A. Crooks, et al. "Ecosystem Engineering in Space and Time." *Ecology Letters* 10, no. 2 (2007): 153–64.

Hawley, S. *Recovering a Lost River: Removing Dams, Rewilding Salmon, Revitalizing Communities*. Boston: Beacon Press, 2011.

Hayden, Cori. *When Nature Goes Public: The Making and Unmaking of Bioprospecting in Mexico*. Princeton, N.J.: Princeton University Press, 2003.

Head, Ian M., D. Martin Jones, and Wilfred F. M. Röling. "Marine Microorganisms Make a Meal of Oil." *Nature Reviews Microbiology* 4 (2006): 173.

Head, Lesley. *Hope and Grief in the Anthropocene*. London: Routledge, 2016.

Hearn, Robert, Charles Watkins, and Ross Balzaretti. "The Cultural and Land Use Implications of the Reappearance of the Wild Boar in North West Italy: A Case Study of the Val Di Vara." *Journal of Rural Studies* 36 (2014): 52–63.

Heck, Heinz. "The Breeding-Back of the Aurochs." *Oryx* 1, no. 3 (1951): 117–22.

Heck, Heinz. "Ueber die Rueckzuechtung des Urs." *Das Tier und Wir* 7, no 10 (1936): 8–13

Heck, Lutz. *Animals, My Adventures*. London: Methuen, 1954.

Heck, Lutz. "Behördliche Landschaftsgestaltung im Osten." *Neues Bauerntum mit Landbaumeister* 34, no. 6 (1942): 213–15.

Heck, Lutz. "Die Neuzüchtung des Auerochsen." *Wild und Hund* 37 (1939): 537–39.

Heck, Lutz. "Über die Neuzüchtung des Ur oder Auerochs." *Berichte der Internationalen Gesellschaft zur Erhaltung des Wisents* 3 (1936): 225–94.

Helm, Dieter. *Green and Prosperous Land: A Blueprint for Rescuing the British Countryside.* New York: HarperCollins, 2019.

Helmreich, Stefan. *Alien Ocean: Anthropological Voyages in Microbial Seas.* Berkeley: University of California Press, 2009.

Helmreich, Stefan. *Sounding the Limits of Life: Essays in the Anthropology of Biology and Beyond.* Princeton, N.J.: Princeton University Press, 2015.

Hennessy, Elizabeth. "The Molecular Turn in Conservation: Genetics, Pristine Nature, and the Rediscovery of an Extinct Species of Galápagos Giant Tortoise." *Annals of the Association of American Geographers* 105, no. 1 (2014): 87–104.

Hennessy, Elizabeth. "Producing 'Prehistoric' Life: Conservation Breeding and the Remaking of Wildlife Genealogies." *Geoforum* 49 (2013): 71–80.

Henrich, Joseph, Steven J. Heine, and Ara Norenzayan. "The Weirdest People in the World?" *Behavioral and Brain Sciences* 33, no. 2–3 (2010): 61–83.

Herf, Jeffrey. *Reactionary Modernism: Technology, Culture, and Politics in Weimar and the Third Reich.* Cambridge: Cambridge University Press, 1984.

Heynen, Nik. *Neoliberal Environments: False Promises and Unnatural Consequences.* New York: Routledge, 2007.

Hinchliffe, Steve. "Cities and Natures: Intimate Strangers." In *Unsettling Cities*, edited by J. E. Allen, 137–80. London: Open University Press, 1999.

Hinchliffe, Steve. "Reconstituting Nature Conservation: Towards a Careful Political Ecology." *Geoforum* 39, no. 1 (2008): 88–97.

Hinchliffe, Steve, John Allen, Stephanie Lavau, Nick Bingham, and Simon Carter. "Biosecurity and the Topologies of Infected Life: From Borderlines to Borderlands." *Transactions of the Institute of British Geographers* 38, no. 4 (2013): 531–43.

Hinchliffe, Steve, Nick Bingham, John Allen, and Simon Carter. *Pathological Lives: Disease, Space and Biopolitics.* London: Blackwell, 2016.

Hinchliffe, Steve, Andrea Butcher, and Muhammad Meezanur Rahman. "The AMR Problem: Demanding Economies, Biological Margins, and Co-producing Alternative Strategies." *Palgrave Communications* 4, no. 1 (2018): 142.

Hinchliffe, Steve, Matt Kearnes, Monica Degen, and Sarah Whatmore. "Urban Wild Things: A Cosmopolitical Experiment." *Environment and Planning D: Society and Space* 23, no. 5 (2005): 643–58.

Hinchliffe, Steve, and Stephanie Lavau. "Differentiated Circuits: The Ecologies of Knowing and Securing Life." *Environment and Planning D: Society and Space* 31, no. 2 (2013): 259–74.

Hinchliffe, Steve, and Kim J. Ward. "Geographies of Folded Life: How Immunity Reframes Biosecurity." *Geoforum* 53 (2014): 136–44.

Hintz, J. "Some Political Problems for Rewilding Nature." *Ethics, Place, and Environment* 10 (2007): 177–216.

Hird, Myra J. "Indifferent Globality: Gaia, Symbiosis and 'Other Worldliness.'" *Theory, Culture, and Society* 27, no. 2–3 (2010): 54–72.

Hird, Myra. *The Origins of Sociable Life: Evolution after Science Studies*. London: Palgrave Macmillan, 2009.

Hobbs, R. J., E. S. Higgs, and C. Hall. *Novel Ecosystems: Intervening in the New Ecological World Order*. New York: Wiley, 2013.

Hodder, K. H., and J. M. Bullock. "Really Wild? Naturalistic Grazing in Modern Landscapes." *British Wildlife* 37 (2009): 37–43.

Hodgetts, Timothy. "Wildlife Conservation, Multiple Biopolitics and Animal Subjectification: Three Mammals' Tales." *Geoforum* 79 (2017): 17–25.

Hodgetts, Timothy, and Jamie Lorimer. "Animals' Mobilities." *Progress in Human Geography* 44, no. 1 (2019): 4–26.

Hodgson, Jarrod C., Shane M. Baylis, Rowan Mott, Ashley Herrod, and Rohan H. Clarke. "Precision Wildlife Monitoring Using Unmanned Aerial Vehicles." *Scientific Reports* 6 (2016): 22574.

Hogarth, Stuart, and Paula Saukko. "A Market in the Making: The Past, Present and Future of Direct-to-Consumer Genomics." *New Genetics and Society* 36, no. 3 (2017): 197–208.

Holling, C. S., and Gary K. Meffe. "Command and Control and the Pathology of Natural Resource Management." *Conservation Biology* 10, no. 2 (1996): 328–37.

Holloway, L., C. Morris, B. Gilna, and D. Gibbs. "Biopower, Genetics and Livestock Breeding: (Re)Constituting Animal Populations and Heterogeneous Biosocial Collectivities." *Transactions of the Institute of British Geographers* 34, no. 3 (2009): 394–407.

Hood, G. *The Beaver Manifesto*. Victoria, Canada: RMB, 2011.

Hooks, Katarzyna B., and Maureen A. O'Malley. "Dysbiosis and Its Discontents." *mBio* 8, no. 5 (2017).

Hootsman, M., and H. Kampf. *Ecological Networks: Experiences in the Netherlands*. The Hague: Ministry of Agriculture, Nature and Food Quality of the Netherlands, 2005.

Hörl, Erich, and James Burton, eds. *General Ecology: The New Ecological Paradigm*. London: Bloomsbury, 2018.

Hotez, Peter. *Forgotten People, Forgotten Diseases: The Neglected Tropical Diseases and Their Impact on Global Health and Development.* Washington, D.C.: ASM Press, 2013.

Hotez, P. "Neglected Diseases amid Wealth in the United States and Europe." *Health Affairs (Milwood)* 28, no. 6 (2009): 1720–25.

Hotez, Peter, David Diemert, Kristina Bacon, et al. "The Human Hookworm Vaccine." *Vaccine* 31, supplement 2 (2013): B227–32.

Hotez, P. J., A. Fenwick, L. Savioli, and D. H. Molyneux. "Rescuing the Bottom Billion through Control of Neglected Tropical Diseases." *Lancet* 373, no. 9674 (2009): 1570–75.

Hotez, P. J., and J. R. Herricks. "Helminth Elimination in the Pursuit of Sustainable Development Goals: A 'Worm Index' for Human Development." Editorial. *PLoS Neglected Tropical Diseases* 9, no. 4 (2015): e0003618.

Hribal, Jason. "'Animals Are Part of the Working Class': A Challenge to Labor History." *Labor History* 44, no. 4 (2003): 435–53.

Hribal, J., and J. S. Clair. *Fear of the Animal Planet: The Hidden History of Animal Resistance.* Chico, Calif.: AK Press, 2010.

Ingold, Tim. "The Architect and the Bee: Reflections on the Work of Animals and Men." *Man* 18, no. 1 (1983): 1–20.

Ingold, Tim. *Hunters, Pastoralists and Ranchers: Reindeer Economies and Their Transformations.* Cambridge: Cambridge University Press, 1988.

Ingold, Tim. *Making: Anthropology, Archaeology, Art and Architecture.* London: Taylor & Francis, 2013.

Ingold, T., and G. Palsson. *Biosocial Becomings: Integrating Social and Biological Anthropology.* Cambridge: Cambridge University Press, 2013.

Ingram, Mrill. "Fermentation, Rot, and Other Human–Microbial Performances." In *Knowing Nature Conversations at the Intersection of Political Ecology and Science Studies*, edited by Mara Goldman and Matthew D. Turner, 99–112. Chicago: University of Chicago Press, 2011.

International Committee on the Management of Large Herbivores in the Oostvaardersplassen (ICMO). *Natural Processes, Animal Welfare, Moral Aspects and Management of the Oostvaardersplassen. Report of the Second International Commission on Management of the Oostvaardersplassen.* The Hague: Wageningen, 2010.

International Committee on the Management of Large Herbivores in the Oostvaardersplassen (ICMO). *Reconciling Nature and Human Interests: Advice of the International Committee on the Management of Large Herbivores in the Oostvaardersplassen.* The Hague: Wageningen, 2006.

Jackson, M. *Asthma: The Biography.* Oxford: Oxford University Press, 2009.

Jacob, Mills, Philip Weinstein, Nicholas J. C. Gellie, Laura S. Weyrich, Andrew J. Lowe, and Martin F. Breed. "Urban Habitat Restoration Provides a Human Health Benefit through Microbiome Rewilding: The Microbiome Rewilding Hypothesis." *Restoration Ecology* 25, no. 6 (2017): 866–72.

Jahiel, René I., and Thomas F. Babor. "Industrial Epidemics, Public Health Advocacy and the Alcohol Industry: Lessons from Other Fields." *Addiction* 102, no. 9 (2007): 1335–39.

Janzen, Daniel H., and Paul S. Martin. "Neotropical Anachronisms: The Fruits the Gomphotheres Ate." *Science* 215, no. 4528 (1982): 19–27.

Jasarevic, Larisa. "The Thing in a Jar: Mushrooms and Ontological Speculations in Post-Yugoslavia." *Cultural Anthropology* 30, no. 1 (2015): 36–64.

Jepson, Paul. "Recoverable Earth: A Twenty-First Century Environmental Narrative." *Ambio* 48, no. 2 (2019): 123–30.

Jepson, Paul. "A Rewilding Agenda for Europe: Creating a Network of Experimental Reserves." *Ecography* 39, no. 2 (2016): 117–24.

Jepson, P., F. Schepers, and W. Helmer. "Governing with Nature: A European Perspective on Putting Rewilding Principles into Practice." *Philosophical Transactions of the Royal Society B: Biological Sciences* 373, no. 1761 (2018): 20170434.

Jha, D. N. *The Myth of the Holy Cow*. New York: Verso, 2004.

Johnson, Elizabeth R. "At the Limits of Species Being: Sensing the Anthropocene." *South Atlantic Quarterly* 116, no. 2 (2017): 275–92.

Johnson, Elizabeth R., and Jesse Goldstein. "Biomimetic Futures: Life, Death, and the Enclosure of a More-than-Human Intellect." *Annals of the Association of American Geographers* 105, no. 2 (2015): 387–96.

Jones, Clive G., John H. Lawton, and Moshe Shachak. "Organisms as Ecosystem Engineers." *Oikos* 69, no. 3 (1994): 373–86.

Jordán, Ferenc. "Keystone Species and Food Webs." *Philosophical Transactions of the Royal Society B: Biological Sciences* 364, no. 1524 (2009): 1733–41.

Jørgensen, Dolly. "Rethinking Rewilding." *Geoforum* 65 (2015): 482–88.

Kalikow, T. J. "Konrad Lorenz's Ethological Theory: Explanation and Ideology, 1938–1943." *Journal of the History of Biology* 16, no. 1 (1983): 39–73.

Kater, Michael H. *Das "Ahnenerbe" der SS, 1935–1945: Ein Beitrag zur Kulturpolitik des Dritten Reiches*. Munich: R. Oldenbourg, 2006.

Katz, S. E. *The Art of Fermentation: An In-depth Exploration of Essential Concepts and Processes from around the World*. White River Junction, Vt.: Chelsea Green, 2012.

Katz, S. E., and S. F. Morell. *Wild Fermentation: The Flavor, Nutrition, and Craft of Live-Culture Foods*. 2nd ed. White River Junction, Vt.: Chelsea Green, 2016.

Kay, Kelly. "Europeanization through Biodiversity Conservation: Croatia's Bid for E.U. Accession and the Natura 2000 Designation Process." *Geoforum* 54 (2014): 80–90.

Keck, F. "Liberating Sick Birds: Poststructuralist Perspectives on the Biopolitics of Avian Influenza." *Cultural Anthropology* 30, no. 2 (2015): 224–35.

Kelly, Ann, and Javier Lezaun. "Urban Mosquitoes, Situational Publics, and the Pursuit of Interspecies Separation in Dar es Salaam." *American Ethnologist* 41, no. 2 (2014): 368–83.

Keulartz, Jozef. "The Emergence of Enlightened Anthropocentrism in Ecological Restoration." *Nature and Culture* 7, no. 1 (2012): 48–71.

Keulartz, Jozef. "Rewilding." In *Oxford Research Encyclopedia of Environmental Science*, edited by Hank Shugart. Oxford: Oxford University Press, 2018.

Keulartz, Jozef, and Henk van den Belt. "DIY-Bio—Economic, Epistemological and Ethical Implications and Ambivalences." *Life Sciences, Society and Policy* 12, no. 1 (2016): 7.

Khan, Sehroon, Sadia Nadir, Zia Ullah Shah, et al. "Biodegradation of Polyester Polyurethane by *Aspergillus tubingensis*." *Environmental Pollution* 225 (2017): 469–80.

Kintisch, Eli. "Born to Rewild." *Science* 350, no. 6265 (2015): 1148.

Kirksey, Eben. *Emergent Ecologies*. Durham, N.C.: Duke University Press, 2015.

Kirksey, Eben. *The Multispecies Salon*. Durham, N.C.: Duke University Press, 2014.

Kirksey, Eben, and Stefan Helmreich. "The Emergence of Multispecies Ethnography." *Cultural Anthropology* 25, no. 4 (2010): 545–76.

Klaver, I., J. Keulartz, H. van den Belt, and B. Gremmen. "Born to Be Wild: A Pluralistic Ethics Concerning Introduced Large Herbivores in the Netherlands." *Environmental Ethics* 24, no. 1 (2002): 3–21.

Klinke, Ian. "Vitalist Temptations: Life, Earth and the Nature of War." *Political Geography* 72 (2019): 1–9.

Kohler, Robert E. *Landscapes and Labscapes: Exploring the Lab–Field Border in Biology*. Chicago: University of Chicago Press, 2002.

Korthals, M., J. Keulartz, H. van den Belt, I. Klaver, and B. Bremmen. "Battles of Nature: The Ethical Side of Grazing with Large Herbivores." *Vakblad Natuurbeheer* 41 (2002): 43–45.

Krohn, Wolfgang, and Johannes Weyer. "Society as a Laboratory: The Social Risks of Experimental Research." *Science and Public Policy* 21, no. 3 (1994): 173–83.

Krzywoszynska, Anna. "Caring for Soil Life in the Anthropocene: The Role of Attentiveness in More-than-Human Ethics." *Transactions of the Institute of British Geographers* 44, no. 4 (2019): 661–75.

Lacoue-Labarthe, Philippe, Jean-Luc Nancy, and Brian Holmes. "The Nazi Myth." *Critical Inquiry* 16, no. 2 (1990): 291–312.

Lake, Philip S. "Resistance, Resilience and Restoration." *Ecological Management and Restoration* 14, no. 1 (2013): 20–24.

Laland, Kevin, Blake Matthews, and Marcus W. Feldman. "An Introduction to Niche Construction Theory." *Evolutionary Ecology* 30, no. 2 (2016): 191–202.

Landecker, Hannah. "Food as Exposure: Nutritional Epigenetics and the New Metabolism." *BioSocieties* 6, no. 2 (2011): 167–94.

Lasanta, T., J. Arnáez, N. Pascual, P. Ruiz-Flaño, M. P. Errea, and N. Lana-Renault. "Space-Time Process and Drivers of Land Abandonment in Europe." *Catena* 149 (2017): 810–23.

Last, Angela. "Experimental Geographies." *Geography Compass* 6, no. 12 (2012): 706–24.

Latour, Bruno. "An Attempt at a 'Compositionist Manifesto.'" *New Literary History* 41, no. 3 (2010): 471–90.

Latour, Bruno. "Bruno Latour Tracks Down Gaia." *L.A. Review of Books*, July 3, 2018.

Latour, Bruno. *Down to Earth: Politics in the New Climatic Regime*. New York: Wiley, 2018.

Latour, Bruno. *Facing Gaia: Eight Lectures on the New Climatic Regime*. New York: Wiley, 2017.

Latour, Bruno. "Fifty Shades of Green." *Environmental Humanities* 7, no. 1 (2016): 219–25.

Latour, Bruno. "'It's Development, Stupid!,' or How to Modernize Modernization." n.d. http://www.bruno-latour.fr/sites/default/files/107-NORDHAU S%26SHELLENBERGER.pdf.

Latour, Bruno. "Love Your Monsters." In *Love Your Monsters: Postenvironmentalism and the Anthropocene*, edited by Michael Shellenberger and Ted Nordhaus. Oakland, Calif.: Breakthrough Institute, 2011.

Latour, Bruno. *Pandora's Hope: Essays on the Reality of Science Studies*. Cambridge, Mass.: Harvard University Press, 1999.

Latour, Bruno. *Politics of Nature: How to Bring the Sciences into Democracy*. Translated by Catherine Porter. Cambridge, Mass.: Harvard University Press, 2004.

Latour, Bruno. "Why Gaia Is Not a God of Totality." *Theory, Culture, and Society* 34, no. 2–3 (2017): 61–81.

Latour, Bruno. "Why Has Critique Run Out of Steam? From Matters of Fact to Matters of Concern." *Critical Inquiry* 30, no. 2 (2004): 225–48.

Laundré, J. W., L. Hernández, and K. B. Altendorf. "Wolves, Elk, and Bison: Reestablishing the 'Landscape of Fear' in Yellowstone National Park, USA." *Canadian Journal of Zoology* 79, no. 8 (2001): 1401–9.

Law, Alan, Martin J. Gaywood, Kevin C. Jones, Paul Ramsay, and Nigel J. Willby. "Using Ecosystem Engineers as Tools in Habitat Restoration and Rewilding: Beaver and Wetlands." *Science of the Total Environment* 605–606 (2017): 1021–30.

Lawton, John. *Making Space for Nature: A Review of England's Wildlife Sites and Ecological Network*. London: DEFRA, 2010.

Leach, H. M. "Human Domestication Reconsidered." *Current Anthropology* 44, no. 3 (2003): 349–68.

Leach, Jeff. "(Re)Becoming Human." Human Food Project, September 30, 2014. http://humanfoodproject.com/.

Leach, Jeff. *Rewild*. CreateSpace Independent Publishing Platform, 2015.

Leiper, Chelsea. "'Re-wilding' the Body in the Anthropocene and Our Ecological Lives' Work." Society and Space, November 14, 2017. https://www.societyandspace.org/.

Lenton, Timothy M., and Bruno Latour. "Gaia 2.0." *Science* 361, no. 6407 (2018): 1066–68.

Lerner, Aaron, Patricia Jeremias, and Torsten Matthias. "The World Incidence and Prevalence of Autoimmune Diseases Is Increasing." *International Journal of Celiac Disease* 3, no. 4 (2015): 151–55.

Leung, Jaqueline M., Sarah A. Budischak, Hao Chung The, et al. "Rapid Environmental Effects on Gut Nematode Susceptibility in Rewilded Mice." *PLoS Biology* 16, no. 3 (2018): e2004108.

Levi, Taal, A. Marm Kilpatrick, Marc Mangel, and Christopher C. Wilmers. "Deer, Predators, and the Emergence of Lyme Disease." *Proceedings of the National Academy of Sciences of the United States of America* 109, no. 27 (2012): 10942–47.

Lieberman, D. *The Story of the Human Body: Evolution, Health and Disease*. London: Penguin, 2013.

Liu, J., R. A. Morey, J. K. Wilson, and W. Parker. "Practices and Outcomes of Self-Treatment with Helminths Based on Physicians' Observations." *Journal of Helminthology* 91, no. 3 (2017): 267–77.

Livingstone, D. N. "The Spaces of Knowledge—Contributions towards a Historical Geography of Science." *Environment and Planning D: Society and Space* 13, no. 1 (1995): 5–34.

Lloyd-Price, Jason, Galeb Abu-Ali, and Curtis Huttenhower. "The Healthy Human Microbiome." *Genome Medicine* 8 (2016): 51.

Lloyd-Price, Jason, Anup Mahurkar, Gholamali Rahnavard, et al. "Strains, Functions and Dynamics in the Expanded Human Microbiome Project." *Nature* 550, no. 7674 (2017): 61–66.

Lock, Margaret. "Comprehending the Body in the Era of the Epigenome." *Current Anthropology* 56, no. 2 (2015): 151–77.

Logan, Alan C., Felice N. Jacka, and Susan L. Prescott. "Immune–Microbiota Interactions: Dysbiosis as a Global Health Issue." *Current Allergy and Asthma Reports* 16, no. 2 (2016): 13.

Loon, Joost van. "Epidemic Space." *Critical Public Health* 15, no. 1 (2005): 39–52.

Lorimer, Hayden. "Scaring Crows." *Geographical Review* 103, no. 2 (2013): 177–89.

Lorimer, Jamie. "The Anthropo-scene: A Guide for the Perplexed." *Social Studies of Science* 47, no. 1 (2016): 117–42.

Lorimer, Jamie. "Counting Corncrakes: The Affective Science of the U.K. Corncrake Census." *Social Studies of Science* 38, no. 3 (2008): 377–405.

Lorimer, Jamie. "Decoupling without Disconnection: Conservation and Democracy in an Urbanized World." Breakthrough Institute, August 17, 2017. https://thebreakthrough.org/.

Lorimer, Jamie. "Gut Buddies: Multispecies Studies and the Microbiome." *Environmental Humanities* 8, no. 1 (2016): 57–76.

Lorimer, Jamie. "Nonhuman Charisma." *Environment and Planning D: Society and Space* 25, no. 5 (2007): 911–32.

Lorimer, Jamie. "Rot." *Environmental Humanities* 8, no. 2 (2016): 235–39.

Lorimer, Jamie. *Wildlife in the Anthropocene: Conservation after Nature*. Minneapolis: University of Minnesota Press, 2015.

Lorimer, Jamie, and Clemens Driessen. "Bovine Biopolitics and the Promise of Monsters in the Rewilding of Heck Cattle." *Geoforum* 48 (2013): 249–59.

Lorimer, Jamie, and Clemens Driessen. "Wild Experiments at the Oostvaardersplassen: Rethinking Environmentalism in the Anthropocene." *Transactions of the Institute of British Geographers* 39, no. 2 (2014): 169–81.

Lorimer, Jamie, Timothy Hodgetts, and Maan Barua. "Animals' Atmospheres." *Progress in Human Geography* 43, no. 1 (2019): 26–45.

Lorimer, Jamie, Chris Sandom, Paul Jepson, Chris Doughty, Maan Barua, and Keith J. Kirby. "Rewilding: Science, Practice, and Politics." *Annual Review of Environment and Resources* 40, no. 1 (2015): 39–62.

Loukas, Alex, Peter J. Hotez, David Diemert, et al. "Hookworm Infection." *Nature Reviews Disease Primers* 2 (2016): 1–18.

Louv, R. *Last Child in the Woods: Saving Our Children from Nature-Deficit Disorder*. Chapel Hill, N.C.: Algonquin Books, 2008.

Louys, Julien, Richard T. Corlett, Gilbert J. Price, Stuart Hawkins, and Philip J. Piper. "Rewilding the Tropics, and Other Conservation Translocations Strategies in the Tropical Asia-Pacific Region." *Ecology and Evolution* 4, no. 22 (2014): 4380–98.

Lovbrand, E., J. Stripple, and B. Wiman. "Earth System Governmentality Reflections on Science in the Anthropocene." *Global Environmental Change-Human and Policy Dimensions* 19, no. 1 (2009): 7–13.

Lovelock, James. *Gaia: A New Look at Life on Earth*. Oxford: Oxford University Press, 1979.

Lovelock, James E., and Lynn Margulis. "Atmospheric Homeostasis by and for the Biosphere: The Gaia Hypothesis." *Tellus* 26, no. 1–2 (1974): 2–10.

Low, T. *Feral Future*. London: Penguin, 2001.

Lowe, Celia. "Viral Clouds: Becoming H5N1 in Indonesia." *Cultural Anthropology* 25, no. 4 (2010): 625–49.

Luke, Timothy W. "Eco-managerialism: Environmental Studies as a Power/Knowledge Formation." In *Living with Nature: Environmental Politics as Cultural Discourse*, edited by Frank Fischer and Maarten Hajer, 103–20. Oxford: Oxford University Press, 1999.

Lukeš, Julius, Roman Kuchta, Tomáš Scholz, and Kateřina Pomajbíková. "(Self-)Infections with Parasites: Re-interpretations for the Present." *Trends in Parasitology* 30, no. 8 (2014): 377–85.

Lulka, David. "Form and Formlessness: The Spatiocorporeal Politics of the American Kennel Club." *Environment and Planning D: Society and Space* 27, no. 3 (2009): 531–53.

Lulka, David. "Stabilizing the Herd: Fixing the Identity of Nonhumans." *Environment and Planning D: Society and Space* 22, no. 3 (2004): 439–63.

Lynas, Mark. *The God Species: How the Planet Can Survive the Age of Humans*. London: Fourth Estate, 2011.

Ma, Yonghui, Jiayu Liu, Catherine Rhodes, Yongzhan Nie, and Faming Zhang. "Ethical Issues in Fecal Microbiota Transplantation in Practice." *American Journal of Bioethics* 17, no. 5 (2017): 34–45.

Mackenzie, D. *An Engine, Not a Camera: How Financial Models Shape Markets*. Cambridge, Mass.: MIT Press, 2008.

MacKinder, Halford. "The Geographical Pivot of History." *Geographical Journal* 23 (1904): 421–37.

Maizels, R. M., H. J. McSorley, and D. J. Smyth. "Helminths in the Hygiene Hypothesis: Sooner or Later?" *Clinical and Experimental Immunology* 177, no. 1 (2014): 38–46.

Maizels, Rick M., and Daniel H. Nussey. "Into the Wild: Digging at Immunology's Evolutionary Roots." *Nature Immunology* 14 (2013): 879.

Maizels, Rick M., and Ursula Wiedermann. "Immunoregulation by Microbes and Parasites in the Control of Allergy and Autoimmunity." In *The Hygiene Hypothesis and Darwinian Medicine*, edited by Graham Rook, 45–75. Basel: Birkhäuser, 2009.

Malhi, Yadvinder, Christopher E. Doughty, Mauro Galetti, Felisa A. Smith, Jens-Christian Svenning, and John W. Terborgh. "Megafauna and Ecosystem Function from the Pleistocene to the Anthropocene." *Proceedings of the National Academy of Sciences of the United States of America* 113, no. 4 (2016): 838–46.

Mansfield, Becky, and Julie Guthman. "Epigenetic Life: Biological Plasticity, Abnormality, and New Configurations of Race and Reproduction." *Cultural Geographies* 22, no. 1 (2015): 3–20.

Marchesi, Julian R., David H. Adams, Francesca Fava, et al. "The Gut Microbiota and Host Health: A New Clinical Frontier." *Gut* 65, no. 2 (2016): 330–39.

Margulis, Lynn, and Dorion Sagan. *Acquiring Genomes: A Theory of the Origins of Species*. New York: Basic Books, 2002.

Marris, E. "Conservation Biology: Reflecting the Past." *Nature* 462, no. 7269 (2009): 30–32.

Marris, Emma. *Rambunctious Garden: Saving Nature in a Post-wild World*. New York: Bloomsbury, 2011.

Martin, Emily. *Flexible Bodies: Tracking Immunity in American Culture from the Days of Polio to the Age of AIDS*. Boston: Beacon Press, 1994.

Martin, Paul S. *Twilight of the Mammoths: Ice Age Extinctions and the Rewilding of America*. Berkeley: University of California Press, 2005.

Marx, Karl. *Capital: Volume 1*. 1867. London: Penguin, 1976.

Marx, Karl. *Economic and Philosophic Manuscripts of 1844*. 1844. New York: Dover, 2012.

Marx, Leo. *The Machine in the Garden: Technology and the Pastoral Ideal in America*. 1964. Oxford: Oxford University Press, 2000.

Massumi, Brian. "National Enterprise Emergency: Steps toward an Ecology of Powers." *Theory, Culture, and Society* 26, no. 6 (2009): 153–85.

Massumi, Brian. *Ontopower: War, Powers, and the State of Perception*. Durham, N.C.: Duke University Press, 2015.

Matless, David. *Landscape and Englishness*. London: Reaktion, 1998.

Maxmen, Amy. "Living Therapeutics: Scientists Genetically Modify Bacteria to Deliver Drugs." *Nature Medicine* 23, no. 1 (2017): 5–7.

McFall-Ngai, Margaret. "Noticing Microbial Worlds: The Post Modern Synthesis in Biology." In *Arts of Living on a Damaged Planet: Ghosts and Monsters of the Anthropocene*, edited by Anna Lowenhaupt Tsing, Heather Anne Swanson, Elaine Gan, and Nils Bubandt, M51–70 Minneapolis: University of Minnesota Press, 2017.

McFall-Ngai, Margaret, Michael G. Hadfield, Thomas C. G. Bosch, et al. "Animals in a Bacterial World, a New Imperative for the Life Sciences."

*Proceedings of the National Academy of Sciences of the United States of America* 110, no. 9 (2013): 3229–36.

McKenna, Megan L., Shannon McAtee, Patricia E. Bryan, et al. "Human Intestinal Parasite Burden and Poor Sanitation in Rural Alabama." *American Journal of Tropical Medicine and Hygiene* 97, no. 5 (2017): 1623–28.

McLachlan, Jason, Jessica Hellmann, and Mark Schwartz. "A Framework for Debate of Assisted Migration in an Era of Climate Change." *Conservation Biology* 21, no. 2 (2007): 297–302.

McMullen, Matthew. "Rewilding Scotlands Highlands: Life, Death and the Labour of Nonhumans." PhD diss., University of Manchester, 2018.

Meloni, Maurizio. "A Postgenomic Body: Histories, Genealogy, Politics." *Body and Society* 24, no. 3 (2018): 3–38.

Merckx, Thomas, and Henrique M. Pereira. "Reshaping Agri-environmental Subsidies: From Marginal Farming to Large-Scale Rewilding." *Basic and Applied Ecology* 16, no. 2 (2015): 95–103.

Miller, Carol. "The Contribution of Natural Fire Management to Wilderness Fire Science." *International Journal of Wilderness* 20, no. 2 (2014): 20–25.

Mitchell, F. J. G. "How Open Were European Primeval Forests? Hypothesis Testing Using Palaeoecological Data." *Journal of Ecology* 93, no. 1 (2005): 168–77.

Mitchell, Piers D. "The Origins of Human Parasites: Exploring the Evidence for Endoparasitism throughout Human Evolution." *International Journal of Paleopathology* 3, no. 3 (2013): 191–98.

Mitman, Gregg. *Breathing Space: How Allergies Shape Our Lives and Landscapes.* New Haven, Conn.: Yale University Press, 2008.

Mitman, Gregg. "Hay Fever Holiday: Health, Leisure, and Place in Gilded-Age America." *Bulletin of the History of Medicine* 77, no. 3 (2003): 600–635.

Monbiot, George. *Feral: Searching for Enchantment on the Frontiers of Rewilding.* London: Penguin, 2013.

Moore, Jason W. *Capitalism in the Web of Life: Ecology and the Accumulation of Capital.* New York: Verso, 2015.

Moos, Christine, Peter Bebi, Massimiliano Schwarz, Markus Stoffel, Karen Sudmeier-Rieux, and Luuk Dorren. "Ecosystem-Based Disaster Risk Reduction in Mountains." *Earth-Science Reviews* 177 (2018): 497–513.

Moran-Thomas, Amy. "A Salvage Ethnography of the Guinea Worm: Witchcraft, Oracles and Magic in a Disease Eradication Program." In *When People Come First: Critical Studies in Global Health*, edited by João Biehl and Adriana Petryna, 207–40. Princeton, N.J.: Princeton University Press, 2013.

Morton, Oliver. *The Planet Remade: How Geoengineering Could Change the World.* Princeton, N.J.: Princeton University Press, 2015.

Muniesa, Fabian, and Michel Callon. "Economic Experiments and the Construction of Markets." In *Do Economists Make Markets? On the Performativity of Economics*, edited by Donald MacKenzie, Fabian Muniesa, and Lucia Siu, 163–89. Princeton, N.J.: Princeton University Press, 2007.

Münster, Daniel. "Performing Alternative Agriculture: Critique and Recuperation in Zero Budget Natural Farming, South India." *Journal of Political Ecology* 25, no. 1 (2018): 748–64.

Myelnikov, Dmitriy. "An Alternative Cure: The Adoption and Survival of Bacteriophage Therapy in the USSR, 1922–1955." *Journal of the History of Medicine and Allied Sciences* 73, no. 4 (2018): 385–411.

Nagy, K., P. D. Johnson, and R. Malamud. *Trash Animals: How We Live with Nature's Filthy, Feral, Invasive, and Unwanted Species*. Minneapolis: University of Minnesota Press, 2013.

Nash, R. *Wilderness and the American Mind*. New Haven, Conn.: Yale University Press, 2001.

Navarro, Severine, Ivana Ferreira, and Alex Loukas. "The Hookworm Pharmacopoeia for Inflammatory Diseases." *International Journal for Parasitology* 43, no. 3–4 (2013): 225–31.

Neel, Brian A., and Robert M. Sargis. "The Paradox of Progress: Environmental Disruption of Metabolism and the Diabetes Epidemic." *Diabetes* 60, no. 7 (2011): 1838–48.

Neimark, B. D., and B. Wilson. "Re-mining the Collections: From Bioprospecting to Biodiversity Offsetting in Madagascar." *Geoforum* 66 (2015): 1–10.

Nelson, Sara Holiday. "Resilience and the Neoliberal Counter-revolution: From Ecologies of Control to Production of the Common." *Resilience* 2, no. 1 (2014): 1–17.

Nelson, Sara, and Bruce Braun. "Autonomia in the Anthropocene: New Challenges to Radical Politics." *South Atlantic Quarterly* 116, no. 2 (2017): 223–35.

Ng, Siew C., Hai Yun Shi, Nima Hamidi, et al. "Worldwide Incidence and Prevalence of Inflammatory Bowel Disease in the 21st Century: A Systematic Review of Population-Based Studies." *Lancet* 390, no. 10114 (2017): 2769–78.

Ngo, S. T., F. J. Steyn, and P. A. McCombe. "Gender Differences in Autoimmune Disease." *Frontiers in Neuroendocrinology* 35, no. 3 (2014): 347–69.

Niazi, Asfandyar Khan, and Sanjay Kalra. "Diabetes and Tuberculosis: A Review of the Role of Optimal Glycemic Control." *Journal of Diabetes and Metabolic Disorders* 11 (2012): 288.

Nibert, D. *Animal Oppression and Human Violence: Domesecration, Capitalism, and Global Conflict*. New York: Columbia University Press, 2013.

Nilsen, Erlend B., E. J. Milner-Gulland, Lee Schofield, Atle Mysterud, Nils Chr. Stenseth, and Tim Coulson. "Wolf Reintroduction to Scotland: Public Attitudes and Consequences for Red Deer Management." *Proceedings of the Royal Society B: Biological Sciences* 274, no. 1612 (2007): 995–1003.

Noske, B. *Humans and Other Animals: Beyond the Boundaries of Anthropology.* London: Pluto Press, 1989.

Oelschlaeger, M. *The Idea of Wilderness: From Prehistory to the Age of Ecology.* New Haven, Conn.: Yale University Press, 1991.

Ogden, Laura A. "The Beaver Diaspora: A Thought Experiment." *Environmental Humanities* 10, no. 1 (2018): 63–85.

Ogden, Laura, Nik Heynen, Ulrich Oslender, Paige West, Karim-Aly Kassam, and Paul Robbins. "Global Assemblages, Resilience, and Earth Stewardship in the Anthropocene." *Frontiers in Ecology and the Environment* 11, no. 7 (2013): 341–47.

Olson, M. *Unlearn, Rewild: Earth Skills, Ideas and Inspiration for the Future Primitive.* Gabriola Island, Canada: New Society Publishers, 2012.

Olson, Valerie. *Into the Extreme: U.S. Environmental Systems and Politics beyond Earth.* Minneapolis: University of Minnesota Press, 2018.

Orland, Barbara. "Turbo-Cows: Producing a Competitive Animal in the Nineteenth and Early Twentieth Centuries." In *Industrializing Organisms*, edited by Susan Schrepfer and Philip Scranton, 191–214. New York: Routledge, 2004.

Osborne, Jari. "Leave It to Beavers." *Nature.* PBS, May 14, 2014. http://www.pbs.org/.

O'Toole, Paul W., Julian R. Marchesi, and Colin Hill. "Next-Generation Probiotics: The Spectrum from Probiotics to Live Biotherapeutics." *Nature Microbiology* 2 (2017): 17057.

Paine, R. T. "A Conversation on Refining the Concept of Keystone Species." *Conservation Biology* 9, no. 4 (1995): 962–64.

Paine, R. T. "A Note on Trophic Complexity and Community Stability." *American Naturalist* 103, no. 929 (1969): 91–93.

Palmer, Clare A. "'Taming the Wild Profusion of Existing Things'? A Study of Foucault, Power, and Human/Animal Relationships." *Environmental Ethics* 23, no. 4 (2001): 339–58.

Palmer, Stevan. *Launching Global Health: The Caribbean Odyssey of the Rockefeller Foundation.* Ann Arbor: University of Michigan Press, 2010.

Palmer, Stevan. "Migrant Clinics and Hookworm Science: Peripheral Origins of International Health, 1840–1920." *Bulletin of the History of Medicine* 83, no. 4 (2009): 676–709.

Panic, Gordana, Urs Duthaler, Benjamin Speich, and Jennifer Keiser. "Repurposing Drugs for the Treatment and Control of Helminth Infections."

*International Journal for Parasitology, Drugs, and Drug Resistance* 4, no. 3 (2014): 185–200.

Papadopoulos, Dimitris. "Insurgent Posthumanism." *Ephemera: Theory and Politics in Organization* 10, no. 2 (2010): 134–51.

Park, Stephen D. E., David A. Magee, Paul A. McGettigan, et al. "Genome Sequencing of the Extinct Eurasian Wild Aurochs, *Bos primigenius*, Illuminates the Phylogeography and Evolution of Cattle." *Genome Biology* 16, no. 1 (2015): 234.

Parker, Melissa, and Tim Allen. "De-politicizing Parasites: Reflections on Attempts to Control the Control of Neglected Tropical Diseases." *Medical Anthropology* 33, no. 3 (2014): 223–39.

Parker, Melissa, Tim Allen, and Julie Hastings. "Resisting Control of Neglected Tropical Diseases: Dilemmas in the Mass Treatment of Schistosomiasis and Soil-Transmitted Helminths in North-West Uganda." *Journal of Biosocial Science* 40, no. 2 (2008): 161–81.

Parker, William. "The 'Hygiene Hypothesis' for Allergic Disease Is a Misnomer." *BMJ* 348 (2014): g5267.

Parker, William, and Jeff Ollerton. "Evolutionary Biology and Anthropology Suggest Biome Reconstitution as a Necessary Approach toward Dealing with Immune Disorders." *Evolution, Medicine, and Public Health* 2013, no. 1 (2013): 89–103.

Parker, William, Sarah E. Perkins, Matthew Harker, and Michael P. Muehlenbein. "A Prescription for Clinical Immunology: The Pills Are Available and Ready for Testing. A Review." *Current Medical Research and Opinion* 28, no. 7 (2012): 1193–202.

Pauly, D. "Anecdotes and the Shifting Baseline Syndrome of Fisheries." *Trends in Ecology and Evolution* 10, no. 10 (1995): 430.

Paxson, Heather. *The Life of Cheese: Crafting Food and Value in America*. Berkeley: University of California Press, 2012.

Paxson, Heather. "Microbiopolitics." In *The Multispecies Salon*, edited by Eben Kirksey, 115–21. Durham, N.C.: Duke University Press, 2014.

Paxson, Heather. "The Naturalization of Nature as Working." Theorizing the Contemporary, Fieldsights, July 26, 2018. https://culanth.org/.

Paxson, Heather. "Post-Pasteurian Cultures: The Microbiopolitics of Raw-Milk Cheese in the United States." *Cultural Anthropology* 23, no. 1 (2008): 15–47.

Paxson, Heather, and Stefan Helmreich. "The Perils and Promises of Microbial Abundance: Novel Natures and Model Ecosystems, from Artisanal Cheese to Alien Seas." *Social Studies of Science* 44, no. 2 (2014): 165–93.

Pearce, Fred. *The New Wild: Why Invasive Species Will Be Nature's Salvation*. London: Icon Books, 2015.

Peet, Richard, and Michael Watts. *Liberation Ecologies: Environment, Development, Social Movements*. London: Routledge, 2004.

Pellis, Arjaan, Martijn Felder, and Rene van der Duim. "The Socio-political Conceptualization of Serengeti Landscapes in Europe: The Case of Western-Iberia." Paper presented at the 10th World Wilderness Congress, Salamanca, Spain, 2013.

Peluso, Nancy Lee. "Whose Woods Are These? Counter-mapping Forest Territories in Kalimantan, Indonesia." *Antipode* 27, no. 4 (1995): 383–406.

Pereira, H. M., and L. Navarro. *Rewilding European Landscapes*. Amsterdam: Springer, 2015.

Perreault, T., G. Bridge, and J. McCarthy. *The Routledge Handbook of Political Ecology*. London: Taylor & Francis, 2015.

Perry, George H. "Parasites and Human Evolution." *Evolutionary Anthropology* 23, no. 6 (2014): 218–28.

Pettorelli, Nathalie, Jos Barlow, Philip A. Stephens, et al. "Making Rewilding Fit for Policy." *Journal of Applied Ecology* 55, no. 3 (2018): 1114–25.

Pettorelli, N., S. M. Durant, and J. T. du Toit. *Rewilding*. Cambridge: Cambridge University Press, 2019.

Phalan, Benjamin. "What Have We Learned from the Land Sparing–Sharing Model?" *Sustainability* 10, no. 6 (2018): 1760.

Phalan, B., M. Onial, A. Balmford, and R. E. Green. "Reconciling Food Production and Biodiversity Conservation: Land Sharing and Land Sparing Compared." *Science* 333, no. 6047 (2011): 1289–91.

Philo, C. "Animals, Geography, and the City: Notes on Inclusions and Exclusions." *Environment and Planning D: Society and Space* 13, no. 6 (1995): 655–81.

Pierotti, R., and B. R. Fogg. *The First Domestication: How Wolves and Humans Coevolved*. New Haven, Conn.: Yale University Press, 2017.

Pires, Mathias Mistretta. "Rewilding Ecological Communities and Rewiring Ecological Networks." *Perspectives in Ecology and Conservation* 15, no. 4 (2017): 257–65.

Podolsky, S. H. *The Antibiotic Era: Reform, Resistance, and the Pursuit of a Rational Therapeutics*. Baltimore, Md.: Johns Hopkins University Press, 2015.

Podolsky, Scott H. "Metchnikoff and the Microbiome." *Lancet* 380, no. 9856 (2012): 1810–11.

Poliquin, R. *Beaver*. London: Reaktion, 2015.

Pollan, Michael. *Cooked: A Natural History of Transformation*. London: Penguin, 2013.

Pollard, T. M. *Western Diseases: An Evolutionary Perspective*. Cambridge: Cambridge University Press, 2008.

Porcher, J. *The Ethics of Animal Labor: A Collaborative Utopia*. Amsterdam: Springer, 2017.

Povinelli, E. A. *Geontologies: A Requiem to Late Liberalism*. Durham, N.C.: Duke University Press, 2016.

Prior, Jonathan, and Kim J. Ward. "Rethinking Rewilding: A Response to Jørgensen." *Geoforum* 69 (2016): 132–35.

Pritchard, David I., and Alan Brown. "Is *Necator americanus* Approaching a Mutualistic Symbiotic Relationship with Humans?" *Trends in Parasitology* 17, no. 4 (2001): 169–72.

Proctor, Michael F., Scott E. Nielsen, Wayne F. Kasworm, et al. "Grizzly Bear Connectivity Mapping in the Canada–United States Trans-border Region." *Journal of Wildlife Management* 79, no. 4 (2015): 544–58.

Puig de la Bellacasa, Maria. "Making Time for Soil: Technoscientific Futurity and the Pace of Care." *Social Studies of Science* 45, no. 5 (2015): 691–716.

Pullan, Rachel L., Jennifer L. Smith, Rashmi Jasrasaria, and Simon J. Brooker. "Global Numbers of Infection and Disease Burden of Soil Transmitted Helminth Infections in 2010." *Parasites and Vectors* 7, no. 1 (2014): 37.

Puttock, Alan, Hugh A. Graham, Andrew M. Cunliffe, Mark Elliott, and Richard E. Brazier. "Eurasian Beaver Activity Increases Water Storage, Attenuates Flow and Mitigates Diffuse Pollution from Intensively-Managed Grasslands." *Science of the Total Environment* 576 (2017): 430–43.

Quan, L. *Rewilded: Saving the South China Tiger*. Godalming, U.K.: Evans Mitchell Books, 2011.

Rabinow, Paul, and Nikolas Rose. "Biopower Today." *BioSocieties* 1, no. 2 (2006): 195–217.

Radkau, Joachim, and F. Uekoetter. *Naturschutz und Nationalsozialismus*. Frankfurt am Main: Campus-Verlag, 2003.

Raimundo, Rafael L. G., Paulo R. Guimarães, and Darren M. Evans. "Adaptive Networks for Restoration Ecology." *Trends in Ecology and Evolution* 33, no. 9 (2018): 664–75.

Rajan, Kaushik Sunder, ed. *Lively Capital: Biotechnologies, Ethics, and Governance in Global Markets*. Durham, N.C.: Duke University Press, 2012.

Rangarajan, M. *Fencing the Forest: Conservation and Ecological Change in India's Central Provinces, 1860–1914*. Oxford: Oxford University Press, 1996.

Ratner, Mark. "Seres's Pioneering Microbiome Drug Fails Mid-stage Trial." *Nature Biotechnology* 34 (2016): 1004.

Reardon, S. "Phage Therapy Gets Revitalized." *Nature* 510, no. 7503 (2014): 15–16.

Redzepi, R., and D. Zilber. *The Noma Guide to Fermentation: Including Koji, Kombuchas, Shoyus, Misos, Vinegars, Garums, Lacto-ferments, and Black Fruits and Vegetables*. Pleasanton, Calif.: Artisan, 2018.

Reichert, H. *Das Nibelungenlied: Nach der St. Galler Handschrift.* Berlin: De Gruyter, 2005.

Renaud, F. G., K. Sudmeier-Rieux, and M. Estrella. *The Role of Ecosystems in Disaster Risk Reduction.* Tokyo: United Nations University Press, 2013.

Rewilding Europe. *Circle of Life: A New Way to Support Europe's Scavengers.* Nijmegen, The Netherlands: Rewilding Europe, 2017.

Rewilding Europe. *Rewilding Europe: A New Beginning, for Wildlife, for Us.* Nijmegen, The Netherlands: ARK, 2010.

Rewilding Europe. *A Vision for a Wilder Europe.* Nijmegen, The Netherlands: Rewilding Europe, 2015.

Rheinberger, Hans-Jörg. *Toward a History of Epistemic Things: Synthesizing Proteins in the Test Tube.* Stanford, Calif.: Stanford University Press, 1997.

Ricciardi, Anthony, and Daniel Simberloff. "Assisted Colonization Is Not a Viable Conservation Strategy." *Trends in Ecology and Evolution* 24, no. 5 (2009): 248–53.

Richardson, David M., Petr Pysek, Marcel Rejmanek, Michael G. Barbour, F. Dane Panetta, and Carol J. West. "Naturalization and Invasion of Alien Plants: Concepts and Definitions." *Diversity and Distributions* 6, no. 2 (2000): 93–107.

Ripple, William J., and Robert L. Beschta. "Wolves and the Ecology of Fear: Can Predation Risk Structure Ecosystems?" *BioScience* 54, no. 8 (2004): 755–66.

Ritvo, Harriet. *The Animal Estate: The English and Other Creatures in the Victorian Age.* Cambridge, Mass.: Harvard University Press, 1987.

Ritvo, Harriet. *Noble Cows and Hybrid Zebras: Essays on Animals and History.* Charlottesville: University of Virginia Press, 2010.

Robbins, P. "Fixed Categories in a Portable Landscape: The Causes and Consequences of Land-Cover Categorization." *Environment and Planning A* 33, no. 1 (2001): 161–79.

Robbins, P. *Political Ecology: A Critical Introduction.* New York: Wiley, 2011.

Robbins, P., and S. A. Moore. "Ecological Anxiety Disorder: Diagnosing the Politics of the Anthropocene." *Cultural Geographies* 20, no. 1 (2013): 3–19.

Robertson, Morgan. "Measurement and Alienation: Making a World of Ecosystem Services." *Transactions of the Institute of British Geographers* 37, no. 3 (2012): 386–401.

Robertson, Morgan. "The Nature that Capital Can See: Science, State, and Market in the Commodification of Ecosystem Services." *Environment and Planning D: Society and Space* 24, no. 3 (2006): 367–87.

Rockström, J., W. Steffen, K. Noone, et al. "Planetary Boundaries: Exploring the Safe Operating Space for Humanity." *Ecology and Society* 14, no. 2 (2009): 32.

Rook, Graham. "Hygiene Hypothesis and Autoimmune Diseases." *Clinical Reviews in Allergy and Immunology* 42, no. 1 (2012): 5–15.

Rook, Graham. "Review Series on Helminths, Immune Modulation and the Hygiene Hypothesis: The Broader Implications of the Hygiene Hypothesis." *Immunology* 126, no. 1 (2009): 3–11.

Rook, Graham, ed. *The Hygiene Hypothesis and Darwinian Medicine*. Basel: Birkhäuser, 2009.

Rook, Graham, Fredrik Bäckhed, Bruce R. Levin, Margaret J. McFall-Ngai, and Angela R. McLean. "Evolution, Human–Microbe Interactions, and Life History Plasticity." *Lancet* 390, no. 10093 (2017): 521–30.

Rook, G. A. W., C. L. Raison, and C. A. Lowry. "Microbial 'Old Friends,' Immunoregulation and Socioeconomic Status." *Clinical and Experimental Immunology* 177, no. 1 (2014): 1–12.

Rose, Nikolas. *The Politics of Life Itself: Biomedicine, Power, and Subjectivity in the Twenty-First Century*. Princeton, N.J.: Princeton University Press, 2006.

Ruddiman, William F. "The Anthropogenic Greenhouse Era Began Thousands of Years Ago." *Climatic Change* 61, no. 3 (2003): 261–93.

Ruddiman, William F. "The Early Anthropogenic Hypothesis: Challenges and Responses." *Reviews of Geophysics* 45, no. 4 (2007): RG4001.

Ruddiman, William F., Erle C. Ellis, Jed O. Kaplan, and Dorian Q. Fuller. "Defining the Epoch We Live In." *Science* 348, no. 6230 (2015): 38–39.

Rutherford, Stephanie. "The Anthropocene's Animal? Coywolves as Feral Cotravelers." *Environment and Planning E: Nature and Space* 1, no. 1–2 (2018): 206–23.

Rynkiewicz, Evelyn C., Amy B. Pedersen, and Andy Fenton. "An Ecosystem Approach to Understanding and Managing Within-Host Parasite Community Dynamics." *Trends in Parasitology* 31, no. 5 (2015): 212–21.

Sachs, Rachel E., and Carolyn A. Edelstein. "Ensuring the Safe and Effective FDA Regulation of Fecal Microbiota Transplantation." *Journal of Law and the Biosciences* 2, no. 2 (2015): 396–415.

Sandom, Chris, and D. MacDonald. "What Next? Rewilding as a Radical Future for the British Countryside." In *Wildlife Conservation on Farmland: Managing for Nature on Lowland Farms*, edited by D. MacDonald and Ruth Feber, 291–316. Oxford: Oxford University Press, 2015.

Sangodeyi, Funke Iyabo. "The Making of the Microbial Body, 1900s–2012." PhD diss., Harvard University, 2014. http://nrs.harvard.edu/urn-3:HUL .InstRepos:12274300.

Saraiva, T. *Fascist Pigs: Technoscientific Organisms and the History of Fascism*. Cambridge, Mass.: MIT Press, 2016.

Sassone-Corsi, Martina, and Manuela Raffatellu. "No Vacancy: How Beneficial Microbes Cooperate with Immunity to Provide Colonization Resistance to Pathogens." *Journal of Immunology* 194, no. 9 (2015): 4081–87.

Sax, Boria. *Animals in the Third Reich: Pets, Scapegoats, and the Holocaust.* New York: Continuum, 2000.

Sax, Boria. "What Is a 'Jewish Dog'? Konrad Lorenz and the Cult of Wildness." *Society and Animals* 5 (1997): 3–21.

Schama, Simon. *Landscape and Memory.* New York: Random House, 1995.

Schölmerich, Jürgen, Klaus Fellermann, Frank W. Seibold, et al. "A Randomised, Double-Blind, Placebo-Controlled Trial of *Trichuris suis* Ova in Active Crohn's Disease." *Journal of Crohn's and Colitis* 11, no. 4 (2017): 390–99.

Schulze, Sylvie, Jana Schleicher, Reinhard Guthke, and Jörg Linde. "How to Predict Molecular Interactions between Species?" *Frontiers in Microbiology* 7 (2016): 442.

Schwartz, Katrina Z. S. *Nature and National Identity after Communism: Globalizing the Ethnoscape.* Pittsburgh, Pa.: University of Pittsburgh Press, 2006.

Schwartz, Katrina Z. S. "Wild Horses in a 'European Wilderness': Imagining Sustainable Development in the Post-communist Countryside." *Cultural Geographies* 12, no. 3 (2005): 292–320.

Schweiger, Andreas H., Isabelle Boulangeat, Timo Conradi, Matt Davis, and Jens-Christian Svenning. "The Importance of Ecological Memory for Trophic Rewilding as an Ecosystem Restoration Approach." *Biological Reviews* 94, no. 1 (2019): 1–15.

Scott, J. C. *Against the Grain: A Deep History of the Earliest States.* New Haven, Conn.: Yale University Press, 2017.

Scout, Urban. *Rewild or Die.* Urban Scout, 2010.

Scranton, R. *Learning to Die in the Anthropocene: Reflections on the End of a Civilization.* San Francisco, Calif.: City Lights, 2015.

Sears, Cynthia L., and Drew M. Pardoll. "Perspective: Alpha-Bugs, Their Microbial Partners, and the Link to Colon Cancer." *Journal of Infectious Diseases* 203, no. 3 (2011): 306–11.

Seddon, Nathalie, Beth Turner, Pam Berry, Alexandre Chausson, and Cécile A. J. Girardin. "Grounding Nature-Based Climate Solutions in Sound Biodiversity Science." *Nature Climate Change* 9, no. 2 (2019): 84–87.

Seddon, Philip J., Christine J. Griffiths, Pritpal S. Soorae, and Doug P. Armstrong. "Reversing Defaunation: Restoring Species in a Changing World." *Science* 345, no. 6195 (2014): 406–12.

Sennett, R. *The Craftsman.* London: Penguin Adult, 2009.

Serpell, J. *The Domestic Dog.* Cambridge: Cambridge University Press, 2016.

Shapin, S. "Placing the View from Nowhere: Historical and Sociological Problems in the Location of Science." *Transactions of the Institute of British Geographers* 23, no. 1 (1998): 5–12.

Sharma, M. *Green and Saffron: Hindu Nationalism and Indian Environmental Politics.* Delhi, India: Permanent Black, 2012.

Shepherd, J. G. "Geoengineering the Climate: An Overview and Update." *Philosophical Transactions of the Royal Society A: Mathematical, Physical and Engineering Sciences* 370, no. 1974 (2012): 4166–75.

Shine, R., and H. W. Greene. *Cane Toad Wars.* Berkeley: University of California Press, 2018.

Shukin, Nicole. *Animal Capital: Rendering Life in Biopolitical Times.* Minneapolis: University of Minnesota Press, 2009.

Singer, Merrill. "Ecosyndemics: Global Warming and the Coming Plagues of the Twenty-First Century." In *Plagues and Epidemics: Infected Spaces Past and Present*, edited by D. A. Herring and A. C. Swedlund, 21–38. Oxford: Berg, 2010.

Skabelund, A. "Breeding Racism: The Imperial Battlefields of the 'German' Shepherd Dog." *Society and Animals* 16, no. 4 (2008): 354–71.

Smallwood, Taylor B., Paul R. Giacomin, Alex Loukas, Jason P. Mulvenna, Richard J. Clark, and John J. Miles. "Helminth Immunomodulation in Autoimmune Disease." *Frontiers in Immunology* 8, no. 453 (2017).

Smit, C., J. L. Ruifrok, R. van Klink, and H. Olff. "Rewilding with Large Herbivores: The Importance of Grazing Refuges for Sapling Establishment and Wood-Pasture Formation." *Biological Conservation* 182 (2015): 134–42.

Smith, Neil. "Nature as an Accumulation Strategy." In *Socialist Register, 2007: Coming to Terms with Nature*, edited by L. Panitch and L. Leys, 16–36. London: Merlin, 2007.

Smyth, Kendra, Claire Morton, Amanda Mathew, et al. "Production and Use of *Hymenolepis diminuta* Cysticercoids as Anti-inflammatory Therapeutics." *Journal of Clinical Medicine* 6, no. 10 (2017): 98.

Song, Yang, Shashank Garg, Mohit Girotra, et al. "Microbiota Dynamics in Patients Treated with Fecal Microbiota Transplantation for Recurrent *Clostridium difficile* Infection." *PLoS One* 8, no. 11 (2013): e81330.

Sonnenburg, Justin, and Erica Sonnenburg. *The Good Gut: Taking Control of Your Weight, Your Mood, and Your Long-Term Health.* London: Penguin, 2015.

Soulé, Michael E. "What Is Conservation Biology?" *BioScience* 35, no. 11 (1985): 727–34.

Soulé, Michael E., James A. Estes, Brian Miller, and Douglas L. Honnold. "Strongly Interacting Species: Conservation Policy, Management, and Ethics." *BioScience* 55, no. 2 (2005): 168–76.

Soulé, Michael, and Reed Noss. "Rewilding and Biodiversity: Complementary Goals for Continental Conservation." *Wild Earth* 22 (1998): 19–28.

Spackman, Christy C. W. "Formulating Citizenship: The Microbiopolitics of the Malfunctioning Functional Beverage." *BioSocieties* 13, no. 1 (2018): 41–63.

Spector, T. *The Diet Myth: The Real Science behind What We Eat.* London: Orion, 2015.

Stallins, J. Anthony, Derek M. Law, Sophia A. Strosberg, and Jarius J. Rossi. "Geography and Postgenomics: How Space and Place Are the New DNA." *GeoJournal* 83, no. 1 (2018): 153–68.

Stein, Miguel, Zalman Greenberg, Mona Boaz, Zeev T. Handzel, Mesfin K. Meshesha, and Zvi Bentwich. "The Role of Helminth Infection and Environment in the Development of Allergy: A Prospective Study of Newly-Arrived Ethiopian Immigrants in Israel." *PLoS Neglected Tropical Diseases* 10, no. 1 (2016): e0004208.

Stengers, Isabelle. "Autonomy and the Intrusion of Gaia." *South Atlantic Quarterly* 116, no. 2 (2017): 381–400.

Stengers, Isabelle. "Gaia, the Urgency to Think (and Feel)." The Thousand Names of Gaia, Rio de Janeiro, 2014. https://osmilnomesdegaia.files.word press.com/2014/11/isabelle-stengers.pdf.

Stilgoe, J. *Experiment Earth: Responsible Innovation in Geoengineering.* London: Taylor & Francis, 2015.

Stillfried, Milena, Pierre Gras, Matthias Busch, Konstantin Börner, Stephanie Kramer-Schadt, and Sylvia Ortmann. "Wild Inside: Urban Wild Boar Select Natural, Not Anthropogenic Food Resources." *PLoS One* 12, no. 4 (2017): e0175127.

Strachan, David. "Hay Fever, Hygiene, and Household Size." *BMJ* 299, no. 6710 (1989): 1259–60.

Strathern, M. *After Nature: English Kinship in the Late Twentieth Century.* Cambridge: Cambridge University Press, 1992.

Sullivan, Sian. "Banking Nature? The Spectacular Financialisation of Environmental Conservation." *Antipode* 45, no. 1 (2013): 198–217.

Sullivan, Sian. "Nature on the Move 3: (Re)Countenancing an Animate Nature." *New Proposals: Journal of Marxism and Interdisciplinary Inquiry* 6, no. 1–2 (2013): 50–71.

Sunder Rajan, Kaushik. *Biocapital: The Constitution of Postgenomic Life.* Durham, N.C.: Duke University Press, 2006.

Sunstein, Cass. "The Idea of a Useable Past." *Columbia Law Review* 601 (1995): 601–8.

Sutherland, W. J. "Conservation Biology—Openness in Management." *Nature* 418, no. 6900 (2002): 834–35.

Svenning, Jens-Christian. "Future Megafaunas: A Historical Perspective on the Scope for a Wilder Anthropocene." In *Arts of Living on a Damaged Planet: Ghosts and Monsters of the Anthropocene*, edited by Anna Lowenhaupt Tsing, Heather Anne Swanson, Elaine Gan, and Nils Bubandt, G67–86. Minneapolis: University of Minnesota Press, 2017.

Svenning, Jens-Christian. "Proactive Conservation and Restoration of Botanical Diversity in the Anthropocene's 'Rambunctious Garden.'" *American Journal of Botany* 105, no. 6 (2018): 963–66.

Svenning, Jens-Christian, Pil B. M. Pedersen, C. Josh Donlan, et al. "Science for a Wilder Anthropocene: Synthesis and Future Directions for Trophic Rewilding Research." *Proceedings of the National Academy of Sciences of the United States of America* 113, no. 4 (2016): 898–906.

Swanson, Heather Anne, Nils Bubandt, and Anna Tsing. "Less than One but More than Many: Anthropocene as Science Fiction and Scholarship-in-the-Making." *Environment and Society* 6 (2015): 149–66.

Swanson, Heather Anne, Marianne Lien, and Gro Ween. *Domestication Gone Wild: Politics and Practices of Multispecies Relations*. Durham, N.C.: Duke University Press, 2018.

Swyngedouw, Erik. "Apocalypse Forever? Post-political Populism and the Spectre of Climate Change." *Theory, Culture, and Society* 27, no. 2–3 (2010): 213–32.

Tack, Jurgen. *Wild Boar* (Sus scrofa) *Populations in Europe: A Scientific Review of Population Trends and Implications for Management*. Brussels: European Landowners Organization, 2018.

Takacs, David. *The Idea of Biodiversity: Philosophies of Paradise*. Baltimore, Md.: Johns Hopkins University Press, 1996.

Tauber, Alfred. "The Immune System and Its Ecology." *Philosophy of Science* 75, no. 2 (2008): 224–45.

Tauber, Alfred. *Immunity: The Evolution of an Idea*. Oxford: Oxford University Press, 2017.

Taylor, J. *Body by Darwin: How Evolution Shapes Our Health and Transforms Medicine*. Chicago: University of Chicago Press, 2015.

Taylor, Peter. *The Spirit of Rewilding: Steps toward a Shamanic Ecology*. ethos-uk .com, 2016.

Terborgh, J., and J. A. Estes. *Trophic Cascades: Predators, Prey, and the Changing Dynamics of Nature*. Washington, D.C.: Island Press, 2010.

Thomas, C. D. *Inheritors of the Earth: How Nature Is Thriving in an Age of Extinction*. London: Penguin, 2017.

Thompson, Helen. "How Oral Vaccines Could Save Ethiopian Wolves from Extinction." *Science News* 193, no. 6 (2018): 20.

Tree, Isabella. *Wilding: The Return of Nature to a British Farm*. London: Pan Macmillan, 2018.

Tsai, Y. L., I. Carbonell, J. Chevrier, and A. L. Tsing. "Golden Snail Opera: The More-than-Human Performance of Friendly Farming on Taiwan's Lanyang Plain." *Cultural Anthropology* 31, no. 4 (2016): 520–44.

Tsing, Anna. "The Buck, the Bull, and the Dream of the Stag: Some Unexpected Weeds of the Anthropocene." *Suomen Antropologi* 42, no. 1 (2017): 3–21.

Tsing, Anna. *The Mushroom at the End of the World: On the Possibility of Life in Capitalist Ruins*. Princeton, N.J.: Princeton University Press, 2015.

Tsing, Anna. "A Threat to Holocene Resurgence Is a Threat to Livability." In *The Anthropology of Sustainability: Beyond Development and Progress*, edited by Marc Brightman and Jerome Lewis, 51–65. London: Palgrave Macmillan, 2017.

Tsing, Anna, and Nils Bubandt, eds. *Arts of Living on a Damaged Planet: Ghosts and Monsters of the Anthropocene*. Minneapolis: University of Minnesota Press, 2017.

Tylianakis, Jason M., Etienne Laliberté, Anders Nielsen, and Jordi Bascompte. "Conservation of Species Interaction Networks." *Biological Conservation* 143, no. 10 (2010): 2270–79.

van den Belt, H. "Networking Nature, or Serengeti behind the Dikes." *History and Technology* 20, no. 3 (2004): 311–33.

van Heel, B. F., A. M. Boerboom, J. M. Fliervoet, H. J. R. Lenders, and R. J. G. van den Born. "Analysing Stakeholders' Perceptions of Wolf, Lynx and Fox in a Dutch Riverine Area." *Biodiversity and Conservation* 26, no. 7 (2017): 1723–43.

van Klink, R., J. L. Ruifrok, and C. Smit. "Rewilding with Large Herbivores: Direct Effects and Edge Effects of Grazing Refuges on Plant and Invertebrate Communities." *Agriculture, Ecosystems and Environment* 234 (2016): 81–97.

van Nood, Els, Anne Vrieze, Max Nieuwdorp, et al. "Duodenal Infusion of Donor Feces for Recurrent *Clostridium difficile*." *New England Journal of Medicine* 368, no. 5 (2013): 407–15.

van Vuure, C. *Retracing the Aurochs: History, Morphology and Ecology of an Extinct Wild Ox*. Sofia, Bulgaria: Pensoft, 2005.

Vasile, Monica. "The Vulnerable Bison: Practices and Meanings of Rewilding in the Romanian Carpathians." *Conservation and Society* 16, no. 3 (2018): 217–31.

Velasquez-Manoff, M. "Among Trillions of Microbes in the Gut, a Few Are Special." *Scientific American*, March 1, 2015.

Velasquez-Manoff, Moises. *An Epidemic of Absence: A New Way of Understanding Allergies and Autoimmune Diseases*. New York: Scribner, 2012.

Vera, Frans. *Grazing Ecology and Forest History*. Wallingford, U.K.: CABI, 2000.

Vera, Frans. "Large-Scale Nature Development—The Oostvaardersplassen." *British Wildlife*, June 2009. https://diaplan.ku.dk/pdf/large-scale_nature_development_the_Oostvaardersplassen.pdf.

von Hertzen, Leena, Ilkka Hanski, and Tari Haahtela. "Natural Immunity: Biodiversity Loss and Inflammatory Diseases Are Two Global Megatrends that Might Be Related." *EMBO Reports* 12, no. 11 (2011): 1089–93.

von Mutius, Erika, and Donata Vercelli. "Farm Living: Effects on Childhood Asthma and Allergy." *Nature Reviews Immunology* 10, no. 12 (2010): 861–68.

Wadiwel, Dinesh. "Chicken Harvesting Machine: Animal Labor, Resistance, and the Time of Production." *South Atlantic Quarterly* 117, no. 3 (2018): 527–49.

Wadiwel, D. *The War against Animals*. Leiden: Brill, 2015.

Wakefield, Stephanie. "Infrastructures of Liberal Life: From Modernity and Progress to Resilience and Ruins." *Geography Compass* 12, no. 7 (2018): e12377.

Wakefield, Stephanie. "Inhabiting the Anthropocene Back Loop." *Resilience* 6, no. 2 (2018): 77–94.

Wakefield, Stephanie, and Bruce Braun. "Oystertecture: Infrastructure, Profanation, and the Sacred Figure of the Human." In *Infrastructure, Environment, and Life in the Anthropocene*, edited by Kregg Hetherington, 193–215. Durham, N.C.: Duke University Press, 2019.

Walker, Jeremy, and Melinda Cooper. "Genealogies of Resilience: From Systems Ecology to the Political Economy of Crisis Adaptation." *Security Dialogue* 42, no. 2 (2011): 143–60.

Wallace, Rob. *Big Farms Make Big Flu: Dispatches on Influenza, Agribusiness, and the Nature of Science*. New York: Monthly Review Press, 2016.

Wallace, R., and R. G. Wallace. "Blowback: New Formal Perspectives on Agriculturally Driven Pathogen Evolution and Spread." *Epidemiology and Infection* 143, no. 10 (2015): 2068–80.

Wammes, Linda, Harriet Mpairwe, Alison Elliott, and Maria Yazdanbakhsh. "Helminth Therapy or Elimination: Epidemiological, Immunological, and Clinical Considerations." *Lancet Infectious Diseases* 14, no. 11 (2014): 1150–62.

Wanderer, Emily Mannix. "Biologies of Betrayal: Judas Goats and Sacrificial Mice on the Margins of Mexico." *BioSocieties* 10, no. 1 (2015): 1–23.

Warde, P., L. Robin, and S. Sörlin. *The Environment: A History of the Idea*. Baltimore, Md.: Johns Hopkins University Press, 2018.

Wark, M. K. *A Hacker Manifesto*. Cambridge, Mass.: Harvard University Press, 2009.

Warner, J. F., A. van Buuren, and J. Edelenbos. *Making Space for the River*. London: IWA Publishing, 2012.

Waterton, C. "From Field to Fantasy: Classifying Nature, Constructing Europe." *Social Studies of Science* 32, no. 2 (2002): 177–204.

Weeks, K. *The Problem with Work: Feminism, Marxism, Antiwork Politics, and Postwork Imaginaries.* Durham, N.C.: Duke University Press, 2011.

White, Sam. "From Globalized Pig Breeds to Capitalist Pigs: A Study in Animal Cultures and Evolutionary History." *Environmental History* 16, no. 1 (2011): 94–120.

Whittaker, R., M. Araujo, P. Jepson, R. Ladle, J. Watson, and K. Willis. "Conservation Biogeography: Assessment and Prospect." *Diversity and Distributions* 11, no. 1 (2005): 3–23.

Wilkie, Rhoda. "Sentient Commodities and Productive Paradoxes: The Ambiguous Nature of Human–Livestock Relations in Northeast Scotland." *Journal of Rural Studies* 21, no. 2 (2005): 213–30.

Williams, Raymond. *The Country and the City.* Oxford: Oxford University Press, 1973.

Wilmers, Christopher C., Barry Nickel, Caleb M. Bryce, Justine A. Smith, Rachel E. Wheat, and Veronica Yovovich. "The Golden Age of Bio-logging: How Animal-Borne Sensors Are Advancing the Frontiers of Ecology." *Ecology* 96, no. 7 (2015): 1741–53.

Woelfle-Erskine, Cleo. "Beavers as Commoners? Invitations to River Restoration Work in a Beavery Mode." *Community Development Journal* 54, no. 1 (2018): 100–118.

Woelfle-Erskine, Cleo, and July Cole. "Transfiguring the Anthropocene: Stochastic Reimaginings of Human–Beaver Worlds." *Transgender Studies Quarterly* 2, no. 2 (2015): 297–316.

Wolf, E. R., T. H. Eriksen, and N. L. Diaz. *Europe and the People without History.* Berkeley: University of California Press, 2010.

Wolfe, Cary. *Before the Law: Humans and Other Animals in a Biopolitical Frame.* Chicago: University of Chicago Press, 2012.

Wolfe, Cary. "Ecologizing Biopolitics, or What Is the 'Bio-' of Biopolitics and Bioart?" In *General Ecology: The New Ecological Paradigm*, edited by Erich Hörl and James Burton, 217–34. London: Bloomsbury, 2017.

Wolf-Meyer, Matthew J. "Normal, Regular, and Standard: Scaling the Body through Fecal Microbial Transplants." *Medical Anthropology Quarterly* 31, no. 3 (2017): 297–314.

Worm, Boris, and Robert T. Paine. "Humans as a Hyperkeystone Species." *Trends in Ecology and Evolution* 31, no. 8 (2016): 600–607.

Worster, Donald. "The Ecology of Order and Chaos." *Environmental History Review* 14, no. 1/2 (1990): 1–18.

Worster, Donald. *Nature's Economy: A History of Ecological Ideas.* 2nd ed. Cambridge: Cambridge University Press, 1994.

Yong, Ed. *I Contain Multitudes: The Microbes within Us and a Grander View of Life.* New York: Random House, 2016.

Youatt, R. "Counting Species: Biopower and the Global Biodiversity Census." *Environmental Values* 17, no. 3 (2008): 393–417.

Young, K. A. "Of Poops and Parasites: Unethical FDA Overregulation." *Food and Drug Law Journal* 69, no. 4 (2014): 555–74, ii.

Yusoff, Kathryn. "Anthropogenesis: Origins and Endings in the Anthropocene." *Theory, Culture, and Society* 33, no. 2 (2016): 3–28.

Yusoff, Kathryn, Elizabeth Grosz, Nigel Clark, et al. "Geopower: A Panel on Elizabeth Grosz's Chaos, Territory, Art: Deleuze and the Framing of the Earth." *Environment and Planning D: Society and Space* 30, no. 6 (2012): 971–88.

Zaiss, M. M., and N. L. Harris. "Interactions between the Intestinal Microbiome and Helminth Parasites." *Parasite Immunology* 38 (2016): 5–11.

Zaiss, Mario M., Alexis Rapin, Luc Lebon, et al. "The Intestinal Microbiota Contributes to the Ability of Helminths to Modulate Allergic Inflammation." *Immunity* 43, no. 5 (2015): 998–1010.

Zimov, S. A. "Pleistocene Park: Return of the Mammoth's Ecosystem." *Science* 308, no. 5723 (2005): 796–98.

Zimov, S. A., N. S. Zimov, A. N. Tikhonov, and F. S. Chapin. "Mammoth Steppe: A High-Productivity Phenomenon." *Quaternary Science Reviews* 57 (2012): 26–45.

Zuckerman, Molly Kathleen, Kristin Nicole Harper, Ronald Barrett, and George John Armelagos. "The Evolution of Disease: Anthropological Perspectives on Epidemiologic Transitions." *Global Health Action* 7 (2014): 10.3402/gha.v7.23303.

# INDEX

(*continued from page ii*)

Jamie Lorimer is associate professor in the School of Geography and the Environment at the University of Oxford. He is the author of *Wildlife in the Anthropocene: Conservation after Nature* (Minnesota, 2015).